Advances in Biochemical Engineering/ Biotechnology

Volume 193

Aims and Scope

This book series reviews current trends in modern biotechnology and biochemical engineering. Its aim is to cover all aspects of these interdisciplinary disciplines, where knowledge, methods and expertise are required from chemistry, biochemistry, microbiology, molecular biology, chemical engineering and computer science. This series supports and advances the UN's sustainable development goals, mainly SDG7 (Affordable and Clean Energy), SDG3 (Good Health and Well Being), and SDG13 (Climate Change).

Volumes are organized topically and provide a comprehensive discussion of developments in the field over the past 3–5 years. The series also discusses new discoveries and applications. Special volumes are dedicated to selected topics which focus on new biotechnological products and new processes for their synthesis and purification.

In general, volumes are edited by well-known guest editors. The series editor and publisher will, however, always be pleased to receive suggestions and supplementary information. Manuscripts are accepted in English.

In references, Advances in Biochemical Engineering/Biotechnology is abbreviated as *Adv. Biochem. Engin./Biotechnol.* and cited as a journal.

Antonina Lavrentieva • Jette Feveile Young
Editors

Cultivated Meat Bioprocessing

An End-to-End Production Chain Prospective

With contributions by

J. Bernfeld · S. Bode · T. Callue · S. Fehlmann · M. P. Hanga ·
C. S. Hee · M. Henkel · M. Kirsch · N. Kleger · T. Krampitz ·
O. N. Kubinyecz · L. Lähteenmäki · N. Lavon · A. Lavrentieva ·
H. E. Lester · J. Liedtke · A. Maor-Shoshani · A. Millbank ·
J. Morales-Dalmau · L. Pasitka · M. J. Post · M. K. Rasmussen ·
F. Rivero-Pino · A. Rossi · D. C. Rueda · T. Ryynänen ·
M. L. Schlieker · S. Shalel-Levanon · E. Theodosiou ·
M. Therkildsen · R. R. Thøgersen · C. Treinen · H. L. Tuomisto ·
M. L. Vieri · D. Vorländer · J. F. Young · S. Zschieschang

Springer

Editors
Antonina Lavrentieva
Institute of Technical Chemistry
Leibniz University of Hannover
Hannover, Niedersachsen, Germany

Jette Feveile Young
Department of Food Science
Aarhus University
Aarhus, Denmark

ISSN 0724-6145 ISSN 1616-8542 (electronic)
Advances in Biochemical Engineering/Biotechnology
ISBN 978-3-032-32099-5 ISBN 978-3-032-32100-8 (eBook)
https://doi.org/10.1007/978-3-032-32100-8

This Springer imprint is published by the registered company Springer Nature Switzerland AG
The registered company address is: Gewerbestrasse 11, 6330 Cham, Switzerland

Preface

In light of a rapidly changing world, new concepts need to be introduced to stakeholders and society through various channels and in diverse formats. Food is a topic that is relevant to everyone as we are all consumers, and in a societal context, an increasing focus on the environmental impact of food has made this topic even more pressing. We need to find different ways of producing food, strike different balances, and even explore a completely new food concept. In this light, cultivated meat, developed from cells cultivated/grown outside the animal, is emerging as part of the solution. For enabling the transition from the concept to technological maturity, scalability, environmental impact, legislation, and consumer acceptance, a global community of diverse experts must collaborate and create the scientific foundation for solutions as well as the environment for constructive information to society and consumers and advice for policymakers.

In this book, we have emphasised the importance of collaboration and co-creation between industry and academia, as reflected in chapters with input from both environments and mixed author combinations within chapters. It contains chapters describing how the fundamental base of cultivated meat – the cells – can be adapted and engineered to meet the criteria of scalability, e.g. the chapter "Fundamentals of Cultivated Meat Technology: Cell Culture and Tissue Engineering". This chapter also addresses the critical interplay between cells, the culture media, and a physical industrial setting, while the chapters "Edible Scaffolds for Cultivated Meat Production" and "Development of Cell Culture Media for Cultivated Meat Production" focus, respectively, on the requirements and solutions of cell-adhering/embedding scaffolds and how different cultivation media requirements drive the development. The chapter "Bioprocess Engineering for Cultivated Meat" covers general concepts of process development and discusses the critical parameters that determine process performance. The chapter "Bioreactor Design and Engineering for Cultivated Meat Manufacturing" focuses specifically on scaling conditions and how bioreactor designs need to be adapted. The chapter "From Muscle Fibres to Functional Foods: Bridging Conventional and Cultivated Approaches" outlines which factors are relevant for developing a quality food product.

All these chapters have an eye for how cultivated meat technology can be applied and adapted from established know-how from tissue engineering, pharma bioprocessing, and meat science to meet the new challenge of producing animal cells at food-relevant scales, costs, and quality.

The finalising chapters address the broader societal perspectives of cultivated meat. Parameters relevant for calculating the environmental impact are given in the chapter "Environmental Sustainability of Cultivated Meat", and fundamental safety concepts and legislation in different parts of the world are presented in the chapter "Safety Aspects of Cell Culture-Derived Food for Human Consumption". Finally, the chapter "Consumer and Social Perspectives to Cultivated Meat" discusses factors influencing consumer and social acceptance and reflects on the future acceptability of cultivated meat. In their chapters, all authors discuss not only current technological, regulatory, and social challenges, but also possible future developments.

The entire book gives an overview of all the aspects of cultivated meat that are required to drive the development through research and innovation, emphasising that all elements must be developed and matured in parallel. This makes the book relevant for researchers, industry, and political decision-makers. We all need a little insight into the elements of the cultivated meat landscape in which we ourselves are not involved – this is precisely where a book like this stands out as a collective introduction and inventory of the situation within the world of cultivated meat.

We are deeply grateful to all contributing authors from academia, start-ups, and industry for generously sharing their knowledge and experience with broader community and helping advance the field of cultivated meat. Thanks to their dedication, enthusiasm, and coordinated efforts, cultivated meat technology has a strong chance of being implemented in the near future and becoming a reality.

Hannover, Germany — Antonina Lavrentieva
Aarhus, Denmark — Jette Feveile Young
May 2026

Contents

Adv Biochem Eng Biotechnol (2026) 193: 1–16
https://doi.org/10.1007/10_2025_289

Published online: 10 December 2025

Fundamentals of Cultivated Meat Technology: Cell Culture and Tissue Engineering (Essentials of Cell Culture, Know-How Transfer from the Red Biotechnology)

Neta Lavon, Sagit Shalel-Levanon, and Ayelet Maor-Shoshani

Contents

Abstract This chapter focuses on the fundamentals of animal cell culture for cultivated meat production. We will explore how cultivated meat companies approach cell cultivation to align with the industry's goals of creating food products that meet consumer expectations for taste, cost, and safety. Building on the foundations established by the "red biotech" industry, we have adapted many of its technologies but have also had to make significant advancements to address the unique requirements of food production. The chapter covers the current state of the art and highlights future developments in the field. We will explain the basics of cell culture, including production facilities, cell lines, cell culture growth conditions, and cell culture media. Special emphasis will be placed on growth media optimization,

N. Lavon (✉), S. Shalel-Levanon, and A. Maor-Shoshani
Research and Development Department, Aleph Farms, Rehovot, Israel
e-mail: neta.lavon@aleph-farms.com

which remains one of the most critical areas of development. We will discuss general approaches for medium optimization including analytical methods for monitoring key quality attributes. Finally, we will present examples of optimization processes we performed in our platform using bovine embryonic stem cells for large-scale cell mass production.

Graphical Abstract

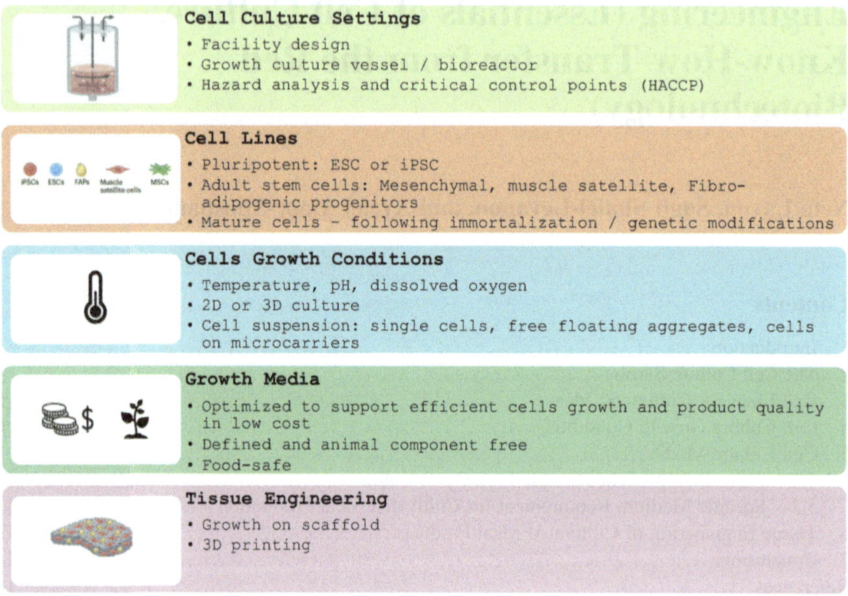

Keywords Cell culture, DoE, Food safety, Medium optimization, Tissue engineering

Abbreviations

ACF	Animal component free
bESC	Bovine embryonic stem cell
BSE	Bovine spongiform encephalopathy
DO	Dissolved oxygen
DoE	Design of experiment
ESC	Embryonic stem cell
FAP	Fibro-adipogenic progenitor
GCR	Glucose consumption rate
GRAS	Generally recognized as safe
HACCP	Hazard analysis and critical control points

iPSC	Induced pluripotent stem cell
MSC	Mesenchymal stem cell
PCR	Polymerase chain reaction
SNP	Single nucleotide polymorphism
STR	Short tandem repeat
TCA	Tricarboxylic acid
TSEs	Transmissible spongiform encephalopathies

1 Introduction

Cultivated meat production is a novel way to produce meat without the need to breed, raise, and slaughter animals. To meet the demand to feed 10B by 2050, we plan to grow only the edible part of the animal and supply the world with nutritional protein-rich food that is produced in a much more sustainable manner. By depleting the need to grow the whole animal, we save resources of water, land, and crops and reduce the amount of methane, manure, and all the waste products of the meat industry. Countries with limited land resources and/or high temperatures with limited capacity to grow animals would be able to grow cultivated meat and increase their food security. Moreover, the process is performed in a controlled manner and allows full traceability from the initial cell that started the culture to the final meat product and increases food safety.

Like all scientists, we learn how to produce cultivated meat by observing the way that nature does that. We explore the requirements to allow animals to grow, or more specifically their body mass increase due to cell proliferation. These conditions are the starting point for our industrial commercial cultivated meat production process, and we optimize these cell culture processes to outperform. Animals were not created to be our food; we made them our food because we could. On top of the ethical burden, they are not optimal for that purpose. By optimization, we can design a process of growing animal cells to become efficient as much as possible. Working in controlled conditions in bioreactors (temperature, pH, dissolved oxygen) with pre-defined cells and feed avoids the inefficiencies related to changing weather conditions, animals' health, or adventitious agents. The whole process is indifferent to external noncontrolled changes and monitored 24/7 to trace any changes if occurs and immediately eliminate their effects.

Most amazing for us cultivated meat scientists is the fact that we are not limited to evolutionary boundaries; we can design our novel process and product to have specific characteristics, as long as we promise its safety. Our goal is to bring the nutrition that is required to feed the growing world population, especially proteins, while promising costs and taste that meet consumers' expectations to promise the mass market repeated purchases. Accordingly, most cultivated meat products today are hybrid products combining plant-based ingredients with animal cells, a product that combines the best of both kingdoms, animals and plants. These products were designed to meet the nutrition, taste, and cost required by the consumers and increase the supply of protein-rich meat products for the needs of the growing population.

Moreover, as designers, we can genetically modify the cells and the ingredients used in the process of growing the animal cells in culture, aiming to increase the efficiency of cultivated meat production. For example, we can design cells to proliferate in an unlimited manner, to grow in single cell suspension, to reduce their sensitivity to shear stress, to produce their own feed, and to reduce the production of waste products. Furthermore, in cases where the mechanism of action is known, we can replace specific native molecules with small molecules, RNA, or enhanced proteins that activate the same mechanism with a much more efficient manner due to higher stability or better receptor binding. Most of these processes and ingredients are still in development and would probably be part of the future cultivated meat products once we can exhibit the reproducibility and the safety of the resulting meat products.

Growing animal cell cultures was first demonstrated by the pioneering work of Ross Granville that grew in vitro frog nerve fibers back in 1906 [1]. Since then, many developments have occurred, and the use of cell cultures to produce commercial products in the biotechnology industry caused a dramatic evolution in the whole field. The first products were part of the "red biotechnology," with the use of biotechnology in the medical and pharmaceutical industries, health preservation, production of vaccines and antibiotics, regenerative therapies, and creation of artificial organs. To support the commercial production of these products, many advancements happened in the last 30 years at all aspects improving cell lines, growth media, carriers, coatings, vessels, bioreactors, perfusion systems, harvesting methods, and more. The same basics of "red biotechnology" in cell cultures are used in the production of cultivated meat, taking into account the specific considerations of making food products and their unique proposition. In the below sections, we will detail the specific requirements for producing cultivated meat in cell cultures.

2 The Cell Culture Setting

Ensuring safe and efficient production of high-quality biological material in cell culture begins with facility design to promise good practices. It requires qualifications of each material included in the process to assure contaminant-free culture. The work performed in a biological safety cabinet needs to provide an aseptic environment for the product while protecting the operator from aerosols. In the R&D phase and in small-scale animal cultures, working in a cabinet of class 2 is adequate. The level of the working environment, room quality and protection gear, is set according to the specific cell line and materials to be handled taking into consideration potential hazards to the product, the operator, and the environment. In larger scales of production, the work is performed in bioreactors in a closed environment. At that production stage, all outputs and inputs are connected in an aseptic manner to promise that no contaminants can enter the bioreactor. The production room air quality, while working in closed systems, should be adequate to support sterile production. Like in "red biotechnology," also with cultivated meat, prior to production, a risk assessment should be performed aiming to ensure the safety of food product to human consumption. In the food industry, this is done using the hazard analysis and critical control points (HACCP) plan to identify potential physical, chemical, or biological hazards during the production process [2].

3 Cell Lines for Cultivated Meat

As the famous chef Thomas Keller once said, "The final product can only be as good as the ingredients you start with"; cultivated meat quality depends on the source of cells that we use to grow the meat from. The quality of your cell line is critical to promise efficient and reproducible process. The properties of the cell line should meet the requirements of the process and product in line with the company's product and strategy. In the first 8 years since the cultivated meat ecosystem evolved, most cultivated meat companies derived their own cell lines. In the last 5 years, we can also find companies that are selling cell lines for cultivated meat producers. Whether you derive the cell line yourself or purchase it from an external supplier, first, you need to characterize them thoroughly for the cell line identity. Techniques routinely used include DNA barcoding, single nucleotide polymorphism (SNP) analysis, isoenzyme analysis, or short tandem repeat (STR) profiling. Then you must establish a hierarchy of master and working cell banks, cryopreserved in liquid- or gas-phase nitrogen, that are demonstrated free from microbial, mycoplasma, and viral contaminations [3]. Several cell types can serve as cell sources for cultivated meat [4]. Most companies use cell lines that promise continuous cultures such as stem cells or transformed cells to promise an almost unlimited availability. The main advantage of pluripotent cells, such as embryonic stem cells (ESCs) [5] and induced pluripotent stem cells (iPSCs) [6], is their indefinite proliferative capabilities and high growth rate, which are essential for the high population doubling levels required throughout the cultured meat production process and to establish cell banks to avoid repeated biopsy collection. However, derivation of ESCs can be challenging, their genetic stability needs to be confirmed, and although they can differentiate into nearly any cell type, differentiation protocols can be complex. On the other hand, adult stem cells (muscle satellite cells [7, 8], mesenchymal stem cells (MSCs) [9], and fibro-adipogenic progenitor (FAP) cells [10, 11]) are multipotent and can differentiate relatively easily into a selected number of cell types, as dictated by the organ or tissue from which they were isolated. However, their proliferation is limited to 50–60 divisions and is dependent on the animal age according to the Hayflick limit [12]. As a result, multiple biopsies must be performed to support a commercial cultured meat product, which introduces variability into the process, requiring repeated costly characterization tests to confirm reproducibility and safety. Mature cells have limited proliferation capacity and can only be used as source for cultured meat if they are genetically modified or immortalized [13]. Cell lines become immortal when they lose control of their cell cycle checkpoints and circumvent the process of senescence. Immortal cell lines can be established by forced expression of catalytic subunit of telomerase (TERT) or by inactivation of P53/P14/RB through the expression of specific viral genes or through a process of spontaneous immortalization [14].

4 Cell Culture Growth Conditions

To produce cultivated meat in scale, we need to find the best conditions that support our process, allowing the cells to proliferate and, in some cases, to further differentiate in culture. We should supply the cells with optimal physical conditions such as temperature, pH, dissolved oxygen, and substrate stiffness, and the right biological signals coming from the culture medium ingredients and materials they adhere to. Some signals come from the neighboring cells, and accordingly control of cell concentration and ratios between cell types while co-culturing further needs to be optimized.

Most animal cell lines, especially mammals, routinely grow in 37°C, pH 7.0–7.4, and 20–60% air saturation. To optimize growth and meet commercial production, it is advised to expand the boundaries of these parameters to find the specific parameters for each cell line. For example, some animals' body temperature is higher than 37°C, and accordingly it is advised to test which temperature enhances the cells' growth.

The characteristics of the cell line will set their mode of culturing methodology. Most animal cells are anchorage dependent, as within the animal, they grow attached to a matrix that supplies them with the microenvironment and the essential physical and biological signals for cell cycle progression. In small scales, it is done in monolayers in tissue culture plates. In larger scales, microcarriers are routinely used and supply the cells with the same tissue culture surface to adhere to and allow their growth in suspension. In some cases, as with bovine embryonic stem cells (bESCs), interactions between the cells themselves support their growth, in free-floating aggregates, replacing the requirement for anchorage dependency [15].

The second category of cells is anchorage independent. These cells can grow in single cell suspension. Such growth allows easier scaling up for commercial applications, allowing high densities of cells, and omits the need for cells' dissociation. In animals, we cannot find cells that proliferate in single cell suspension. Even if blood cells only circulate in the bloodstream but do not divide in suspension, they divide in organs such as bone marrow and spleen. Accordingly, we need to actively change the cells that we have isolated from the animals by either transformation or adaptation, to create such cell lines.

The cells' growth medium is composed of many nutrients, and accordingly, it is part of the environment that can mostly influence the cells' state and the process efficiency. There are various commercially defined basal media types. Originally, balanced salt solutions were used to maintain mammalian cells. Animal serum was added to the salt solutions, as a rich source of nutrients to support the cell growth. Since then, many advancements have happened in the field, and today, we can completely define the media formulation to specifically support each cell needs. Due to the major influence of the growth media composition on the cultivated meat production and the unique requirements of its composition to meet cost and safety measures of cultivated meat products, we have allocated more focus on this aspect in this chapter.

5 Cell Culture Media

5.1 General Approaches for Medium Optimization

Media optimization can solve the two major challenges of the cultivated meat industry: increased production yield and decrease in production costs. Media optimization includes approaches focusing on removing components that are not essential for cell growth, increasing the concentration of nutrients that contribute to cell growth, and reducing the production of waste products that inhibit the growth of cells [16]. In other words, the main challenge is to identify the right balance of metal ions, vitamins, amino acids, lipids, proteins, etc. in culture medium, which is essential for controlling the desired cell metabolism, growth, and phenotype.

It is widely recognized that medium composition can affect different performance of various cell lines and clones [17, 18]. This includes cell lines from the same species and type. Therefore, platform medium formulation for all cell lines is an effective method only when the platform medium formulation is used during clone selection, where a few clones with the highest growth and productivity are selected among thousands of clones. In the cultivated meat industry, isolating clonal embryonic stem cell lines poses a challenge, typically resulting in a single cell line per embryo, and accordingly, such screening platform is not relevant. As a result, optimizing the medium remains the preferred and recommended method in process development within the cultured meat industry. Moreover, various cell types, such as ESC, iPSC, MSC, FAPs, and muscle satellite cells, can be utilized in the production of cultivated meat [4, 19, 20], necessitating a medium that supports the growth and characteristics of these specific cells. Additionally, different media formulations are required for various stages of the process, like cell propagation or differentiation.

In biomanufacturing, to optimize a medium effectively, it is essential to have a comprehensive understanding of key metabolic pathways, particularly those influencing growth, productivity, and product quality. For instance, the quality of monoclonal antibodies can affect their potency, efficacy, and antibody-dependent cellular cytotoxicity [21]. Therefore, medium components that improve critical quality attributes might be prioritized, even if it means sacrificing higher productivity [22]. In the cultivated meat industry, critical quality attributes could include pluripotency and differentiation states. Maintaining the cells at a pluripotent stage is essential for process scale-up, as only when the cells are pluripotent can they propagate indefinitely while maintaining their ability to differentiate to a specific lineage, once given the right conditions. Consequently, in the production of cultivated meat product, components that enhance pluripotency and differentiation to mesodermal lineage might be preferred, even if they result in lower yields (see example: Pluripotency Monitoring).

Example: Pluripotency Monitoring
Pluripotency can be assessed by various analytical methods, including flow cytometry and real-time polymerase chain reaction (PCR) for pluripotent markers [15]. Aleph Farms developed a platform to assess the impact of materials on the quality attributes of bESC and their differentiated derivatives in aggregates grown in suspension culture. We test pluripotency in bESC and for expression of mesodermal markers in differentiated cells. This platform involves immunostaining with antibodies targeting pluripotency or early mesoderm markers, followed by fluorescent confocal imaging and quantitative analysis of the marker's fluorescent signal normalized to the DAPI signal (for cell nuclei) in whole aggregates. The platform analyzes several dozen aggregates to ensure reliable results, enabling an accurate assessment of the culture state. The benefit of this platform over other protein or gene expression quantification methods, such as flow cytometry or real-time PCR, is that it does not require aggregate disaggregation. Moreover, analyzing whole aggregates provides important information about the distribution of marker expression within aggregates.

As can be seen in Fig. 1a, in a culture grown in conditions which support pluripotency, OCT4 (pluripotency marker) expression is high, as indicated by the red fluorescent signal obtained for aggregates formed in this condition. When the conditions are less favorable for pluripotency, the OCT4 red signal decreases, while the DAPI staining (for cells' nuclei) becomes more prominent. OCT4 signal normalized to DAPI provides quantitative analysis for the culture state, as shown in Fig. 1b.

The initial step of each medium optimization process involves a thorough analysis of the conditioned cell line medium (i.e., the supernatant of the cultivated cells, also known as spent medium). The analysis should encompass components that play roles in metabolic pathways including glycolysis, tricarboxylic acid (TCA) cycle, pentose phosphate pathway, amino acid metabolism, lipid and fatty acid metabolism, polyamine metabolism, nucleotide metabolism, and essential signaling pathways that regulate pluripotency and differentiation. For instance, pluripotent stem cells are noted for having underdeveloped mitochondria; hence, during their proliferation, glycolytic genes are upregulated, while pyruvate dehydrogenase is downregulated, restricting pyruvate's entry into the TCA cycle [23]. Consequently, optimizing pluripotent stem cell growth should prioritize energy sources other than the TCA cycle, such as glycolysis, glutaminolysis, and pentose phosphate pathway.

Based on spent medium analysis and in-depth screening of the dominant metabolic pathways, studies should be designed to evaluate the effect of different components on cell growth and phenotype by manipulating the rates of different

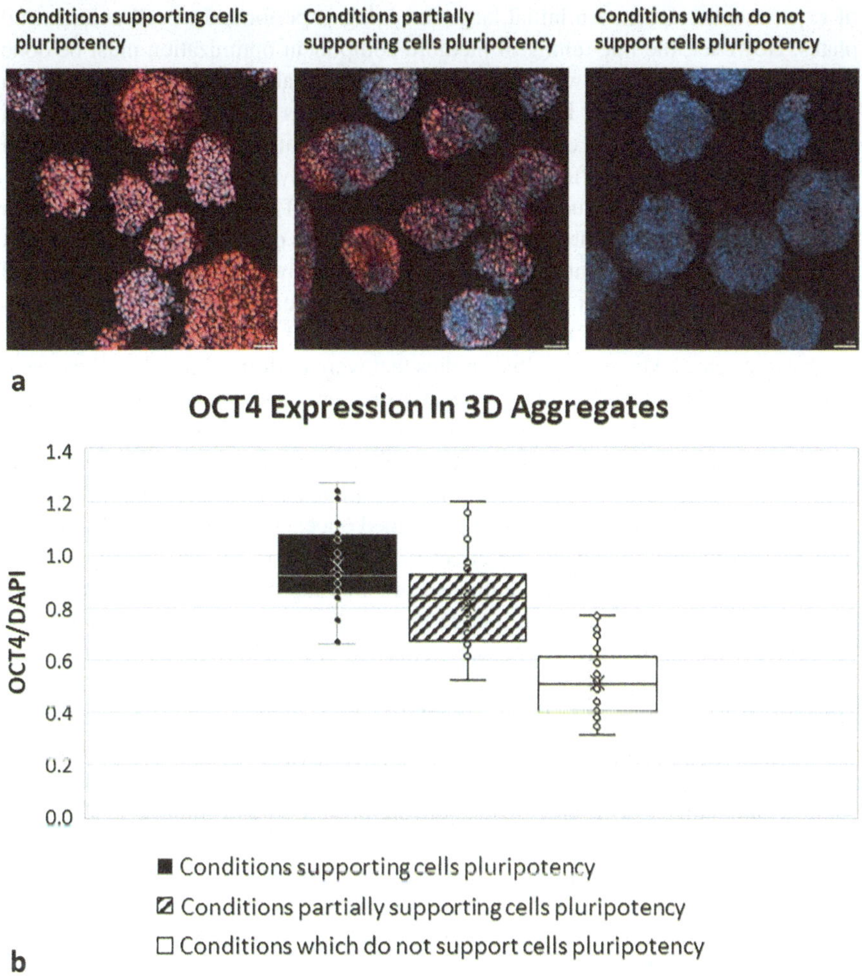

Fig. 1 OCT4 expression in aggregates. Whole-aggregate analysis for OCT4 expression by fluorescent confocal analysis. (**a**) One field of aggregates imaged by confocal microscope. Red – OCT4 signal; blue – DAPI signal. (**b**) Quantitative analysis of OCT4 fluorescent signal normalized to DAPI signal for about 30 aggregates from each condition

metabolic pathways. This can be achieved by increasing the concentration of substrates, increasing enzyme activity (by supplementing, for example, cofactors), and decreasing or removing the effect of inhibitors.

The effects of different components can be studied using one factor at a time. However, the most efficient method is to study the effect of multiple components using a design of experiment (DoE) approach [24]. The use of DoE approach enables the evaluation of the effect of each component alone and the interaction between additives. Statistical software such as Design-Expert, JMP, and Minitab is typically used to design and analyze DoE studies, ensuring robust and reliable interpretation of experimental results. An initial large screening is performed in cell culture well plates. However, the fine-tuning of medium component optimization must be done in small-scale bioreactors, which are used as a downscale model for the production bioreactor system. This is because process parameters such as pH and dissolved oxygen, which are tightly controlled in bioreactors but are not controlled in well plates, may significantly affect the outcome of the study.

An example of a DoE study is presented in Table 1. The purpose of the study was to determine the effect of three amino acids identified as depleted during spent media analysis and thus as potential candidates for cell growth improvement. The full factorial design experiment was conducted in minibioreactors (Ambr® 250, Sartorius), testing two levels of each amino acid (low level, 0; and high level, 1). The levels were selected based on initial studies that were performed in six well plates in suspension culture, using animal component-free and chemically defined medium. Figure 2 presents an example of data analysis (using JMP 18.1.1) that evaluates the effect of these three amino acids on glucose consumption rate (GCR), which is correlated with cell growth. An increase in GCR indicates an increase in cell growth. In this DoE study, each condition was performed once; however, as shown in Fig. 2, this was sufficient to evaluate the effect of each amino acid and their combined effect. As shown in Fig. 2, when amino acid #1 was supplemented at a low level (0), the GCR levels were low regardless of the levels of the other two amino acids. However, when amino acid #1 was supplemented at a high level (1), an interaction was observed between this amino acid and amino acid #2. At the lowest level of amino acid #2 (0), the GCR levels were low, similar to the levels obtained at the lowest level of amino acid #1. However, when amino acid #2 was supplemented at a high level (1), a significant increase in GCR levels was observed. These results suggest that to increase the GCR level, both amino acid #1 and amino acid #2 should be supplemented at a high level (1).

The results of the study also emphasize the benefit of using a DoE study over a one-factor-at-a-time study for medium optimization, as a one-factor-at-a-time study would have resulted in the conclusion that neither amino acid #1 nor amino acid #2 had an effect on GCR, since only when these amino acids are supplemented together at the highest concentrations was an effect on GCR level observed.

Table 1 Full factorial design experiment, for the evaluation of the effect of three amino acids

Condition #	Amino acid #1	Amino acid #2	Amino acid #3
1	0	0	0
2	0	0	1
3	0	1	0
4	0	1	1
5	1	0	0
6	1	0	1
7	1	1	0
8	1	1	1

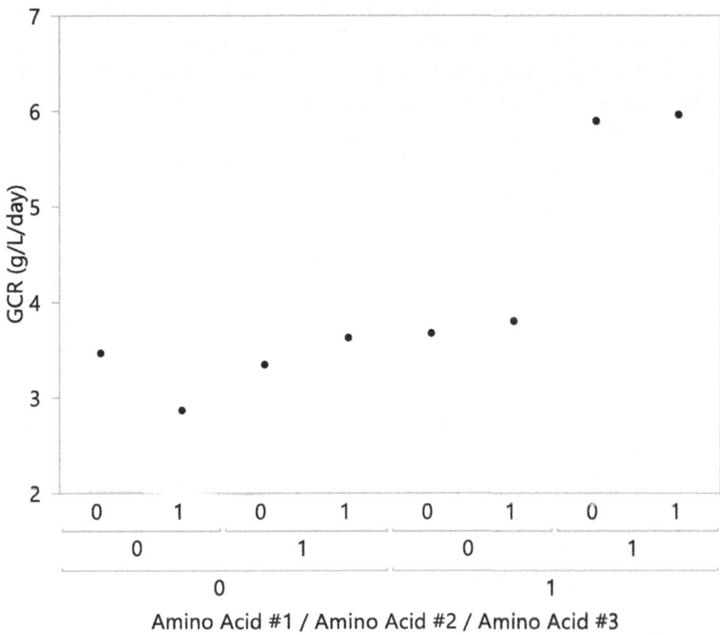

Fig. 2 The effect of three amino acids on GCR. *X*-axis: external scale – amino acid #1; middle scale – amino acid #2; internal scale – amino acid #3. 0 low level; 1 – high level

5.2 *Specific Medium Requirement for Cultivated Meat Production*

There are several unique medium requirements for cultivated meat production to meet the food industry costs, safety requirements, and consumer acceptance.

Firstly, the cost of the media must be 100 to 1,000 times lower than that used in "red biotechnology" to ensure that the resulting meat product is affordable and comparable in price to conventional meat [25]. To achieve this, strategies include minimizing the number and concentration of medium components necessary for cell growth and characteristics, particularly by reducing the use of recombinant proteins. Since recombinant proteins significantly impact the cost of cultivated meat media, obtaining them from companies that have optimized and scaled up their production processes for the cultivated meat sector can greatly reduce costs [25, 26]. These companies have achieved a 10- to 5,000-fold reduction in production costs compared to pharmaceutical-grade recombinant protein production, approaching the low-cost targets set by the GFI paper in 2020 [25]. Another approach is to substitute costly materials that are used in relatively high concentration, with lower cost alternatives that can fulfil the same role in the medium [17].

Another distinctive aspect of cultivated meat medium is the involvement of ethical considerations reflecting consumer acceptance. Avoiding the use of animal-derived components is aligned with cultivated meat goals to reduce the need for growing animals for human consumption. This has prompted many cultivated meat companies to develop serum-free [6, 27] or an animal component-free (ACF) and/or chemically defined medium, as we use in Aleph Farms. Instead of using animal source ingredients, ACF medium is composed of ingredients such as recombinant proteins and synthetic or plant-based alternatives. On the top of the ethical issue, ACF medium brings another value to the cultivated meat production – it is considered safer because it does not include animal-sourced materials that can introduce adventitious agents into the process. However, ACF medium can present difficulties for cell culture, particularly when serum or serum albumin is removed, as this increases the cells' sensitivity to trace materials whose negative effect is otherwise masked by the serum.

Finally, materials used in cultivated meat production must be food-safe [2, 18, 28]. This involves ensuring that media components are recognized as food ingredients or generally recognized as safe (GRAS) for the food industry, using food-grade components produced in food-grade facilities, and complying with food safety standards, such as allergen, transmissible spongiform encephalopathies (TSE)/bovine spongiform encephalopathy (BSE), and heavy metal testing. Consequently, several media companies have started the process of replacing components of DMEM/F12, a basal medium most frequently used for many cultivated meat processes, with food-grade components, and have begun producing media in food-grade facilities to achieve food-grade certification [29]. Additionally, many recombinant protein manufacturers have transitioned their production processes to food-grade facilities and adopted food quality methods to certify their recombinant proteins as food-grade products.

As discussed thoroughly, media materials should support efficient cell growth and characteristics. Food-grade materials should also meet these requirements. Verification that certain contaminants that might be present in them like metals, heavy metals, and endotoxins, which are safe for food but could potentially be harmful to the cell culture, is required.

Consequently, extensive testing is necessary when transitioning from pharmaceutical- or R&D-grade materials, which are often used for cell culture, to food-grade materials, to ensure a consistent and effective process. This evaluation can be done as discussed in Sect. 5.1. Whether a single or a group of components were tested, it is highly recommended to eventually test all the food-grade components collectively, to verify that there is no cumulative effect of contaminants negatively affecting cell growth or characteristics.

6 Tissue Engineering of Cultivated Meat Products

When the cultivated meat field first emerged, it evolved from research labs focused on muscle tissue formation for medical applications [30]. Naturally, early efforts assumed that producing meat, which consists primarily of muscle, would require similar tissue engineering processes, and most research initially concentrated on mimicking these processes using animal cells [31] and edible scaffolds [32]. Over the years, various groups have developed edible scaffolds for cultivated meat using diverse fabrication methods that yield a range of stiffness and structural properties to support muscle and fat cell growth. Materials such as gelatin, collagen, soymilk, peanut protein, alginate, and chitosan, used individually or in combination, have not only facilitated tissue development but also contributed to the nutritional and organoleptic qualities of the final product [33–35]. To avoid the need for de novo scaffold fabrication, researchers have proposed using decellularized plant tissues and autoclaved vegetables as naturally structured, edible scaffolds [36, 37]. Aleph Farms' first product is a tissue-engineered cultivated meat that incorporates a porous scaffold made of textured vegetable protein, supporting a mass of bovine cells growing within it. It is the first cultivated beef product in the world to receive regulatory approval for marketing (https://www.gov.il/en/pages/17012024-02).

As the cultivated meat ecosystem developed, however, the focus shifted. The industry refined its approach to meet the overarching goal of providing the nutrition, particularly protein, needed to feed the growing global population, while also achieving the cost and taste profiles necessary to drive widespread consumer adoption. As a result, most cultivated meat products today are hybrid products: they combine plant-based ingredients with animal cells and typically do not involve growing cells on scaffolds to form structured tissues. These hybrids are designed to deliver the required nutrition, taste, and affordability, helping to expand the supply of protein-rich foods to meet increasing demand.

Nevertheless, it remains essential to explore tissue engineering approaches for cultivated meat and explore potential advantages. Growing structured tissues could enable further cost reductions through the cross talk between different cell types and the extracellular matrix in co-cultures [4]. Tissue formation could also introduce new organoleptic experiences thanks to the variety of cells involved.

Currently, one major challenge of tissue engineering is the difficulty of growing thick tissues, primarily due to limitations in nutrient and oxygen diffusion. More recently, 3D printing technologies have enabled the creation of thicker cuts [38]. Furthermore, additive printing enables precise placement of different cell types, such as muscle and fat, within a single construct, customizing the organization according to specific design needs [39].

As these methods advance toward higher levels of automation and production scalability, it is expected that they will lower costs to meet food industry standards. Integrating such technologies will not only expand the cultivated meat product portfolio but also offer consumers new and unique organoleptic experiences.

7 Conclusion

Extensive knowledge accumulated in the "red biotechnology" regarding cell culture techniques and media optimization has established the basis for the cultured meat industry. However, in each of these cell culture techniques, there are unique challenges to the cultured meat industry that require further development and adjustment. This includes choosing the right cell type for the production process, media optimization, and tissue engineering considerations.

The main requirement from cells used for cultivated meat production is to reach a high biomass. Several cell types are commonly used for cultivated meat production, each cell type has its pros and cons, and the cultivated meat manufacturer chooses the right cell type for its process and goals. While ESCs have potentially indefinite division capability which is required to reach the high cell mass cultivated meat process dictates, their derivation and differentiation protocols are relatively complicated. On the other hand, mesenchymal, muscle satellite, and FAPs require easier derivation and differentiation protocols, but their proliferation is limited to a certain number of divisions.

Regarding media optimization, although the basis for cultivated meat growth medium can be common to certain cell types and processes, each species, cell type, cell line, as well as process stage requires specific optimization. Medium optimization is essential to reach maximal biomass formation while supporting cell characteristics to allow scale-up at a reasonable commercial cost. In addition, media for cultivated meat production should be food-safe, and preferably animal component free, to align with cultivated meat goals to reduce the need for growing animals for human consumption.

Finally, tissue engineering of cultivated meat products could be replaced with a hybrid product that maintains nutritional value and organoleptic experience at a simpler and at a lower cost process.

Ongoing advancements in red biotechnology are poised to enhance multiple dimensions of cultivated meat production, from cellular efficiency to bioprocessing scalability; conversely, progress within the cultivated meat sector, particularly in driving down costs and improving production efficiencies, stands to reciprocally inform and accelerate developments across the broader red biotechnology domain.

References

1. Harrison RG (1906) Observations on the living developing nerve fiber. Proc Soc Exp Biol Med 4:140–143
2. Ong KJ, Johnston J, Datar I, Sewalt V, Holmes D, Shatkin JA (2021) Food safety considerations and research priorities for the cultured meat and seafood industry. Compr Rev Food Sci Food Saf 20:5421
3. Sigma-Aldrich (2016) Fundamental techniques in cell culture: laboratory handbook. 3rd edn. Sigma-Aldrich Co. LLC

4. David S, Tsukerman A, Safina D, Maor-Shoshani A, Lavon N, Levenberg S (2023) Co-culture approaches for cultivated meat production. Nat Rev Bioeng 1:817
5. Bogliotti YS et al (2018) Efficient derivation of stable primed pluripotent embryonic stem cells from bovine blastocysts. Proc Natl Acad Sci USA 115:2090
6. Stout AJ et al (2022) Immortalized bovine satellite cells for cultured meat applications. bioRxiv. https://doi.org/10.1101/2022.12.02.518927
7. Ding S, Swennen GNM, Messmer T, Gagliardi M, Molin DGM, Li C, Zhou G, Post MJ (2018) Maintaining bovine satellite cells stemness through p38 pathway. Sci Rep 8:10808
8. Li B, Li P, Huang R, Sun W, Wang H, Li Q, Chen J, Wu W, Liu H (2015) Isolation, culture and identification of porcine skeletal muscle satellite cells. Asian Australas J Anim Sci 28:1171
9. Musina RA, Bekchanova ES, Belyavskii AV, Sukhikh GT (2006) Differentiation potential of mesenchymal stem cells of different origin. Bull Exp Biol Med 141:147
10. Biferali B, Proietti D, Mozzetta C, Madaro L (2019) Fibro-adipogenic progenitors cross-talk in skeletal muscle: the social network. Front Physiol 10:1074
11. Uezumi A, Fukada S, Yamamoto N, Takeda S, Tsuchida K (2010) Mesenchymal progenitors distinct from satellite cells contribute to ectopic fat cell formation in skeletal muscle. Nat Cell Biol 12:143
12. Hayflick L (1965) The limited in vitro lifetime of human diploid cell strains. Exp Cell Res 37: 614
13. Maqsood MI, Matin MM, Bahrami AR, Ghasroldasht MM (2013) Immortality of cell lines: challenges and advantages of establishment. Cell Biol Int 37:1038
14. Soice E, Johnston J (2021) Immortalizing cells for human consumption. Int J Mol Sci 22:11660
15. Zehorai E, Maor-Shoshani A, Molotski N, Dorojkin A, Marelly N, Dvash T, Lavon N (2023) From fertilised oocyte to cultivated meat – harnessing bovine embryonic stem cells in the cultivated meat industry. Reprod Fertil Dev 36:124
16. Gomez Romero S, Boyle N (2023) Systems biology and metabolic modeling for cultivated meat: a promising approach for cell culture media optimization and cost reduction. Compr Rev Food Sci Food Saf 22:3422
17. Combe M, Sokolenko S (2021) Quantifying the impact of cell culture media on CHO cell growth and protein production. Biotechnol Adv 50:107761
18. O'Neill EN, Cosenza ZA, Baar K, Block DE (2021) Considerations for the development of cost-effective cell culture media for cultivated meat production. Compr Rev Food Sci Food Saf 20:686
19. Reiss J, Robertson S, Suzuki M (2021) Cell sources for cultivated meat: applications and considerations throughout the production workflow. Int J Mol Sci 22:7513
20. Post MJ (2012) Cultured meat from stem cells: challenges and prospects. Meat Sci 92:297
21. Goetze AM, Schenauer MR, Flynn GC (2010) Assessing monoclonal antibody product quality attribute criticality through clinical studies. mAbs 2:500
22. Torkashvand F, Vaziri B (2017) Main quality attributes of monoclonal antibodies and effect of cell culture components. Iran Biomed J 21:131
23. Tanosaki S, Tohyama S, Kishino Y, Fujita J, Fukuda K (2021) Metabolism of human pluripotent stem cells and differentiated cells for regenerative therapy: a focus on cardiomyocytes. Inflamm Regener 41:5
24. Mandenius C, Brundin A (2008) Bioprocess optimization using design-of-experiments methodology. Biotechnol Prog 24:1191
25. Specht L (2020) An analysis of culture medium costs and production volumes for cultivated meat. White Paper. The Good Food Institute
26. Venkatesan M, Semper C, Skrivergaard S, Di Leo R, Mesa N, Rasmussen MK, Young JF, Therkildsen M, Stogios PJ, Savchenko A (2022) Recombinant production of growth factors for application in cell culture. iScience 25:105054
27. Messmer T, Klevernic I, Furquim C, Ovchinnikova E, Dogan A, Cruz H, Post MJ, Flack JE (2022) A serum-free media formulation for cultured meat production supports bovine satellite cell differentiation in the absence of serum starvation. Nat Food 3:74

28. Manning L (2024) Responsible innovation: mitigating the food safety aspects of cultured meat production. J Food Sci 89:4638
29. Kanayama T, Tanaka K, Shimada H, Ichiyama K, Higuchi S, Kawashima I (2022) Improving the safety of cultured meat using basal medium prepared using food ingredients. bioRxiv. https://doi.org/10.1101/2022.03.28.486069
30. Martins ALL, Giorno LP, Santos AR (2022) Tissue engineering applied to skeletal muscle: strategies and perspectives. Bioengineering 9:744
31. Langelaan MLP, Boonen KJM, Polak RB, Baaijens FPT, Post MJ, Van Der Schaft DWJ (2010) Meet the new meat: tissue engineered skeletal muscle. Trends Food Sci Technol 21:59
32. Ben-Arye T, Shandalov Y, Ben-Shaul S, Landau S, Zagury Y, Ianovici I, Lavon N, Levenberg S (2020) Textured soy protein scaffolds enable the generation of three-dimensional bovine skeletal muscle tissue for cell-based meat. Nat Food 1:210
33. Li C-H, Yang I-H, Ke C-J, Chi C-Y, Matahum J, Kuan C-Y, Celikkin N, Swieszkowski W, Lin F-H (2022) The production of fat-containing cultured meat by stacking aligned muscle layers and adipose layers formed from gelatin-soymilk scaffold. Front Bioeng Biotechnol 10:875069
34. Zheng Y-Y, Chen Y, Zhu H-Z, Li C-B, Song W-J, Ding S-J, Zhou G-H (2022) Production of cultured meat by culturing porcine smooth muscle cells in vitro with food grade peanut wire-drawing protein scaffold. Food Res Int 159:111561
35. Li L et al (2022) Chitosan-sodium alginate-collagen/gelatin three-dimensional edible scaffolds for building a structured model for cell cultured meat. Int J Biol Macromol 209:668
36. Young JF, Skrivergaard S (2020) Cultured meat on a plant-based frame. Nat Food 1:195
37. Liu Y et al (2025) Growing meat on autoclaved vegetables with biomimetic stiffness and micro-patterns. Nat Commun 16:161
38. Zagury Y, Ianovici I, Landau S, Lavon N, Levenberg S (2022) Engineered marble-like bovine fat tissue for cultured meat. Commun Biol 5:927
39. Ianovici I, Zagury Y, Afik N, Hendel M, Lavon N, Levenberg S (2024) Embedded three-dimensional printing of thick pea-protein-enriched constructs for large, customized structured cell-based meat production. Biofabrication 16:045023

Adv Biochem Eng Biotechnol (2026) 193: 17–62
https://doi.org/10.1007/10_2025_292

Published online: 15 October 2025

Bioprocess Engineering for Cultivated Meat

Julia Bernfeld, Simon Zschieschang, Marline Kirsch, Jordi Morales-Dalmau, Jasper Liedtke, and Antonina Lavrentieva

Contents

J. Bernfeld, S. Zschieschang, and J. Liedtke
Institute of Technical Chemistry, Leibniz University of Hannover, Hannover, Germany

Cultimate Foods GmbH, Berlin/Hannover, Germany

M. Kirsch and J. Morales-Dalmau
Cultimate Foods GmbH, Berlin/Hannover, Germany

A. Lavrentieva (✉)
Institute of Technical Chemistry, Leibniz University of Hannover, Hannover, Germany
e-mail: lavrentieva@iftc.uni-hannover.de

Abstract Cultivated meat (CM) has emerged as a promising solution to the environmental, ethical, and food security concerns associated with conventional meat production. However, realising its full potential depends on developing robust, scalable, and cost-effective bioprocessing strategies. This chapter explores the entire CM bioprocessing pipeline, from upstream challenges such as the development of robust cell lines, scaling up cell cultures, and evaluating different operating modes like batch, fed-batch, and perfusion to downstream processes such as biomass harvesting, purification, and product structuring. Key considerations include ensuring sterility, monitoring critical process parameters, and supporting effective cell proliferation and differentiation. Economically, the high cost of media and capital expenditures remain a major barrier to scale. Strategies such as media recycling, bulk ingredient sourcing, and in-house production are being explored to reduce costs. Emerging technologies like artificial intelligence, machine learning, and digital twins offer new tools for optimising operations, though adoption is still in early stages. As the industry progresses toward commercialisation, continued innovation in bioprocess engineering, guided by techno-economic modelling, will be crucial to achieving scalable and sustainable meat alternatives.

Graphical Abstract

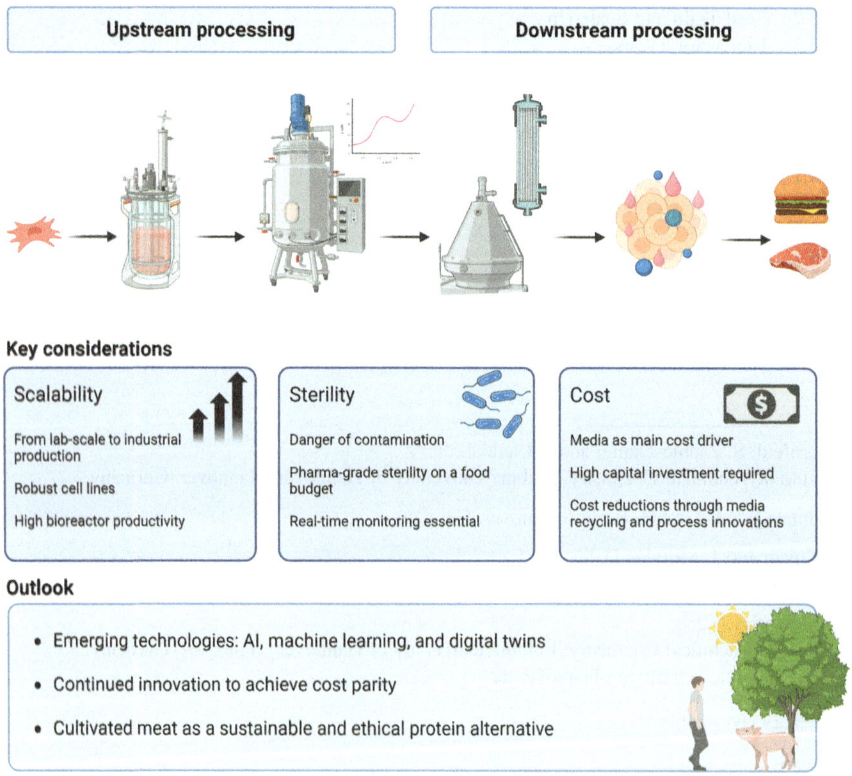

Keywords Cell culture, Cultivated meat, Downstream processing, Process control, Scale-up, Upstream processing

Abbreviations

2D	2-Dimensional
3D	3-Dimensional
7-AAD	7-Aminoactinomycin D
AI	Artificial intelligence
ATF	Alternating tangential flow
BODIPY	Boron-dipyrromethene
CAPEX	Capital expenditure
CFD	Computational fluid dynamics
cGMP	Current good manufacturing practice
CHO	Chinese hamster ovary
CIP	Clean-in-place
CM	Cultivated meat
CO_2	Carbon dioxide
CRE	Carbapenem-resistant Enterobacteriaceae
dCO_2	Dissolved carbon dioxide
DO	Dissolved oxygen
DSP	Downstream processing
E&L	Extractables and leachables
ELISA	Enzyme-linked immunosorbent assay
ERK	Extracellular signal-regulated kinase
ESCs	Embryonic stem cells
FATS	Fast Adipogenesis Tracking System
FBS	Foetal bovine serum
FDA	Food and Drug Administration
FGF2	Fibroblast growth factor 2
GRAS	Generally Recognized as Safe
HFB	Hollow-fibre bioreactor
hMSC	Human mesenchymal stem cell
HPTS	Hydroxypyrene trisulfonic acid
IBMX	3-Isobutyl-1-methylxanthine
iPSCs	Induced pluripotent stem cells
kLA	Volumetric mass transfer coefficient
LOD	Limit of detection
MEK	Mitogen-activated protein kinase kinase
ML	Machine learning
mL	Millilitre
MRF4	Myogenic regulatory factor
MRSA	Methicillin-resistant *Staphylococcus aureus*
MSCs	Mesenchymal stem cells

MYF5	Myogenic factor 5
MYH1	Myosin 1
MYH2	Myosin 2
MYOD	Myoblast determination protein 1
NIR	Near-infrared
NJS	Just-suspended speed
OD	Optical density
ORO	Oil Red O
P/V	Power per unit volume
PAT	Process Analytical Technology
PAX7	Paired box 7
PBR	Packed-bed bioreactor
PCR	Polymerase chain reaction
PID	Proportional–integral–derivative
PPARγ	Peroxisome proliferator-activated receptor gamma
PTFE	Polytetrafluoroethylene
QbD	Quality by design
R&D	Research and development
RPMs	Revolutions per minute
RTDs	Resistance temperature detectors
RT-qPCR	Reverse transcription quantitative polymerase chain reaction
RXR	Retinoid X receptor
SIP	Steam-in-place
STR	Stirred-tank bioreactor
TEAs	Techno-economic analyses
TFF	Tangential flow filtration
TFPP	Meso-tetra(pentafluorophenyl)porphyrin
VRE	Vancomycin-resistant Enterococcus

1 Introduction

Bioprocessing is defined as the use of living sources, such as cells or their components, to produce desired products at scale through a carefully controlled process [1]. It is crucial in many industries, including pharmaceuticals, biofuels, food, and beverages, but its application in the field of cultivated meat (CM) presents a new and unique challenge. CM involves growing cells outside of an animal to produce muscle and fat tissues that mimic conventional meat products, without the need for animal breeding, husbandry, and slaughter [2].

The potential benefits of CM are powerful and include reduced emission of greenhouse gases, improved food security, and elimination of animal welfare concerns [3–7]. However, these promises rely on solving a critical bottleneck: scaling up the production from laboratory to industrial level. The laboratory-scale production of CM has been technically feasible for over a decade, as has been famously demonstrated by Mark Post in 2013. Yet, to have a real impact on the food industry and reap the environmental, ethical, and health benefits, it is necessary to translate lab-based processes into a much larger, industrial scale. For that, robust, food-safe, and cost-effective bioprocessing strategies are required. In contrast to the pharmaceutical industry, where the high margins facilitate complex and sterile production, CM must meet stringent regulatory and safety standards while remaining economically viable for mass consumption [4, 7].

This chapter explores the distinct bioprocessing challenges and opportunities in the production of CM. It delves into the considerations of upstream and downstream processing, process monitoring, cost reduction strategies, and emerging innovations. We aim to create a road map for engineering scalable and sustainable bioprocessing systems that will enable the transition of CM from the lab to the plate.

2 Key Challenges

Scaling CM from laboratory prototypes to industrial food products involves complex, interdependent challenges that are not typically encountered in pharmaceutical or microbial fermentation processes.

2.1 Scalability

Mammalian cells, particularly muscle and fat progenitors, grow far more slowly and are far more sensitive than bacterial, fungal, or yeast cells. While microorganisms are generally robust, tolerating wide variations in temperature, pH, and osmolarity, and can be cultivated under both anaerobic and aerobic conditions, animal cells are highly sensitive to their microenvironment and require precise control of pH, temperature, oxygen, and CO_2 levels. Therefore, scaling up these cultures requires the use of bioreactors that can maintain the optimal culture conditions over extended periods of time and larger volumes, without compromising the health, viability, or differentiation potential of the cells. Instead of scaling up, alternative strategies such as scale-out (using parallel bioreactors of the same size) may be more effective in preserving consistency. To date, the highest degree of large-scale cell culture has been achieved with Chinese hamster ovary (CHO) cells cultivated in suspension with stirred-tank bioreactors, particularly for the production of recombinant monoclonal antibodies. Industrial facilities, such as Lonza in Visp, Switzerland, routinely operate bioreactors with working volumes up to 20,000 L for this purpose [8].

2.2 Sterility

Sterility is essential in CM production for two primary reasons: (1) maintaining aseptic cultivation conditions, since most microorganisms proliferate much more rapidly than animal cells and can easily outcompete cultures, and (2) ensuring the sterility of the final product to guarantee food safety and regulatory compliance. Due to its food status, the final CM product must be antibiotic-free to meet food safety regulations [9]. This requires a robust aseptic process design and continuous, real-time monitoring for pathogens. Mycoplasma, a common contaminant in cell culture, is particularly difficult to detect and eliminate [10].

2.3 Cost Constraints

Unlike pharmaceuticals, CM needs to be economically competitive with conventional, subsidised meat products. Major cost drivers include growth media and bioreactor operation and maintenance [11]. At bench scale, the price of cell culture medium for microorganisms can be as low as 1–5 US dollars/L, depending on the formulation and additives. In contrast, animal cell culture media range from 50 to 200 US dollars/L. Furthermore, serum-free, stem cell, or specialty media can reach prices of up to 600 US dollars/L. Media formulated for CM or other advanced applications containing growth factors can cost 300–400 US dollars/L or more at a small scale. However, this cost could drop dramatically to $0.24/L or even lower, with targeted strategies such as bulk recombinant protein production, in-house formulation using food-grade ingredients, and medium recycling [12]. Reducing medium cost is therefore a central challenge, requiring the removal of expensive ingredients like foetal bovine serum (FBS) and replacing them with food-grade, animal-free, and cost-effective alternatives.

2.4 Oxygen and Nutrient Transport

For efficient and optimal cell growth, a continuous supply of sufficient oxygen and nutrients to the cells is essential. Oxygen is critical not only for ATP production through oxidative phosphorylation but also for maintaining enzyme activity, supporting various biosynthetic pathways, and facilitating cellular signalling [13, 14]. Therefore, oxygen supply must be maintained throughout the entire upstream bioprocess to support cellular metabolism and proliferation. As cell cultures become denser, increase in volume, or create 3D structures, it is increasingly difficult for nutrients and oxygen to be supplied sufficiently. This can lead to hypoxic conditions, ATP depletion, phenotype changes, or even cell death. Bioreactors use various oxygen supply systems, including spargers, hollow-fibre membranes, and silicone tubing, to efficiently deliver oxygen to cultured cells (see also chapter "Bioreactor Design and Engineering").

2.5 *Shear Stress and Suspension Culture*

Adapting adherent cells to suspension culture is a challenge by itself. Additionally, transitioning from static to dynamic culture conditions introduces shear stress. Most animal cells are highly sensitive to their mechanical microenvironment, converting mechanical forces such as shear stress into biological signals that regulate key cellular processes, including differentiation, proliferation, and gene expression, and even cause cell death [15]. It is essential to strike a balance between agitation (required for efficient mass transfer and prevention of cell sedimentation) and minimising mechanical stress to protect sensitive animal cells. Key factors to optimise include bioreactor geometry, impeller design, and media viscosity, all of which influence the hydrodynamic environment and extent of shear forces experienced by the cells.

2.6 *Food-Grade Process*

The entire bioprocess of CM production must be designed not only to achieve scalability and produce sufficient biomass, but also to comply with strict food safety standards. Therefore, certain standard pharma-grade reagents may not be allowed in food applications and should be eliminated. Although animal cells are not currently included in FDA's Generally Recognized as Safe (GRAS) list – unlike certain microorganisms used in precision fermentation – all ingredients, process aids, and derivatives used in the production of CM must comply with GRAS or food-grade standards to ensure food safety [16, 17].

3 Upstream Processing

Upstream processing includes all steps that lead to biomass generation, such as (1) cell (line) selection, (2) cell expansion, and (3) process monitoring and control. This stage is crucial for establishing optimal conditions that ensure high yields and reproducibility between batches.

3.1 *Cells*

Cells are the foundational component of CM, serving as the building blocks that ultimately determine the quality and structure of the final product. Although native meat is a complex tissue composed of multiple cell types and extracellular components, skeletal muscle fibres and adipocytes are generally regarded as the minimal essential cellular components necessary for recreating meat in vitro [18]. Therefore, selecting the appropriate cell type is a critical decision in CM bioprocess development. An industry survey revealed that producers commonly utilise a diverse range of cell sources, including muscle satellite cells, fibroblasts, mesenchymal stem cells (MSCs), induced pluripotent stem cells (iPSCs), and embryonic stem cells (ESCs) [19]. While adult stem cells, such as satellite cells and MSCs, are more frequently used due to their tissue specificity, ease of isolation, and fewer ethical concerns, pluripotent stem cells (iPSCs and ESCs) offer theoretical advantages in scalability and versatility due to their unlimited self-renewal capacity [18, 20]. However, challenges remain in establishing continuous, stable cell lines for a wide range of livestock species, which limits broader adoption and increases production complexity [19].

From a bioprocessing standpoint, the ideal starting cells must meet stringent criteria. High proliferative potential is essential to achieve biomass accumulation at scale. Primary cells, which are directly isolated from animal tissues, often have limited proliferative capacity and can undergo senescence after a finite number of divisions, making it challenging to achieve the large cell quantities required for commercial CM production.

In contrast, tailored or immortalised cell lines like ISP-4 porcine preadipocytes and the FaTTy mesenchymal stem cell line demonstrate robust expansion while maintaining full differentiation capacity [21, 22], offering process efficiency for industrial applications. Efficient and reliable differentiation into muscle or fat lineages is equally important for CM production. Muscle cells are typically obtained by expanding satellite cells, which are multipotent adult progenitors already committed to muscle lineage, making their differentiation more straightforward [23, 24]. Similarly, fat cells are usually derived from MSCs, which can be isolated from adipose tissue and stimulated to differentiate into mature adipocytes using established adipogenic protocols [25]. In contrast, pluripotent stem cells require more complex and tightly controlled differentiation protocols to direct them into specific lineages, as they are less committed to a specific lineage and have the potential to develop into any cell type [26]. Genetic stability during expansion is also critical – cells must retain a consistent phenotype and avoid genomic abnormalities that could affect their safety or quality [20, 27]. Furthermore, adaptability to large-scale suspension cultures and food-grade, animal component-free media is necessary for cost-effective biomanufacturing. Engineered cell lines, such as the bovine satellite cells that autonomously produce fibroblast growth factor 2 (FGF2), offer a promising solution to reduce reliance on costly media supplements [28]. Ultimately, the development of cell lines for CM must balance high proliferative capacity, differentiation efficiency, genetic stability, and bioprocess compatibility to enable scalable and economically viable production [20, 27].

This chapter provides only a brief overview of the importance of cell selection for the CM bioprocess.

3.2 Seed Train and Scale-Up

CM production depends on large, factory-scale bioreactors to generate sufficient biomass for future demand [29]. The seed train is the critical biological and logistical link between a cryopreserved cell bank and production-scale bioreactors (see Fig. 1). Starting from a small vial of frozen cells, the seed train progressively amplifies cell numbers through multiple proliferation stages in vessels of increasing size, typically scaling by factors of 5–10, until the seeding cell densities for production-scale bioreactors are reached [30, 31]. As cells are thawed and adapted to growth conditions, they undergo characteristic phases: a lag phase (where cells adjust to the new environment), followed by exponential proliferation (log phase), and eventually a stationary phase (where growth slows due to factors such as nutrient depletion and contact inhibition). The cell type, initial seeding density, and volume strongly influence this trajectory, affecting nutrient access and time to reach log phase [32–34]. Seed train progression is typically guided by cell viability and density measurements, which inform when to transfer cultures into larger vessels, ranging from T-flasks and shake flasks to progressively larger bioreactors [31].

Maintaining cell quality during early passages is essential, as it directly impacts performance in later stages [31]. This is particularly crucial because the final cell yield is determined by exponential, not linear, growth, meaning that even small improvements in early-stage cell health can result in substantially higher expansion factors and overall cell numbers. Unlike pharmaceutical processes where product titres are the focus, CM aims to maximise biomass via cell proliferation, making seed

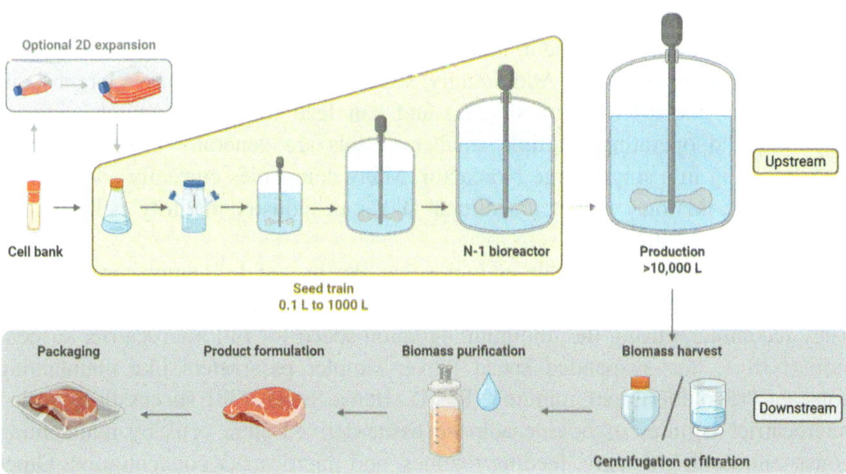

Fig. 1 The bioprocessing workflow for cultivated meat production. The upstream processes begin with the thawing and expansion of cryopreserved cells through a series of progressively larger vessels in the seed train, until the desired seeding densities for production-scale bioreactors are reached. Following this, downstream processing includes the harvesting and purification of biomass, as well as the formulation and packaging of the final product. (Created with BioRender)

train efficiency a cornerstone of process success [30]. According to a survey from the Good Food Institute, most companies seed small-scale cultures at 1×10^5 to 5×10^5 cells/mL, with some spanning from 1×10^4 to 2×10^6. Production bioreactors are inoculated at ~1–20 × 10^6 cells/mL, with final harvest targets exceeding 5×10^7 and reaching up to 4×10^8 cells/mL in some high-density perfusion systems [30].

Optimisation strategies such as N-1 perfusion (where the penultimate bioreactor in the seed train – the "N-1" stage – is operated in perfusion mode to generate a high-density inoculum for the production bioreactor), high-density cryopreservation, and early-stage process intensification can significantly reduce seed train duration and capital requirements [35, 36]. For example, directly thawing high-density cryo-bags into small, controlled bioreactors may eliminate intermediate steps, saving weeks of processing time [36]. Minimising passages also helps reduce senescence and differentiation drift in progenitor cells, preserving proliferation potential [35]. The ultimate goal is to maintain cells in exponential growth while preventing premature differentiation, enabling expansion from ~10^4 to ~10^{13} cells – enough to generate 1 ton of CM [18].

To achieve the necessary biomass for commercial-scale CM production, it is crucial to plan economic and technical scale-up strategies from the earliest stages. Companies pursue either scale-up (expanding working volume) or scale-out, where production is increased by adding more parallel units of a fixed bioreactor size. Scale-up offers consistency and capital efficiency by reducing batch-to-batch variation, but it introduces risk from contamination and increased technical demands for agitation and aeration [29, 30]. Scale-out reduces these operational risks and supports flexibility, including the use of single-use systems and varied product types within a single facility. However, single-use bioreactors are typically limited to volumes of up to 5,000 L due to material stability, risk of leaks at higher volumes, and engineering constraints. Additionally, scale-out strategies require a larger facility footprint (increased required area) and can lead to higher capital costs, as acquiring and operating multiple smaller vessels are generally more expensive than investing in a single large bioreactor. Many companies currently adopt scale-up, although this may reflect transitional stages in industry maturity rather than a fixed preference [30].

Several strategies guide scale-up in practice. Pajčin et al. [37] emphasise that poor performance at larger volumes often results from inappropriate scale-up criteria. They recommend using the minimum agitation speed for full microcarrier suspension (NJS – "just-suspended speed") over simpler parameters like maintaining constant "revolutions per minute" (RPM). Hanga et al. [38] successfully scaled microcarrier cultures of bovine adipose tissue-derived stem cells by maintaining consistent seeding density, feeding regimes, and microcarrier concentration. Other key parameters include constant power per unit volume (P/V), which reflects mixing and shear conditions [39], and sufficient oxygen transfer, typically tracked via the volumetric mass transfer coefficient (kLa) [40]. Maintaining a constant P/V during scale-up is crucial because it helps ensure that cells in larger bioreactors experience similar mixing intensity and shear forces as in smaller vessels, thereby supporting consistent cell growth and product quality across scales. kLa is not inherently dependent on scale-up alone; rather, it is primarily influenced by process parameters (such as agitator speed and gassing rate), equipment design (vessel, impeller, sparger), and physicochemical properties of the medium (viscosity, density, surface tension, etc.) [41].

3.2.1 Emerging Technologies

Due to the unique biological and economic demands of CM production, the design of seed trains is advancing rapidly. Novel technologies are focused on improving efficiency, ensuring reproducibility, and lowering costs in early-stage cell expansion.

Single-Use Bioreactors

Transitioning from standard stainless-steel or glass vessels to single-use bioreactors is becoming increasingly common due to reduced costs, lower microbial contamination and cross-contamination risk, and shorter turnaround times between batches [42]. These disposable vessels require minimal preparation and maintain sterility in antibiotic-free bioprocesses. However, there is a risk of introducing leachables into the culture medium or even the final product. Leachables are chemical substances that can migrate from the surface of processing equipment materials, such as single-use bioreactors, tubing, filters, or storage bags, into the process stream. These substances include plasticisers, stabilisers, additives, or residual monomers from polymeric materials [43]. They are concerning because they can negatively affect the health and viability of cells, thereby compromising the bioprocess [44, 45]. To mitigate this risk, extractables and leachables (E&L) studies are conducted. Extractables are compounds that can be extracted under the most aggressive conditions, representing a worst-case scenario for potential leachables. In contrast, leachables are compounds that migrate into the product under normal bioprocessing conditions. By identifying and quantifying these substances, manufacturers can assess their cytotoxicity and determine their effect on the quality of the final product [46, 47].

Perfusion Operation Mode

An alternative approach to improve the seed train process is the introduction of perfusion technology. Traditionally, a seed train involves a series of discrete steps or batch cultures, each with a progressively larger working volume, to generate sufficient biomass for the inoculation of the production-scale bioreactors. In contrast, perfusion systems continuously supply fresh medium and remove waste products while the cells are retained within the bioreactor. This approach, applied at the N-1 stage or earlier, can enable higher cell densities, which can directly inoculate the final-stage bioreactor. It can also reduce the number of scale-up steps and improve the overall efficiency of the process (elimination of the lag phase) [48–50]. By simplifying seed train operations and decreasing the number of required bioreactors, perfusion technology has the potential to accelerate scale-up. However, optimising the perfusion process is essential to ensure that high cell densities do not adversely affect cell viability or quality of the final product. It should be noted that the perfusion process requires additional, often costly equipment, such as cell retention devices, and can generate larger amounts of waste media due to continuous media exchange. Furthermore, the continuous nature of perfusion processes might pose a higher risk of contamination compared to conventional batch processes.

3.3 Bioreactor Types

A bioreactor is a fundamental element of the bioprocess. It can be defined as a vessel or device used to carry out and optimise biological processes. It provides a sterile and controlled environment for the growth and activity of biological systems such as microorganisms, enzymes, animal or plant cells, and tissues. A bioreactor allows precise regulation of key parameters like oxygen levels, nutrient availability, temperature, pH, and mechanical stress to ensure optimal conditions for the desired biological reaction. By maintaining these conditions, bioreactors support high-efficiency bioprocesses and are commonly used in applications such as fermentation and enzymatic reactions [20, 51–53].

While there are many types of bioreactors and a dedicated chapter in this book explains them in depth (see chapter "Bioreactor Design and Engineering"), we will only briefly describe the most common ones. Different bioreactor architectures can be classified according to the type of energy input: mechanically driven systems (such as stirred tanks, wave bags, and rotary wheel reactors) ensure constant mixing, while hydraulically driven systems (such as packed bed and hollow-fibre reactors) use pumped media to support cell growth on solid, cell-seeded scaffolds.

The stirred tank bioreactor (STR) is one of the most conventional and widely used bioreactor designs, particularly in fermentation and large-scale bioprocessing. It typically consists of a cylindrical vessel equipped with a centrally mounted, motor-driven impeller or agitator, which facilitates essential functions such as mixing, aeration, and mass and heat transfer [53–55]. STRs are favoured for their effective fluid mixing, high oxygen transfer efficiency, good temperature control, and adaptability through various impeller designs. These features support easy scale-up and compliance with current good manufacturing practices (cGMPs). However, STRs also present challenges, including high power consumption, elevated shear forces, and mechanical issues related to shaft sealing and stability in tall reactors. These limitations are particularly critical when culturing shear-sensitive cells like animal cells, prompting efforts to optimise impeller design to balance efficient mixing with minimal cellular damage. Additionally, the geometric configuration of the reactor – such as impeller size, baffle design, sparger position, and ratio of liquid height to tank diameter – plays an important role in its performance and has been extensively studied to enhance bioreactor efficiency and reliability [53, 55].

Hollow-fibre bioreactors (HFBs) are a type of perfusion reactor in which cells grow on or around bundles of semi-permeable hollow fibres rather than being mixed mechanically. Cells can be inoculated inside the fibre lumens or on their extracapillary surfaces, while culture medium is continuously perfused through the extra- or intracapillary compartments, respectively. This flow-based mixing strategy generates very low shear stress and enables precise control over the nutrients and gases that are delivered through the fibre walls. As a result, HFBs achieve cell densities exceeding 10^8 cells/mL, support both anchorage-dependent and suspension cultures for extended periods, and foster three-dimensional cell interactions and physiologic behaviour ideal for tissue engineering or high-value protein secretion [56, 57].

The modular design of HFBs allows customisation of membrane pore size, fibre material, and chamber geometry to suit diverse applications. Inoculation and perfusion modes can be chosen to optimise mass transfer and retention of particular particles [58, 59]. However, as the densely packed fibres can create microscale gradients in nutrient concentrations, oxygen levels, and pH, careful design and monitoring are required to avoid spatial inhomogeneity – especially when, like in CM, the cells themselves are the desired final product [60]. Furthermore, cells grow within the dense network of hollow fibres, making it technically difficult to efficiently detach and harvest all cells or tissues from the bioreactor.

Packed-bed bioreactors (PBRs), also known as column bioreactors, are systems designed for the immobilisation of cells on microcarriers or within porous scaffolds arranged inside a cylindrical chamber. These reactors are especially valued for their ability to support high-density cell cultures in a compact format, making them well suited for both research and large-scale production settings [61, 62]. Culture medium is circulated through the packed bed in various flow patterns, such as upward, downward, or radial, to ensure consistent distribution of nutrients and gases while facilitating waste removal. This flow strategy creates a stable, low-shear environment that supports cell survival and functionality [63]. PBRs can also be integrated with a separate conditioning vessel, allowing for precise control of medium oxygenation and temperature before it is perfused through the cell bed [64]. Owing to their high surface area-to-volume ratio and capacity to sustain cell densities exceeding 10^8 cells/mL, these systems are increasingly utilised in good manufacturing practice (GMP) environments for the production of biopharmaceuticals [62].

3.4 Sterility and Antibiotic Avoidance

The overuse of antibiotics in animal agriculture poses a major public health risk by driving the emergence of antibiotic-resistant pathogens, such as methicillin-resistant *Staphylococcus aureus* (MRSA), carbapenem-resistant *Enterobacteriaceae* (CRE), and vancomycin-resistant *Enterococcus* (VRE), which threaten both human and animal health [65–67]. In most regions, antibiotics are still routinely used not only to prevent disease but also to promote growth, despite growing awareness of the global antibiotic resistance crisis. If current practices persist, we may face a post-antibiotic era in which common infections become untreatable [66]. Indeed, CM offers a promising alternative by eliminating the routine use of antibiotics through controlled aseptic production, thereby potentially reducing the selective pressure that drives antibiotic resistance. Still, preventing microbial contamination is critical to avoid batch loss, compromised cell behaviour, and food safety concerns [9]. The most common contaminants in animal cell culture include bacteria and fungi, which are usually visible, as well as mycoplasmas and viruses, which are harder to detect and require molecular biological assays such as polymerase chain reaction (PCR) or enzyme-linked immunosorbent assay (ELISA) for identification [68, 69]. Although antibiotics are sometimes used in early research and development (R&D) stages, their continued presence in large-scale production can negatively affect cell health by altering proliferation, gene expression, and differentiation potential [70, 71]. To maintain sterility without antibiotics, CM production employs a combination of strategies, including aseptic processing, positive-pressure environments, moist-heat sterilisation of bioreactors, filtration of culture media (typically using 0.2 μm filters), ultraviolet or heat treatment, gamma irradiation of select equipment and materials, and strict adherence to cleanroom protocols [9, 30, 72]. Air sparging systems are also sterilised, and closed-loop systems help minimise exposure to airborne contaminants. Real-time monitoring tools such as spectroscopic sensors and automated contamination detection further support sterility. Antimicrobial peptides, which can target a broad range of pathogens with a lower risk of inducing resistance, are emerging as a potential alternative to traditional antibiotics [73, 74]. Encouragingly, the CM products approved for sale to date are produced entirely without antibiotics, supporting the feasibility of antibiotic-free bioprocessing even in food-grade environments [30]. This success is further supported by the success of antibiotic-free CHO cell cultivation in the biopharmaceutical industry, where strict aseptic techniques and process controls have enabled the large-scale, antibiotic-free production of therapeutic proteins for years. The experience of the biopharmaceutical industry provides a strong precedent for the reliability and scalability of antibiotic-free processes in CM production.

3.5 Mode of Operation

Choosing an operating mode – batch, fed-batch, semi-continuous, or perfusion/continuous – is a first-order decision in CM bioprocess design because it sets the balance between productivity, media use, equipment complexity, and operating cost (see Fig. 2) [30, 37]. Since culture medium is one of the major cost drivers of CM bioprocessing, operation modes that reduce its usage can dramatically improve cost-effectiveness [30].

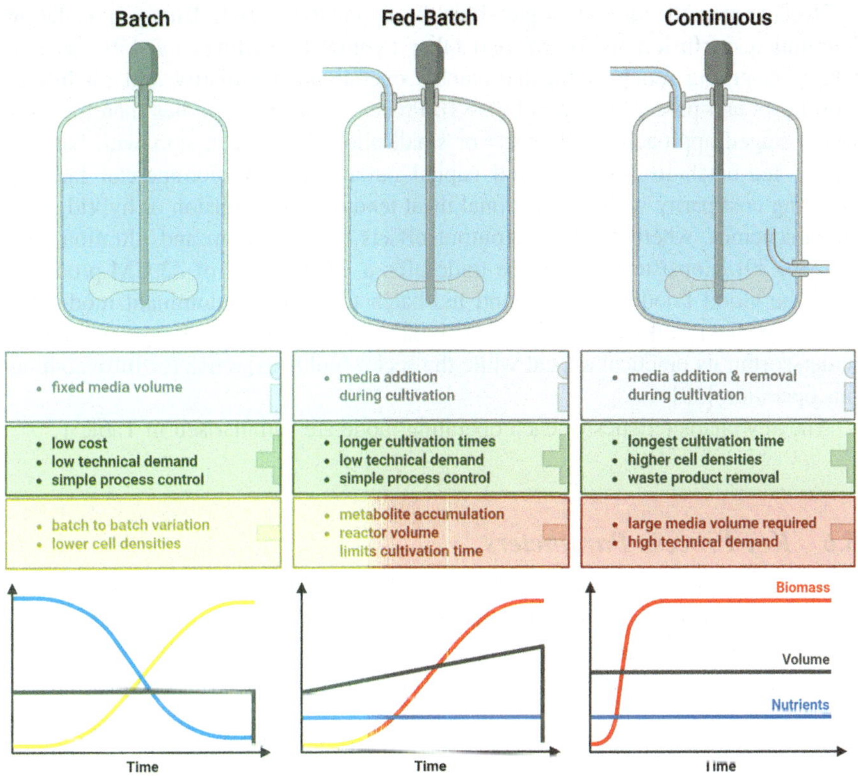

Fig. 2 The bioprocess can be operated in three modes: batch, fed-batch, or continuous/perfusion. Each mode differs in terms of media use, cultivation time, and biomass yield. (Created with BioRender)

In batch culture, a fixed medium volume supports growth until nutrients are exhausted. The broth is then harvested and the vessel cleaned, a simplicity that minimises risk but limits peak cell density and volumetric output [75, 76]. Fed-batch overcomes some of these constraints by adding concentrated feeds during the process, prolonging culture duration and raising viable cell density. However, waste products and CO_2 still accumulate and ultimately arrest cell growth [75, 76]. Semi-continuous and perfusion approaches intermittently or continuously replace spent medium with fresh medium. In full perfusion, cells are retained in the bioreactor while fresh medium is continuously supplied and waste removed, enabling sustained high cell densities. While most cost analyses are based on monoclonal antibody production, studies suggest that at sufficiently high productivities, media costs per gram in perfusion systems can rival or even undercut those of fed-batch processes [77–79]. These gains, however, come with added hardware, such as filters, acoustic settlers, or alternating tangential flow (ATF) devices, as well as tighter process control, all of which raise capital and operating complexity [30, 76].

Recent intensification strategies blur the boundaries. High-density inoculation, intermittent-perfusion fed-batch, and other hybrids boost titres by 45% or more while preserving much of the fed-batch's operational familiarity and established workflows and process controls [80–82]. Process economics studies therefore suggest a staged approach: early R&D or small pilot plants often start with batch or simple fed-batch to curb risk and capital outlay, whereas commercial facilities targeting cost parity with conventional meat tend toward perfusion or hybrid intensified schemes, where the higher output offsets extra medium and filtration costs [30, 76, 79]. Consistent with these trade-offs, a 2024 survey of 22 CM producers from the Good Food Institute found fed-batch to be the predominant mode used (13 companies), followed by batch (10) and continuous/perfusion systems (10/8), underscoring its practical appeal while the sector builds expertise for fully continuous operation [30].

The key characteristics of each operating mode are summarised in Table 1.

3.6 Key Process Parameters

CM production occurs within a bioreactor where cells are cultured in liquid medium and exposed to a gas phase, forming a complex environment in which solid (cells and scaffolds), liquid (medium), and gas (oxygen, carbon dioxide) phases interact.

Table 1 Comparison of bioprocess operation modes

Aspect	Batch	Fed-batch	Perfusion/continuous
Medium management	One fixed volume of medium; no further additions	Concentrated nutrient feeds added (glucose, glutamine, amino acids)	Fresh medium supplied and spent broth simultaneously removed; cells retained in the vessel
Typical cell density and volumetric output	Low-moderate; growth stops once nutrients are depleted	Higher than batch; limited by toxic metabolite build-up and bioreactor volume	Highest; steady nutrient supply and metabolite removal keep cell proliferating
Media use vs. product cost	High, because productivity is low	Moderate; lower than batch	Most media volume used, but cost per gram can equal or beat fed-batch at high productivity
Process complexity and capital	Simplest, minimal hardware	Moderate; requires controlled nutrient feeds	Highest; needs filters/ATF or other retention devices, plus tighter automation
Typical application stage	Early R&D and small pilots (low risk, low capital)	Commercially favoured start-up mode; used by most cultivated meat producers today	Large-scale plants or thick-tissue/high cell density formats aiming for maximal output

These interactions require careful monitoring and control to maintain optimal conditions for cell proliferation and differentiation [83, 84].

There are many variables that influence bioprocess performance, including temperature, dissolved oxygen, pH, nutrient concentrations (e.g. glucose), cell density, and metabolic by-products [85, 86]. The main parameters that need to be controlled in bioprocesses can be divided into three groups: (1) the physical parameters (temperature, viscosity, pressure, flow rate, agitation rate, foaming, etc.), (2) the chemical parameters (pH, dissolved oxygen and CO_2, etc.), and (3) the biological/biochemical parameters (cell morphology, viability, growth rate, biomass concentration, concentration of nutrients and metabolites, differentiation) [84]. Maintaining a tightly controlled culture environment by regulating these variables, along with nutrient delivery, waste removal, and mechanical conditions such as shear stress, is essential for supporting consistent cell growth. Proper optimisation of these factors not only enhances process reproducibility but is also critical for achieving the scalability, efficiency, and quality required for commercially viable CM manufacturing.

Real-time tracking of critical variables within bioreactors is essential to maintain optimal culture conditions during CM production. The FDA *Process Analytical Technology* (*PAT*) framework encourages the implementation of advanced in-process analytics to improve process understanding and control [87]. PAT enables real-time monitoring and automated control of critical process parameters – such as nutrient concentrations, cell density, and product quality – helping to ensure consistent bioprocess performance, reduce manual sampling, and support the principles of quality by design (QbD). However, applying PAT principles to biological systems such as CM presents specific challenges, as analytical tools must deliver accurate, real-time data despite potential interference from culture media and must also withstand harsh bioprocess conditions, including sterilisation via steam-in-place (SIP) and clean-in-place (CIP) procedures [83, 88].

Optical sensors, particularly Raman and infrared spectroscopy, have emerged as promising solutions due to their non-invasive, sterilisable design and ability to monitor multiple analytes simultaneously [88]. Sensors can be integrated in various formats: (1) in-line, where probes are embedded directly in the bioreactor; (2) online, where samples are diverted and analysed in real time; (3) at-line, involving near-process sampling; or (4) offline, where samples are manually analysed in a remote lab [83, 84, 86]. In-line and online systems offer continuous monitoring and are well suited for closed-loop control, but in-line sensors must be highly durable and built to operate reliably without being clogged or coated by proteins or cells. At-line and offline methods can enable more advanced analyses but are subject to manual handling and delays.

As the CM industry advances, a trend toward increased automation and real-time analytics is evident. Survey data from 23 companies showed that most use a combination of online and offline monitoring, with a growing preference for continuous, automated systems [30]. Ultimately, adopting a hybrid strategy tailored to the unique demands of each process, while aligning with PAT guidelines, can enhance precision, streamline production, and reduce reliance on labour-intensive sampling.

3.6.1 Temperature Control

Temperature regulation is a critical parameter in bioprocessing for CM, as it directly affects cell viability, metabolic activity, and production outcomes. Optimal temperature conditions vary depending on the species and cell type being cultured; for instance, mammalian cells typically require incubation around 37°C [37, 84], whereas fish and marine cell cultures, which reflect their native environments, are maintained at lower ranges between 15 and 30°C [89]. Temperature influences enzyme kinetics, metabolic rates, and cell differentiation, making precise control essential for robust cell growth and protein expression.

Bioreactors are generally equipped with automated temperature control systems, including sensors for real-time monitoring and heat exchange elements like water jackets or heating blankets, which respond to deviations from preset values [37]. Commonly used temperature sensors include thermocouples, valued for their low cost but limited by low sensitivity, and resistance temperature detectors (RTDs), which offer higher accuracy and faster response times by measuring changes in metal resistance [84]. More recently, optical thermometry has gained traction for its non-invasive nature, broad detection range, and compatibility with remote sensing, operating by detecting changes in luminescent intensity as a function of temperature variation [90].

3.6.2 pH Control

Accurate pH control is essential in bioprocessing, as even minor deviations from the optimal value (about 7.4 for most animal cells) can significantly impair cell viability, metabolism, and productivity [84]. pH serves as a key indicator of cell health, with decreases often reflecting the accumulation of acidic metabolic by-products like lactate and carbonic acid. Large-scale and high-density cultures exacerbate pH regulation challenges due to intensified metabolite build-up, necessitating frequent addition of pH-regulating agents, which can in turn lead to osmolality imbalances [37]. To maintain stable pH levels, modern bioreactors employ closed-loop feedback systems that adjust pH automatically using acid, base, or CO_2, often in conjunction with bicarbonate-buffered media for mammalian cells [91]. pH is traditionally monitored using electrochemical sensors, which operate based on potentiometric measurements between ion-selective electrodes housed in glass membranes. While widely used, these sensors are fragile, prone to drift and fouling, and require regular recalibration [83, 84, 92]. As an alternative, optical pH sensors have gained prominence due to their compact design, robustness, and non-invasive "through-the-wall" operation. These sensors rely on indicator dyes, such as cresol red, phenol red, and bromophenol blue, immobilised in polymer matrices like PVA or sol-gel, which respond to pH changes via measurable shifts in absorbance or fluorescence [84, 90, 93, 94]. Optical sensing techniques may also exploit changes in refractive index, emission intensity, or surface plasmon resonance to achieve high sensitivity [90, 95] (Steinegger et al. 2020). These advances have positioned optical pH sensors as standard components in single-use bioreactor systems, where they are typically integrated via fibre optics connected to disposable sensor patches containing immobilised dyes behind an optical window, enabling real-time, non-invasive monitoring [93].

3.6.3 Dissolved Oxygen Control

Oxygen is a critical nutrient in mammalian cell culture, essential for cellular metabolism and bioenergetic processes [96–98]. To support optimal cell growth and function, dissolved oxygen (DO) levels in bioreactors are commonly maintained at 40–60% air saturation, corresponding to a pO_2 of approximately 84–126 mmHg (11–17 kPa) [90, 99, 100]. Oxygen solubility in water is naturally low, and at physiological temperature (37°C), a typical cell culture medium contains only about 190 μM dissolved oxygen at air saturation, making oxygen supply a critical challenge for cell cultivation. Variations in oxygen concentration can markedly influence cell physiology: hyperoxia may impair proliferation and metabolic activity, while hypoxia can suppress metabolism, arrest the cell cycle, maintain stemness, and induce pro-survival signalling pathways [101]. Continuous oxygen delivery is necessary to meet the specific metabolic demands of different cell lines, with DO typically introduced through headspace aeration or direct sparging into the culture medium [37, 84]. Since overhead aeration alone is often insufficient in high-density cultures, sparging and agitation play a central role in maintaining adequate DO levels, which are also affected by temperature, pressure, and medium composition [37, 83]. DO monitoring is usually achieved via electrochemical, optical, or paramagnetic sensors [84, 90]. Clark-type electrodes, the most common electrochemical sensors, consist of an anode and cathode in an electrolyte solution encased by an oxygen-permeable membrane, where oxygen reduction generates a current proportional to DO concentration [83]. While electrochemical sensors are compact and effective at high oxygen concentrations, their performance can decline due to membrane degradation [84]. Moreover, Clark-type electrodes consume oxygen by measurement. In contrast, optical sensors, based on fluorescence quenching of immobilised dyes like ruthenium complexes, enable non-invasive, chemically inert, and miniaturised DO measurements suitable for small-scale and diffusion-limited systems [90, 92]. These sensors operate by detecting changes in emission intensity or fluorescence lifetime caused by oxygen-dependent quenching, using components such as dichroic mirrors, LEDs, and photodetectors [92]. Their advantages over Clark electrodes include no oxygen consumption during measurement and compatibility with both liquid and gas phases, though photobleaching remains a limitation [92].

3.6.4 Carbon Dioxide Control

Monitoring dissolved carbon dioxide (dCO_2) is a critical component of CM bioprocessing due to its substantial effects on cellular physiology and bioprocess performance. As a natural by-product of respiration, CO_2 is consistently produced during cell culture, and its accumulation is closely linked to increasing cell density and metabolic activity [84, 102]. Once dissolved in the culture medium, CO_2 readily diffuses across cell membranes, leading to intracellular acidification. This pH perturbation can adversely impact enzymatic reactions, protein folding, and overall cell viability [103]. Elevated dCO_2 levels, particularly those exceeding 20%, have been shown to inhibit cell proliferation, disrupt metabolite consumption, and alter the glycosylation patterns of recombinant proteins [102, 104, 105]. This underscores the importance of dCO_2 monitoring and control strategies, especially during large-scale bioprocessing, where conventional CO_2 removal methods such as aeration, sparging, or agitation face limitations due to mass transfer inefficiencies and shear sensitivity [106].

A variety of sensor technologies are employed for real-time dCO_2 monitoring in bioprocesses. The Severinghaus electrode is one of the most established electrochemical sensors, consisting of a pH electrode in contact with a carbonate buffer behind a CO_2-permeable membrane, such as polytetrafluoroethylene (PTFE) or silicone [84, 107]. As CO_2 diffuses into the buffer, it alters the pH, which is measured electrochemically to infer CO_2 concentration. While widely used, these sensors are limited by slow response times, baseline drift, and challenges associated with sterilisation. To overcome some of these limitations, ion-selective electrodes based on lithium lanthanum titanate ceramics have been proposed, offering improved mechanical stability and thermal resistance [92].

Optical CO_2 sensors have gained increasing attention due to their potential for miniaturisation, fast response times, and suitability for real-time, non-invasive monitoring. These systems often mimic the Severinghaus principle but rely on fluorometric or colourimetric detection of pH changes in indicator dyes, such as hydroxypyrene trisulfonic acid (HPTS) or ruthenium complexes, embedded in a bicarbonate buffer behind a gas-permeable membrane [93, 104, 108]. While these sensors offer the advantage of autoclavability and compatibility with harsh bioprocessing conditions, they still face challenges related to membrane lifespan, slow CO_2 diffusion through the membrane, and temperature-dependent signal drift [84, 109]. Nonetheless, the ongoing development of robust, responsive, and sterilisable optical CO_2 sensors remains essential for enabling precise environmental control in CM production.

Recently, a new approach was introduced that uses exhaust gas analysis to monitor oxygen and carbon dioxide concentrations in the exhaust stream, enabling real-time assessment of cellular metabolism and respiration activity in bioreactor cultures [110]. This technique complements the measurement of dissolved oxygen and dissolved carbon dioxide and provides additional insights for process optimisation in the production of mammalian cells and CM.

3.6.5 Agitation and Shear Stress Control

Agitation monitoring is critically important in CM bioprocessing as it directly influences mixing, oxygen transfer, nutrient distribution, and hydrodynamic conditions cells experience [111, 112]. Proper agitation maintains uniform nutrient availability, temperature, and pH control throughout the bioreactor, ensuring optimal growth conditions [37]. It also facilitates oxygen transfer and helps maintain homogenous cell distribution. This is particularly important in systems using microcarriers or cell aggregates, where uniform suspension supports efficient cell proliferation [111, 113]. Agitation and hydrodynamic mixing in bioreactors are strongly influenced by impeller geometry, with common designs including Rushton turbines, pitched-blade, marine, and cupped impellers, each imparting distinct flow patterns and mixing efficiencies (see chapter "Bioreactor Design and Engineering").

However, agitation must be precisely selected, as inadequate stirring results in poor mixing, nutrient gradients, and oxygen depletion, while excessive agitation exposes cells to harmful shear forces, reducing viability and altering cellular metabolism and morphology [111, 114–116]. The addition of non-ionic surfactants such as poloxamers (e.g. Pluronic F-68) can shield cells from these shear forces by coating the cell membrane and dampening the disruptive impact of bubble bursting [117]. Additionally, vigorous agitation, especially combined with gas sparging, can cause foam formation, further contributing to potential cell damage by trapping cells inside the foam layer. Because dissolved oxygen delivery, foam formation, CO_2 stripping, and pH stability all rise or fall with impeller geometry and speed, adjustments to agitation must be coordinated with gas sparging rates to keep high-density cultures healthy [118]. Optimising agitation rate is also crucial for controlling aggregate size in mammalian stem cell cultures, thus preventing nutrient-limited necrosis and enhancing viability and cell density [112].

Traditional methods used to measure fluid dynamics and resulting shear stress, like computational fluid dynamics (CFD) modelling and polymerase chain reaction (PCR) assays of shear-responsive genes, have limitations, including technical complexity and slow turnaround times. Recently, cell-based shear-stress sensors have been developed that directly map shear levels across diverse bioreactor geometries, guiding the design of low-shear operating windows [119–121]. Routine tracking of impeller geometry and speed and dissolved oxygen concentration, supplemented by such shear sensors, forms a foundation for a robust, energy-efficient, and scalable CM production.

3.6.6 Nutrient and Metabolite Control

Another important aspect in the bioprocess for CM production is the accurate monitoring of nutrients and metabolites, as metabolic demands can significantly vary across cell types and culture phases [85]. Glucose serves as the primary carbon and energy source, and maintaining it within an optimal range is critical for cell proliferation and product quality [122]. As cells consume glucose and glutamine during exponential growth, lactate and ammonia accumulate, respectively, which can inhibit cell growth. For instance, hybridoma cultures show a 50% growth reduction at 55 mM lactate, and CHO cells decrease growth by 25% at 60 mM [123, 124], while ammonia concentrations as low as 5–10 mM impair cell growth and metabolism [125–127].

Traditionally, glucose, lactate, and ammonia are measured via daily offline sampling using multifunctional analysers that employ enzymatic, photometric, or ion-selective assays [128, 129]. More recently, online, in situ spectroscopic techniques, such as near-infrared (NIR) and Raman spectroscopy, have been applied with chemometric models to correlate spectral features to metabolite concentrations, enabling real-time glucose, lactate, and ammonia profiling without manual sampling [88, 90, 130, 131]. For instance, Domján et al. [132] used Raman spectroscopy to monitor 20 amino acids in CHO cultures in real time, and Kozma et al. [133] demonstrated NIR-based glucose monitoring from 1 to 5,000 L bioreactors. Complementary fibre-optic and fluorescent optical ammonia sensors, utilising dyes such as aza-BODIPYs or TFPP embedded in hydrophobic matrices, offer fast (10–15 s) and sensitive (LOD ~0.03 mg/L) detection, with high stability and minimal pH interference, making them well suited for continuous ammonia monitoring in CM systems [90]. Together, these non-destructive, in-line technologies enhance process control in CM bioreactors by enabling timely adjustments to feeding strategies and preventing accumulation of growth-inhibitory by-products such as lactate and ammonia [134–138].

3.6.7 Biomass Monitoring

Monitoring biomass, cell concentration, viability, and morphology is essential in CM bioprocessing, as these parameters directly affect process efficiency, product consistency, and scalability. Cell concentration reflects proliferation efficiency and informs harvest timing and nutrient needs, while cell viability signals the proportion of functional cells and indicates potential stress from suboptimal conditions. Cell morphology provides insight into cell health, differentiation status, and adaptation to scale-up. Various technologies have been developed to assess these parameters, moving beyond traditional offline methods toward more advanced online systems. Offline measurements, which require manual or pump-based sampling, utilise many traditional and equipment-assisted methods used in animal cell culture, including trypan blue exclusion, microscopy, flow cytometry, automated cell counters such as NucleoCounter and Cedex, and dye exclusion assays with stains like propidium iodide or 7-AAD [139].

Optical density (OD) is one of the simplest and most widely used approaches for biomass estimation, but it suffers from limitations such as sensitivity to bubbles and inability to distinguish viable from dead cells [83, 92]. In contrast, dielectric and impedance spectroscopy offer label-free, non-invasive online alternatives, where only viable cells with intact membranes influence the measurement [92, 93,

140]. Capacitance-based sensors, a specific application of impedance spectroscopy, leverage these properties to monitor viable biomass in real time and are effective in industrial fermentations due to their robustness and ability to correlate with key indicators such as viable cell concentration and volume [92, 140–144]. Radio-frequency impedance has also been shown to be accurate and reliable across various bioreactor formats, including cGMP processes, especially for perfusion cultures and systems where traditional methods fail [145]. Optical methods like Raman spectroscopy [138], NIR spectroscopy [146], and scattered light sensors [147, 148] are increasingly adopted for real-time, label-free monitoring. Soft sensors that combine mathematical models with real-time data (e.g. turbidity, oxygen uptake, or permittivity) are gaining traction for estimating viable and lysed cell populations and metabolic state without the need for extensive training data [139, 149, 150]. In situ microscopy systems, which include high-resolution cameras and image analysis algorithms, provide rich data on cell morphology, density, and aggregation, with the added benefit of non-destructive online monitoring [83, 84, 90]. The integration of artificial intelligence with in situ microscopy has enabled real-time classification of viable, necrotic, and apoptotic cells in complex environments, offering a powerful analytical tool for process control [151]. Finally, automated flow cytometry provides high-resolution multiparameter analysis of live cell populations, including size, viability, and cycle data, though it requires periodic sampling and staining [152]. Together, these methods represent a rapidly advancing toolkit for comprehensive biomass monitoring in CM production.

3.6.8 Foam Control

Foam formation is a common challenge in bioprocessing, arising primarily from gas sparging and agitation, which introduce gas into the culture medium. The resulting micro- and macrobubbles, stabilised by proteins and lipids in the medium, accumulate at the liquid surface and can compromise process performance by trapping cells, causing mechanical shear stress, and altering local chemical conditions such as pH and nutrient availability. Moreover, foam hinders heat transport in the upper part of the reactor and blocks exhaust gas filters. This can result in increased pressure and compromise system sterility, as escaping foam may ultimately lead to process termination. The conditions within the foam layer reduce cell viability and productivity and may lead to eventual process failure if not properly managed [37, 153–156]. Traditionally, foam is controlled by chemical antifoam agents – surface-active compounds like silicone oils or polypropylene glycols that suppress foam by lowering surface tension. However, these chemicals can negatively impact cell metabolism, interfere with downstream operations, and require careful optimisation to balance efficacy and biocompatibility [37, 153, 155, 157, 158].

Monitoring systems for foam control are typically based on in situ sensors positioned above the liquid level in the bioreactor, which detect foam through changes in electrical properties such as conductivity, capacitance, or ultrasonic signals. Although these systems allow for real-time foam detection and automated antifoam dosing, they are susceptible to fouling [37, 153, 159]. An early approach by Brown et al. [159] used conductance-based sensors to monitor foam height and structure in real time, enabling estimation of the foam's liquid content and training of neural networks to predict foaming behaviour for improved control. More recently, machine vision technologies employing convolutional neural networks were reported to offer a non-invasive, image-based approach to monitor foam levels with high accuracy. These systems can distinguish between different foam states and support better upstream control by minimising unnecessary antifoam use and reducing foam-related failures [153, 160].

3.6.9 Differentiation Process

Differentiation is a critical step in CM bioprocessing, enabling the generation of certain cell types or structured tissues that closely resemble conventional meat. While muscle tissue constitutes the majority of meat biomass, fat contributes up to 10% and is essential for achieving authentic meaty flavour, juiciness, and mouthfeel [27]. Most research to date has focused on muscle cells, but successful commercial products must replicate both muscle and fat structures to deliver sensory and nutritional equivalence to conventional meat [23, 161]. Differentiation typically follows the proliferation phase and may occur in suspension, as cell aggregates or on microcarriers, or in adherent formats using edible porous scaffolds or hydrogels that support cell attachment and maturation [30, 162]. The process requires careful bioprocess planning, as differentiation may be integrated with or separated from proliferation. The differentiation phase often involves modifications to the bioreactor environment, including changes in media composition, mechanical environment via flow conditions, dissolved oxygen concentrations, or addition of growth factors [4, 30, 163].

Co-culturing myogenic and adipogenic cells is of interest due to its potential to recreate native tissue complexity. However, this approach remains technically challenging, particularly due to conflicting differentiation cues that may impair one lineage while promoting another [4, 164]. As noted by Sugii et al. [165], successful co-culture using meat-relevant cell types has been limited so far, highlighting the need for further optimisation. Given these difficulties, it may be more practical in many cases to differentiate muscle and fat cells separately under optimised conditions and then combine them during downstream processing to achieve the desired product composition and structure.

Myogenic differentiation in CM involves the coordinated exit of satellite cells from the cell cycle and the activation of transcriptional regulators of myogenesis, such as MYOD, MYF5, MRF4, and myogenin, that drive the formation and maturation of muscle fibres [166, 167]. Traditionally, differentiation is induced through serum starvation, which triggers remodelling of the transcriptional landscape and leads to myotube formation by fusion of multiple individual cells, mimicking adult muscle strands [168, 169]. Reportedly, high myogenic differentiation efficiencies have also been reached with serum-free media formulations

[170]. Melzener et al. [167] demonstrated that targeting key signalling pathways such as MEK/ERK, NOTCH, and RXR improves myotube fusion and actomyosin accumulation, enhancing muscle tissue formation in both 2D and 3D cultures.

Adipogenic differentiation is a tightly regulated process involving numerous transcription factors and signalling pathways. It is typically induced in vitro using a cocktail of insulin, dexamethasone, 3-isobutyl-1-methylxanthine (IBMX), and a peroxisome proliferator-activated receptor gamma (PPARγ) agonist, such as thiazolidinediones or indomethacin, which activate adipogenesis across a variety of species, including livestock and seafood [165, 171]. In addition to these standard agents, some protocols supplement media with free fatty acids, such as oleic, palmitoleic, and myristoleic acids, to promote lipid accumulation in bovine preadipocytes [168]. Beyond biochemical cues, the mechanical properties of the cell microenvironment also play an important role. For example, scaffold stiffness can modulate the degree of adipogenic or myogenic differentiation when optimised separately for each cell type. By adjusting polymer crosslinking density, it is possible to influence spatial patterns of differentiation within engineered tissues [4, 161].

Monitoring differentiation is critical to assess the success of CM bioprocesses and to ensure consistent product quality. Myogenic differentiation is typically evaluated using morphological metrics such as myotube formation and fusion index, which quantifies the proportion of nuclei within multinucleated fibres [161, 167, 168]. Molecular assessments commonly include reverse transcription quantitative PCR (RT-qPCR) and immunostaining for myogenic markers such as *Myod*, *Myog*, *Myhl/2*, and *Pax7* [172, 173]. Real-time impedance-based assays also enable dynamic, label-free monitoring of myogenesis and myotube responsiveness throughout the differentiation process [174].

Adipogenic differentiation is usually assessed through a combination of lipid staining, gene expression profiling, and protein analysis. Stains such as Oil Red O (ORO) are widely used but can suffer from non-specific background staining [175]. Fluorescent dyes such as BODIPY, Nile Red, and LipidTOX provide brighter, more specific staining and are especially effective in 3D cultures, where their reduced non-specific binding to complex structures like scaffolds and spheroids enhances signal clarity and quantification [161, 165, 171]. Gene expression analysis of conserved markers such as *Pparg*, *Fabp4*, and *Cebpa* is routinely performed using qPCR [165, 168, 171, 176]. Protein-level verification is commonly achieved via western blotting, although the availability of cross-reactive antibodies for livestock species remains limited [165]. To improve consistency and throughput, automated tools such as the Fast Adipogenesis Tracking System (FATS) have been developed for image-based quantification of lipid accumulation [177]. Looking forward, real-time, non-invasive monitoring through media analysis or spectroscopic techniques could enhance process control for adipogenic differentiation.

4 Downstream Processing

Downstream processing (DSP) in CM production refers to the suite of operations that follow cell expansion and are essential for converting biomass into a structured, safe, and palatable food product (see Fig. 1). Unlike biopharmaceutical manufacturing, where DSP focuses on extracting and purifying specific molecules, CM DSP aims to preserve whole cells or tissues at high yields while meeting food safety and sensory standards [178]. The main stages of DSP include cell harvesting, product formulation, and packaging. Harvesting, the first and arguably most critical step, involves separating cells from the culture medium, typically using centrifugation or filtration, although innovative approaches like thermo-responsive surfaces, dissolvable microcarriers, and edible scaffolds are also under development. The appropriate harvesting technique depends on cell type, culture system, and scalability needs and may differ significantly between systems producing intact tissues versus those generating more unstructured products. Subsequent steps may involve the recovery and recycling of spent medium components, blending of differentiated cell types (e.g. muscle and fat), and structuring of biomass to mimic conventional meat texture and flavour. Final product preservation and packaging must ensure food safety and shelf stability. No single downstream method has yet emerged as standard, and scalability, efficiency, and regulatory compliance will shape the optimal choices. Continued innovation in downstream processing is critical for improving cost-efficiency and supporting large-scale, high-quality production of CM.

4.1 Cell Harvest

Cell harvesting is a pivotal step in the downstream processing of CM, marking the transition from cell cultivation to product formulation. This step involves separating cells from the spent culture medium while preserving their structural integrity, which is particularly important when the harvested biomass consists of differentiated tissues intended for structured CM products. Established techniques such as centrifugation, dead-end filtration, and tangential flow filtration (TFF) are commonly used in biopharmaceuticals, but they require adaptation for CM due to the shear sensitivity of animal cells and the need to retain tissue structure [179, 180]. While these methods have historically been employed to recover extracellular proteins or viruses, the focus in CM is on recovering whole cells or tissues in large volumes and with minimal damage. Designing harvesting processes requires an understanding of the properties of both the culture medium and the cells or microcarriers being processed [179]. Emerging approaches from cell therapy, where high viability and cell quality are crucial, may offer useful strategies for CM applications. However, harvesting at the industrial scale remains a key bottleneck: existing systems for microbial cultures are not yet fully suitable, and technologies must be adapted or developed with scalability, automation, and food safety in mind [30, 162]. The timing of harvest is typically guided by culture duration or cell density, whereas high final viability is not always essential, as cells in the final food product are no longer living [30]. As the field progresses, innovations in scalable, low-shear, and food-grade harvesting technologies will be essential to ensure product quality and economic feasibility.

4.2 *Separating Cells from Scaffolding*

In CM bioprocessing, harvesting anchorage-dependent cells from microcarriers or scaffolds presents unique challenges due to the need to detach adherent cells without compromising their functional integrity. Enzymatic dissociation, typically using proteolytic enzymes such as trypsin or collagenase, is a widely adopted method that cleaves the adhesion proteins anchoring cells to the microcarrier surface, thereby releasing them into suspension. While this technique is effective and standard at the laboratory scale, its industrial application raises costs and regulatory concerns, including the need to remove or inactivate residual enzymes [181]. Scalable harvesting strategies have been developed to address these issues. For example, combining enzymatic treatment with a brief, high-intensity agitation step can efficiently detach cells from microcarriers while preserving key quality attributes, as demonstrated across various bioreactor volumes and cell-microcarrier systems [182–184]. Once cells are in suspension, separation from microcarriers is commonly achieved by size-exclusion filtration, with filters typically sized between 40 and 125 μm to retain the larger carriers while allowing cells to pass [185, 186]. More advanced techniques, such as TFF and continuous centrifugation, may be necessary at larger scales to address issues like filter clogging [181]. In addition to physical separation, alternative strategies include the use of degradable or edible microcarriers. Degradable carriers simplify downstream processing by eliminating the need for separation steps, although enzymatic treatment may still be required to achieve a single-cell suspension. Edible microcarriers, on the other hand, can remain in the final product but must comply with food safety regulations and be evaluated for their impact on sensory qualities such as texture and taste. Depending on the final product format, less stringent separation methods, like sedimentation, may suffice when some residual microcarrier content is acceptable [181].

4.3 *Filtration*

Filtration-based harvesting methods, including microfiltration, depth filtration, and tangential flow filtration, offer gentle and scalable options for separating cultivated cells from their growth medium. Dead-end filtration involves passing the culture directly through a mesh or membrane, which retains cells on its surface. While this approach is simple and cost-effective at moderate volumes (3–200 L), it becomes less practical at larger scales due to rapid fouling from accumulated cells and debris

[179, 186]. Filtration of high-density cultures can also suffer from increased resistance and flow reduction, especially in systems using fine membrane filters [179].

Filtration is especially relevant for adherent cell systems using microcarriers. For instance, cells can be detached enzymatically and then separated from their carriers via size-exclusion mesh filtration. This method takes advantage of the size difference between microcarriers and cells, allowing cells to pass through while retaining the larger carriers. Multi-step filtration protocols have also been applied, starting with coarse sieving to isolate intact microcarriers and followed by finer mesh filtration to collect the detached cells. Such approaches have proven effective at both laboratory and pilot scales, enabling the recovery of large quantities of viable cells for downstream processing [185, 186].

TFF, also called cross-flow filtration, circulates culture media parallel to the membrane surface, reducing clogging and enabling continuous operation. TFF is commonly used in perfusion bioreactors at small scale and is suitable for concentrating cells and removing soluble waste with relatively low shear stress. However, increasing viscosity at high cell densities can limit its throughput and effectiveness [179]. Although originally developed for protein purification, TFF has been adapted for use in clinical human mesenchymal stem cell (hMSC) manufacturing by maintaining low shear rates and pressure, preserving cell viability and phenotype [187]. This demonstrates its potential applicability to CM processes where maintaining whole-cell integrity is important. Further advancements, such as alternating tangential flow (ATF) systems, offer scalable solutions for biomass concentration, with devices available for processing volumes from 1 L to over 5,000 L [186]. If TFF uses a unidirectional flow across the membrane to separate cells from smaller molecules, ATF differs by periodically reversing the flow using a diaphragm pump, which helps minimise filter fouling and enhances process efficiency. While filtration technologies provide advantages in terms of cell viability and continuous processing, their successful deployment at industrial scale requires careful consideration of fouling control, sterility, and system complexity.

4.4 Centrifugation

Centrifugation is a widely used method in both laboratory and industrial settings for separating cells based on density [188]. It is commonly employed after cell expansion to concentrate biomass by sedimenting cells from the culture medium. In industrial bioprocessing, continuous disc-stack centrifuges are well established for their ability to handle large volumes efficiently and are commonly used in biopharmaceutical manufacturing, though they require rigorous cleaning and validation procedures [189]. In CM bioprocessing, continuous centrifugation is a commonly used harvesting method due to its suitability for large-scale operations [30]. Originally developed for biopharmaceutical applications, these systems have demonstrated the capacity to process large volumes and cell loads, although some particles may remain in the supernatant and complicate downstream processing

[186, 190]. Some continuous systems can operate in a fluidised bed mode to gently concentrate cells while continuously removing supernatant, which reduces shear stress and preserves viability, an approach particularly relevant for sensitive cell types like mesenchymal stem cells [187]. However, optimising operational parameters is crucial: excessive centrifugal force or duration can damage fragile muscle or fat cells, potentially releasing intracellular contents that affect product quality [179, 191, 192]. Precise tuning of speed, time, and conditions like temperature or osmolarity is necessary to maximise yield while minimising cell loss or damage [192]. While low g-forces may be sufficient for some cell lines, others require tailored protocols to avoid lysis or clumping [179]. Despite these limitations, centrifugation remains a practical and scalable option for cell harvesting in CM manufacturing.

4.5 Purification

Following the cell harvesting stage in CM production, a biomass washing step is often employed to remove residual culture media components such as growth factors, enzymes, or microcarrier fragments. This purification process, though not held to pharmaceutical-grade standards, is critical for ensuring food safety and consumer acceptance. Residual hormone-related substances present in the medium may pose food safety risks if they persist in the final product and enter the human food chain [193]. Moreover, the presence of additives like hormones or antibiotics has been shown to negatively affect consumer perception of CM [194]. Washing can be conducted either before or after cell harvesting, using simple buffers such as phosphate-buffered saline [30]. In the cell therapy field, washing is often achieved through TFF or centrifugation to remove unwanted substances while concentrating cells to the high densities needed for downstream processing [187]. Scalable and cost-effective approaches must be developed and refined to satisfy both regulatory requirements and economic feasibility in CM production.

4.6 Final Product Structuring Considerations

The final structuring of CM products spans a wide spectrum of complexity, ranging from processed items like nuggets and sausages to highly structured formats such as steaks and fillets. Less structured products are easier to manufacture, while replicating the intricate architecture of whole cuts requires advanced strategies from tissue engineering and biomaterials science. Scaffolds intended to remain in the final product must be safe for consumption and should emulate the biomechanical and sensory properties of native extracellular matrix, particularly during cooking [162]. Achieving realistic texture, colour, and flavour is essential for consumer acceptance, and these properties are strongly influenced by the degree of differentiation and distribution of muscle and adipose cells, as well as media composition and culture conditions [2, 27, 195]. Adipose tissue can enhance juiciness and mouthfeel, with differentiation protocols controlling fatty acid profiles to match high-end meats such as Wagyu beef [196]. Emerging experimental approaches are exploring how scaffold design and controlled cell differentiation can enhance the sensory properties of CM. For example, some techniques aim to replicate flavour development through cooking by incorporating compounds that release aroma when heated, while others focus on assembling separately cultured muscle and fat tissues to better mimic the structure and taste of conventional meat [161, 197]. However, current cultured muscle tissues show reduced umami and altered amino acid composition compared to conventional meat, suggesting that further media optimisation is needed [198]. The complexity of co-culturing muscle and fat cells at scale also raises questions about feasibility and cost [195]. Advanced manufacturing methods such as 3D bioprinting, using either scaffold-based or scaffold-free (composed entirely of cells) bioinks, offer additional pathways for building structured tissues with controlled architecture and cellular arrangement [179, 199]. Ultimately, aligning structural, sensory, and nutritional characteristics with consumer expectations remains a central challenge for CM development.

5 Emerging Technologies

Emerging technologies such as artificial intelligence (AI), machine learning (ML), and digital twins are poised to transform CM bioprocessing by enabling faster, more efficient, and scalable production systems. These technologies offer the potential to accelerate R&D, improve process control, and support scalability, all of which are critical for commercial viability. ML is particularly promising for streamlining tasks traditionally dependent on iterative and labour-intensive experimentation, such as optimising cell growth conditions, media composition, and bioreactor performance. ML can be utilised to optimise mammalian cell cultivation processes by analysing historical and experimental data to identify conditions that enhance cell growth and productivity [200, 201]. By uncovering hidden patterns in experimental data, ML models can accurately predict outcomes, enabling rapid design of experiments and improved process decision-making [202, 203]. Techniques such as artificial neural networks, support vector machines, and probabilistic optimisation have been used for diverse applications ranging from optimising serum-free media [204] and reducing environmental impact [205] to predicting real-time bioreactor conditions and ensuring quality assurance through NIR spectroscopy [206].

Beyond upstream applications, ML also supports innovations in downstream product structuring. It has been applied to predict the printability of bioinks in 3D bioprinting [207], automate extrusion-based printing [208], and improve flavour design through modelling of Maillard reaction profiles and sensory prediction [201, 209]. As CM products evolve from ground forms to structured cuts, these data-driven methods will be critical for refining sensory attributes such as texture, aroma, and juiciness.

In parallel, digital twins (virtual replicas of physical bioreactors or production systems that incorporate real-time sensor data and computational models) are emerging as key enablers of smart biomanufacturing [210]. Digital twins facilitate predictive control by simulating system behaviour under various conditions, allowing operators to anticipate issues, test optimisations virtually, and reduce experimental burden [33, 211]. Their integration with ML further enhances dynamic monitoring and control of complex, non-linear biological processes. This may replace static, rule-based systems such as proportional–integral–derivative (PID) controllers with adaptive, model-driven approaches that are more suitable for biological variability [201]. As CM processing scales, the automation of closed systems will become increasingly important to minimise contamination risk and operational failures. Real-time monitoring will also be essential to enable dynamic adjustments and ensure consistent product quality [212].

Despite the promise of these technologies, their adoption in the CM field remains in the early stages. Key barriers include limited standardisation of datasets, need for domain-specific ML models, and challenge of aligning predictive tools with regulatory and food safety standards [203, 213]. Initiatives like the Cultivated Meat Modeling Consortium aim to foster collaborative model development while protecting proprietary knowledge, highlighting the importance of shared infrastructure and interdisciplinary cooperation to unlock the full potential of computational approaches in CM production [214].

6 Economic Challenges and Cost Reduction Strategies

CM presents a promising solution to the environmental and ethical challenges of conventional meat production, but achieving economic viability remains a critical hurdle. Much of the early development of CM was based on biopharmaceutical mammalian cell culture systems, which were designed for small-batch, high-value products under strict regulatory conditions. In contrast, food production requires large volumes at much lower cost and with different regulatory standards [11]. As a result, substantial innovation is needed to transition CM from the lab to industrial-scale food production. Techno-economic analyses (TEAs) have estimated that to be cost-competitive with conventional beef, CM must achieve production costs below $9/kg [11]. While current estimates remain well above this threshold, ongoing research is making important strides toward reducing these costs.

Among all inputs, the culture medium has consistently emerged as the dominant cost driver in CM production. TEAs show that media can account for 41–83% of operating expenditure, depending on reactor scale and media formulation assumptions [11]. Other studies also highlight media as a leading cost driver, with Garrison et al. [215] reporting it as 31% of total cost per kg. Strategies to reduce media costs include sourcing ingredients through bulk commodity markets and improving production efficiency for key components like amino acids and growth factors [12, 216]. Several targeted innovations have been proposed, including engineering plants or microorganisms to produce recombinant growth factors, developing more potent and stable variants that reduce required concentrations, and bringing production in-house to reduce reliance on high-cost pharmaceutical supply chains [217, 218]. Medium recycling strategies, such as in situ adsorption of metabolic by-products and co-culturing with nutrient-rich microalgae, offer promising ways to extend media lifespan and reduce costs. Although still under development, these approaches could improve media efficiency while enhancing the nutritional quality of CM [193].

Capital expenditure (CAPEX) represents another major challenge. Building facilities capable of producing just 1% of the US beef supply (~121 million kg/year) is estimated to require $1.6–5.5 billion in infrastructure investment, depending on reactor size, cleanroom requirements, and facility design assumptions [11, 216, 217, 219]. However, as production technology evolves, opportunities for CAPEX reduction are emerging. These include simplifying reactor design using less expensive materials and integrating continuous strategies such as perfusion systems, which may improve yields and reduce downtime [30, 218].

Further cost savings in CM production are projected through the scale-up of reactor volumes. Negulescu et al. [11] estimate that utilising 262,000 L airlift reactors could reduce production costs to as low as $17/kg. While perfusion systems offer advantages in cell density and efficiency, their implementation at a large scale remains limited [217]. To support commercialisation, future systems must be designed to balance performance, cost, and scalability for various product types. This may involve the use of computational modelling to guide the creation of economically feasible reactor designs [219].

Advancements in cell line development are also critical. Most current studies rely on primary muscle cells, which are difficult to scale. Therefore, there is a pressing need for robust, scalable, publicly available cell lines that can thrive in suspension culture and achieve high densities in bioreactors while maintaining their capacity to differentiate [217].

Despite these challenges, there is strong momentum and a clear path forward. Strategies such as hybrid products (combining CM with plant-based proteins), low inclusion ratios, and production of undifferentiated cell mass offer near-term solutions that reduce complexity while maintaining sensory appeal [218]. Additionally, refining bioreactor operations and integrating automation are expected to streamline manufacturing, reduce operational complexity, and lower production costs [30]. Lastly, the application of advanced techno-economic modelling ensures that ongoing research is directed toward the most impactful areas, supporting the long-term goal of CM achieving price parity with conventional meat [217, 220].

7 Conclusion

Bioprocess engineering lies at the core of the CM industry's transition from laboratory experiments to industrial-scale food production. This review has examined the full bioprocessing pipeline, highlighting both the challenges and opportunities that shape the development of scalable, safe, and economically viable CM products. Key upstream challenges include developing robust, fast-growing cell lines, achieving large-scale proliferation in suspension cultures, and designing bioreactors that can balance efficiency, sterility, and cost. We explored seed train design, scale-up strategies, and various bioreactor types. Sterility, a major concern due to the vulnerability of animal cells, must be carefully maintained throughout production to prevent contamination and ensure consistent product quality at industrial scale. Precise process monitoring of key parameters such as pH, temperature, dissolved oxygen, nutrient levels, and metabolite accumulation is essential to maintain optimal culture conditions and support successful cell expansion and differentiation.

Downstream processing is equally critical, yet still underdeveloped. Efficient, food-grade methods for harvesting, purifying, and structuring the final product must be advanced to ensure safety and consumer acceptance. Technologies such as filtration and centrifugation need to be optimised for scale and cost. Final product quality remains central, requiring that sensory and nutritional properties match or exceed those of conventional meat. Emerging technologies such as artificial intelligence, machine learning, and digital twins show promise in optimising bioprocess control and accelerating R&D, although their application in CM is still in early stages due to data limitations and regulatory challenges.

Economically, the industry must overcome significant barriers before CM can compete with conventional meat. Techno-economic analyses consistently identify media costs and CAPEX as primary cost drivers. Media alone can account for up to 83% of operating expenses, depending on production scale and formulation. Innovations such as in-house media production and media recycling strategies like in situ adsorption and co-culture with microalgae are promising avenues to reduce these costs. Simultaneously, scaling reactor volumes and adopting continuous production strategies such as perfusion can improve yields and reduce downtime. CAPEX can be further mitigated through simpler reactor designs, food-grade infrastructure, and automation. As the industry grows, techno-economic modelling will remain vital in guiding resource allocation and innovation efforts.

Despite these challenges, the path forward is clear. Already, CM has made remarkable progress, transitioning from a €250,000 prototype burger in 2013 to regulatory approvals and pilot-scale products in several countries [218, 221]. As bioprocess technologies continue to mature, the CM industry is well positioned to scale production, reduce costs, and deliver a sustainable, ethical, and economically viable alternative to conventional meat.

Acknowledgments We would like to thank Eugenia Sagué and George Zheleznyi for their valuable support. This study was funded by the German Federal Ministry for Economic Affairs and Climate Action (BMWK) through projects KK5344003 ML4 and 5653301ML4, "Development of High-Quality Fat Cell Cultures as the Key to Authentic Flavor in Meat Alternatives", as well as by the Adalbert Raps Foundation. This study was also funded by the Investitionsbank Berlin (IBB) on behalf of the State of Berlin within the ProFIT program, supported by the European Regional Development Fund (ERDF 2021–2027) and state funds, through projects 10210753 and 10210754, "CultiFat".

Conflict of Interest Julia Bernfeld, Simon Zschieschang, Marline Kirsch, Jordi Morales-Dalmau, and Jasper Liedtke are employees of Cultimate Foods GmbH (Berlin/Hannover). However, the authors declare that they wrote this review objectively and without bias. The company did not influence manuscript preparation. No other conflicts of interest exist.

References

1. Cossar D (2011) Chapter 1.51 – Bioprocessing techniques. In: Moo-Young M (ed) Comprehensive biotechnology, 2nd edn. Academic Press, pp. 679–690. https://doi.org/10.1016/B978-0-08-088504-9.00078-7
2. Rubio NR, Xiang N, Kaplan DL (2020) Plant-based and cell-based approaches to meat production. Nat Commun 11(1):6276. https://doi.org/10.1038/s41467-020-20061-y
3. Kim S, Beier A, Schreyer HB, Bakshi BR (2022) Environmental life cycle assessment of a novel cultivated meat burger patty in the United States. Sustainability 14(23):16133. https://doi.org/10.3390/su142316133
4. Kirsch M, Morales-Dalmau J, Lavrentieva A (2023) Cultivated meat manufacturing: technology, trends, and challenges. Eng Life Sci 23(12):e2300227. https://doi.org/10.1002/elsc.202300227
5. Mattick CS, Landis AE, Allenby BR, Genovese NJ (2015) Anticipatory life cycle analysis of in vitro biomass cultivation for cultured meat production in the United States. Environ Sci Technol 49(19):11941–11949. https://doi.org/10.1021/acs.est.5b01614
6. Sinke P, Swartz E, Sanctorum H, van der Giesen C, Odegard I (2023) Ex-ante life cycle assessment of commercial-scale cultivated meat production in 2030. Int J Life Cycle Assess 28(3):234–254. https://doi.org/10.1007/s11367-022-02128-8
7. Stephens N, Di Silvio L, Dunsford I, Ellis M, Glencross A, Sexton A (2018) Bringing cultured meat to market: technical, socio-political, and regulatory challenges in cellular agriculture. Trends Food Sci Technol 78:155–166. https://doi.org/10.1016/j.tifs.2018.04.010
8. Lonza (2025) Biologics. https://www.lonza.com/annualreport/2024/our-businesses/biologics. Accessed 23 Jun 2025
9. McNamara E, Bomkamp C (2022) Cultivated meat as a tool for fighting antimicrobial resistance. Nat Food 3(10):791–794. https://doi.org/10.1038/s43016-022-00602-y
10. Young L, Sung J, Stacey G, Masters JR (2010) Detection of mycoplasma in cell cultures. Nat Protoc 5(5):929–934. https://doi.org/10.1038/nprot.2010.43
11. Negulescu PG, Risner D, Spang ES, Sumner D, Block D, Nandi S, McDonald KA (2023) Techno-economic modeling and assessment of cultivated meat: impact of production bioreactor scale. Biotechnol Bioeng 120(4):1055–1067. https://doi.org/10.1002/bit.28324
12. Specht L (2020) An analysis of culture medium costs and production volumes for cultivated meat. The Good Food Institute
13. Csete M (2005) Oxygen in the cultivation of stem cells. Ann N Y Acad Sci 1049(1):1–8. https://doi.org/10.1196/annals.1334.001
14. Wilson JW, Shakir D, Batie M, Frost M, Rocha S (2020) Oxygen-sensing mechanisms in cells. FEBS J 287(18):3888–3906. https://doi.org/10.1111/febs.15374
15. Vining KH, Mooney DJ (2017) Mechanical forces direct stem cell behaviour in development and regeneration. Nat Rev Mol Cell Biol 18(12):728–742. https://doi.org/10.1038/nrm.2017.108
16. Center for Veterinary Medicine (2025) Current animal food GRAS Notices Inventory. FDA. https://www.fda.gov/animal-veterinary/generally-recognized-safe-gras-notification-program/current-animal-food-gras-notices-inventory
17. Human Foods Program (2025) Human food made with cultured animal cells. FDA. https://www.fda.gov/food/food-ingredients-packaging/human-food-made-cultured-animal-cells
18. Post MJ, Levenberg S, Kaplan DL, Genovese N, Fu J, Bryant CJ, Negowetti N, Verzijden K, Moutsatsou P (2020) Scientific, sustainability and regulatory challenges of cultured meat. Nat Food 1(7):403–415. https://doi.org/10.1038/s43016-020-0112-z

19. Ravikumar M, Powell D (2023) Cell line development and utilisation trends in the cultivated meat industry. The Good Food Institute
20. Reiss J, Robertson S, Suzuki M (2021) Cell sources for cultivated meat: applications and considerations throughout the production workflow. Int J Mol Sci 22(14):7513. https://doi.org/10.3390/ijms22147513
21. Cheng Y-M, Hong P-C, Song M-M, Zhu H-N, Qin J, Zhang Z-D, Chen H, Ma X-Z, Tian M-Y, Zhu W-Y, Huang Z (2023) An immortal porcine preadipocyte cell strain for efficient production of cell-cultured fat. Commun Biol 6(1):1–13. https://doi.org/10.1038/s42003-023-05583-7
22. Thrower T, Riley SE, Lee S, Esteves CL, Xavier Donadeu F (2025) A unique spontaneously immortalised cell line from pig with enhanced adipogenic capacity. npj Sci Food 9(1):52. https://doi.org/10.1038/s41538-025-00413-y
23. Kadim IT, Mahgoub O, Baqir S, Faye B, Purchas R (2015) Cultured meat from muscle stem cells: a review of challenges and prospects. J Integr Agric 14(2):222–233. https://doi.org/10.1016/S2095-3119(14)60881-9
24. Post MJ (2012) Cultured meat from stem cells: challenges and prospects. Meat Sci 92(3):297–301. https://doi.org/10.1016/j.meatsci.2012.04.008
25. Ozhava D, Lee K, Bektas C, Jackson A, Patel K, Mao Y (2024) Optimized adipogenic differentiation and delivery of bovine umbilical cord stem cells for cultivated meat. Gels (Basel, Switzerland) 10(8):488. https://doi.org/10.3390/gels10080488
26. Fowler JL, Ang LT, Loh KM (2020) A critical look: challenges in differentiating human pluripotent stem cells into desired cell types and organoids. WIREs Dev Biol 9(3):e368. https://doi.org/10.1002/wdev.368
27. Fish KD, Rubio NR, Stout AJ, Yuen JSK, Kaplan DL (2020) Prospects and challenges for cell-cultured fat as a novel food ingredient. Trends Food Sci Technol 98:53–67. https://doi.org/10.1016/j.tifs.2020.02.005
28. Stout AJ, Zhang X, Letcher SM, Rittenberg ML, Shub M, Chai KM, Kaul M, Kaplan DL (2024) Engineered autocrine signaling eliminates muscle cell FGF2 requirements for cultured meat production. Cell Rep Sustain 1(1). https://doi.org/10.1016/j.crsus.2023.100009
29. Santos ACA, Camarena DEM, Roncoli Reigado G, Chambergo FS, Nunes VA, Trindade MA, Stuchi Maria-Engler S (2023) Tissue engineering challenges for cultivated meat to meet the real demand of a global market. Int J Mol Sci 24(7):6033. https://doi.org/10.3390/ijms24076033
30. Harsini F, Swartz E (2024) Trends in cultivated meat scale-up and bioprocessing. The Good Food Institute
31. Kern S, Platas-Barradas O, Pörtner R, Frahm B (2016) Model-based strategy for cell culture seed train layout verified at lab scale. Cytotechnology 68(4):1019–1032. https://doi.org/10.1007/s10616-015-9858-9
32. Frahm B (2014) Seed train optimization for cell culture. In: Pörtner R (ed) Animal cell biotechnology: methods and protocols. Humana Press, pp 355–367. https://doi.org/10.1007/978-1-62703-733-4_22
33. Hernández Rodríguez T, Frahm B (2020) Design, optimization, and adaptive control of cell culture seed trains. In: Pörtner R (ed) Animal cell biotechnology: methods and protocols. Springer, pp 251–267. https://doi.org/10.1007/978-1-0716-0191-4_14
34. Wang Z (2021) Regulation of cell cycle progression by growth factor-induced cell signaling. Cells 10. https://doi.org/10.3390/cells10123327
35. Allan SJ, De Bank PA, Ellis MJ (2019) Bioprocess design considerations for cultured meat production with a focus on the expansion bioreactor. Front Sustain Food Syst 3. https://doi.org/10.3389/fsufs.2019.00044
36. Heidemann R, Mered M, Wang DQ, Gardner B, Zhang C, Michaels J, Henzler H-J, Abbas N, Konstantinov K (2002) A new seed-train expansion method for recombinant mammalian cell lines. Cytotechnology 38(1):99–108. https://doi.org/10.1023/A:1021114300958
37. Pajčin I, Knežić T, Azoulay IS, Vlajkov V, Djisalov M, Janjušević L, Grahovac J, Gadjanski I (2022) Bioengineering outlook on cultivated meat production. Micromachines 13(3):402. https://doi.org/10.3390/mi13030402

38. Hanga MP, de la Raga FA, Moutsatsou P, Hewitt CJ, Nienow AW, Wall I (2021) Scale-up of an intensified bioprocess for the expansion of bovine adipose-derived stem cells (bASCs) in stirred tank bioreactors. Biotechnol Bioeng 118(8):3175–3186. https://doi.org/10.1002/bit.27842
39. Ascanio G, Castro B, Galindo E (2004) Measurement of power consumption in stirred vessels – a review. Chem Eng Res Des 82(9):1282–1290. https://doi.org/10.1205/cerd.82.9.1282.44164
40. Alavijeh MK, Baker I, Lee YY, Gras SL (2022) Digitally enabled approaches for the scale up of mammalian cell bioreactors. Digit Chem Eng 4:100040. https://doi.org/10.1016/j.dche.2022.100040
41. Aroniada M, Maina S, Koutinas A, Kookos IK (2020) Estimation of volumetric mass transfer coefficient (kLa) – review of classical approaches and contribution of a novel methodology. Biochem Eng J 155:107458. https://doi.org/10.1016/j.bej.2019.107458
42. Shukla AA, Gottschalk U (2013) Single-use disposable technologies for biopharmaceutical manufacturing. Trends Biotechnol 31(3):147–154. https://doi.org/10.1016/j.tibtech.2012.10.004
43. Cuadros-Rodríguez L, Lazúen-Muros M, Ruiz-Samblás C, Navas-Iglesias N (2020) Leachables from plastic materials in contact with drugs. State of the art and review of current analytical approaches. Int J Pharm 583:119332. https://doi.org/10.1016/j.ijpharm.2020.119332
44. Budde D, Jurkiewicz E (2021) Risk analysis of leachables in cell and gene therapy using a CAR-T model process. Int J Pharm 607:121015. https://doi.org/10.1016/j.ijpharm.2021.121015
45. Kelly PS, McSweeney S, Coleman O, Carillo S, Henry M, Chandran D, Kellett A, Bones J, Clynes M, Meleady P, Barron N (2016) Process-relevant concentrations of the leachable bDtBPP impact negatively on CHO cell production characteristics. Biotechnol Prog 32(6):1547–1558. https://doi.org/10.1002/btpr.2345
46. USP (2025) Extractables and leachables. https://www.usp.org/impurities/extractables-and-leachables. Accessed 27 Apr 2025
47. Li K, Rogers G, Nashed-Samuel Y, Lee H, Mire-Sluis A, Cherney B, Forster R, Yeh P, Markovic I (2015) Creating a holistic extractables and leachables (E&L) program for biotechnology products. PDA J Pharm Sci Technol 69(5):590–619. https://doi.org/10.5731/pdajpst.2015.01073
48. Bielser J-M, Wolf M, Souquet J, Broly H, Morbidelli M (2018) Perfusion mammalian cell culture for recombinant protein manufacturing – a critical review. Biotechnol Adv 36(4):1328–1340. https://doi.org/10.1016/j.biotechadv.2018.04.011
49. Wong HE, Chen C, Le H, Goudar CT (2022) From chemostats to high-density perfusion: the progression of continuous mammalian cell cultivation. J Chem Technol Biotechnol 97(9):2297–2304. https://doi.org/10.1002/jctb.6841
50. Yang WC, Lu J, Kwiatkowski C, Yuan H, Kshirsagar R, Ryll T, Huang Y-M (2014) Perfusion seed cultures improve biopharmaceutical fed-batch production capacity and product quality. Biotechnol Prog 30(3):616–625. https://doi.org/10.1002/btpr.1884
51. Chemistry (IUPAC), T. I. U. of P. and A. (1992) IUPAC – bioreactor (B00662). https://doi.org/10.1351/goldbook.B00662
52. Naing MW, Williams DJ (2011) Three-dimensional culture and bioreactors for cellular therapies. Cytotherapy 13(4):391–399. https://doi.org/10.3109/14653249.2011.556352
53. Wang S-J, Zhong J-J (2007) Chapter 6 – Bioreactor engineering. In: Yang S-T (ed) Bioprocessing for value-added products from renewable resources. Elsevier, pp 131–161. https://doi.org/10.1016/B978-044452114-9/50007-4
54. Martin I, Wendt D, Heberer M (2004) The role of bioreactors in tissue engineering. Trends Biotechnol 22(2):80–86. https://doi.org/10.1016/j.tibtech.2003.12.001
55. Zhong J-J (2011) Chapter 2.14 – Bioreactor engineering. In: Moo-Young M (ed) Comprehensive biotechnology2nd edn. Academic Press, pp 165–177. https://doi.org/10.1016/B978-0-08-088504-9.00097-0

56. Bettahalli NMS, Steg H, Wessling M, Stamatialis D (2011) Development of poly(l-lactic acid) hollow fiber membranes for artificial vasculature in tissue engineering scaffolds. J Membr Sci 371(1):117–126. https://doi.org/10.1016/j.memsci.2011.01.026
57. Roberts I, Baila S, Rice RB, Janssens ME, Nguyen K, Moens N, Ruban L, Hernandez D, Coffey P, Mason C (2012) Scale-up of human embryonic stem cell culture using a hollow fibre bioreactor. Biotechnol Lett 34(12):2307–2315. https://doi.org/10.1007/s10529-012-1033-1
58. Godara P, McFarland CD, Nordon RE (2008) Design of bioreactors for mesenchymal stem cell tissue engineering. J Chem Technol Biotechnol 83(4):408–420. https://doi.org/10.1002/jctb.1918
59. Kumar A, Starly B (2015) Large scale industrialized cell expansion: producing the critical raw material for biofabrication processes. Biofabrication 7(4):044103. https://doi.org/10.1088/1758-5090/7/4/044103
60. Eghbali H, Nava MM, Mohebbi-Kalhori D, Raimondi MT (2016) Hollow fiber bioreactor technology for tissue engineering applications. Int J Artif Organs 39(1):1–15. https://doi.org/10.5301/ijao.5000466
61. Krishania M, Sindhu R, Binod P, Ahluwalia V, Kumar V, Sangwan RS, Pandey A (2018) Chapter 5 – Design of bioreactors in solid-state fermentation. In: Pandey A, Larroche C, Soccol CR (eds) Current developments in biotechnology and bioengineering. Elsevier, pp 83–96. https://doi.org/10.1016/B978-0-444-63990-5.00005-0
62. Meuwly F, Ruffieux P-A, Kadouri A, von Stockar U (2007) Packed-bed bioreactors for mammalian cell culture: bioprocess and biomedical applications. Biotechnol Adv 25(1): 45–56. https://doi.org/10.1016/j.biotechadv.2006.08.004
63. Moritz MSM, Verbruggen SEL, Post MJ (2015) Alternatives for large-scale production of cultured beef: a review. J Integr Agric 14(2):208–216. https://doi.org/10.1016/S2095-3119(14)60889-3
64. Pörtner R, Platas OB, Fassnacht D, Nehring D, Czermak P, Märkl H (2007) Fixed bed reactors for the cultivation of mammalian cells: design, performance and scale up. Open Biotechnol J. https://doi.org/10.2174/1874070700701010041
65. Low CX, Tan LT-H, Ab Mutalib N-S, Pusparajah P, Goh B-H, Chan K G, Letchumanan V, Lee L-H (2021) Unveiling the impact of antibiotics and alternative methods for animal husbandry: a review. Antibiotics (Basel, Switzerland) 10(5):578. https://doi.org/10.3390/antibiotics10050578
66. Mann A, Nehra K, Rana JS, Dahiya T (2021) Antibiotic resistance in agriculture: perspectives on upcoming strategies to overcome upsurge in resistance. Curr Res Microb Sci 2:100030. https://doi.org/10.1016/j.crmicr.2021.100030
67. Martin MJ, Thottathil SE, Newman TB (2015) Antibiotics overuse in animal agriculture: a call to action for health care providers. Am J Public Health 105(12):2409–2410. https://doi.org/10.2105/AJPH.2015.302870
68. Malik P, Mukherjee S, Mukherjee TK (2023) Microbial contamination of mammalian cell culture. In: Mukherjee TK, Malik P, Mukherjee S (eds) Practical approach to mammalian cell and organ culture. Springer Nature, pp 187–231. https://doi.org/10.1007/978-981-19-1731-8_5-1
69. Powell DJ, Li D, Smith B, Chen WN (2025) Cultivated meat microbiological safety considerations and practices. Compr Rev Food Sci Food Saf 24(1):e70077. https://doi.org/10.1111/1541-4337.70077
70. Ryu AH, Eckalbar WL, Kreimer A, Yosef N, Ahituv N (2017) Use antibiotics in cell culture with caution: genome-wide identification of antibiotic-induced changes in gene expression and regulation. Sci Rep 7:7533. https://doi.org/10.1038/s41598-017-07757-w
71. Skubis A, Gola J, Sikora B, Hybiak J, Paul-Samojedny M, Mazurek U, Łos MJ (2017) Impact of antibiotics on the proliferation and differentiation of human adipose-derived mesenchymal stem cells. Int J Mol Sci 18(12):2522. https://doi.org/10.3390/ijms18122522
72. Clarke KG (2013) Chapter 10 – Bioprocess asepsis and sterility. In: Clarke KG (ed) Bioprocess engineering. Woodhead Publishing, pp 189–208. https://doi.org/10.1533/9781782421689.189

73. Satchanska G, Davidova S, Gergova A (2024) Diversity and mechanisms of action of plant, animal, and human antimicrobial peptides. Antibiotics 13(3):202. https://doi.org/10.3390/antibiotics13030202
74. Wang J, Dou X, Song J, Lyu Y, Zhu X, Xu L, Li W, Shan A (2019) Antimicrobial peptides: promising alternatives in the post feeding antibiotic era. Med Res Rev 39(3):831–859. https://doi.org/10.1002/med.21542
75. Martínez-Monge I, Comas P, Triquell J, Lecina M, Casablancas A, Cairó JJ (2018) A new strategy for fed-batch process control of HEK293 cell cultures based on alkali buffer addition monitoring: comparison with O.U.R. dynamic method. Appl Microbiol Biotechnol 102(24):10469–10483. https://doi.org/10.1007/s00253-018-9388-4
76. Pollock J, Ho SV, Farid SS (2013) Fed-batch and perfusion culture processes: economic, environmental, and operational feasibility under uncertainty. Biotechnol Bioeng 110(1):206–219. https://doi.org/10.1002/bit.24608
77. Lin H, Leighty RW, Godfrey S, Wang SB (2017) Principles and approach to developing mammalian cell culture media for high cell density perfusion process leveraging established fed-batch media. Biotechnol Prog 33(4):891–901. https://doi.org/10.1002/btpr.2472
78. Walther J, Lu J, Hollenbach M, Yu M, Hwang C, McLarty J, Brower KP (2018) Perfusion cell culture decreases process and product heterogeneity in a head-to-head comparison with fed-batch. Biotechnol J 14. https://doi.org/10.1002/biot.201700733
79. Xu S, Gavin J, Jiang R, Chen H (2017) Bioreactor productivity and media cost comparison for different intensified cell culture processes. Biotechnol Prog 33(4):867–878. https://doi.org/10.1002/btpr.2415
80. Schulze M, Niemann J, Wijffels RH, Matuszczyk J, Martens DE (2022) Rapid intensification of an established CHO cell fed-batch process. Biotechnol Prog 38(1):e3213. https://doi.org/10.1002/btpr.3213
81. Xiang S, Zhang J, Yu L, Tian J, Tang W, Tang H, Xu K, Wang X, Cui Y, Ren K, Cao W, Su Y, Zhou W (2024) Developing an ultra-intensified fed-batch cell culture process with greatly improved performance and productivity. Biotechnol Bioeng 121(2):696–709. https://doi.org/10.1002/bit.28605
82. Yongky A, Xu J, Tian J, Oliveira C, Zhao J, McFarland KS, Borys MC, Li Z (2019) Process intensification in fed-batch production bioreactors using non-perfusion seed cultures. MAbs 11:1502–1514. https://doi.org/10.1080/19420862.2019.1652075
83. Biechele P, Busse C, Solle D, Scheper T, Reardon K (2015) Sensor systems for bioprocess monitoring. Eng Life Sci 15(5):469–488. https://doi.org/10.1002/elsc.201500014
84. Djisalov M, Knežić T, Podunavac I, Živojević K, Radonic V, Knežević NŽ, Bobrinetskiy I, Gadjanski I (2021) Cultivating multidisciplinarity: manufacturing and sensing challenges in cultured meat production. Biology 10(3):204. https://doi.org/10.3390/biology10030204
85. O'Neill EN, Ansel JC, Kwong GA, Plastino ME, Nelson J, Baar K, Block DE (2022) Spent media analysis suggests cultivated meat media will require species and cell type optimization. npj Sci Food 6(1):46. https://doi.org/10.1038/s41538-022-00157-z
86. Zhao L, Fu H, Zhou W, Hu W (2015) Advances in process monitoring tools for cell culture bioprocesses. Eng Life Sci 15(5):459–468. https://doi.org/10.1002/elsc.201500006
87. Center for Drug Evaluation and Research (2004) PAT – a framework for innovative pharmaceutical development, manufacturing, and quality assurance. FDA. https://www.fda.gov/regulatory-information/search-fda-guidance-documents/pat-framework-innovative-pharmaceutical-development-manufacturing-and-quality-assurance
88. Abu-Absi NR, Kenty BM, Cuellar ME, Borys MC, Sakhamuri S, Strachan DJ, Hausladen MC, Li ZJ (2011) Real time monitoring of multiple parameters in mammalian cell culture bioreactors using an in-line Raman spectroscopy probe. Biotechnol Bioeng 108(5):1215–1221. https://doi.org/10.1002/bit.23023
89. Rubio N, Datar I, Stachura D, Kaplan D, Krueger K (2019) Cell-based fish: a novel approach to seafood production and an opportunity for cellular agriculture. Front Sustain Food Syst 3. https://doi.org/10.3389/fsufs.2019.00043

90. Pawar D, Lo Presti D, Silvestri S, Schena E, Massaroni C (2023) Current and future technologies for monitoring cultured meat: a review. Food Res Int 173:113464. https://doi.org/10.1016/j.foodres.2023.113464
91. Xu S, Chen H (2016) High-density mammalian cell cultures in stirred-tank bioreactor without external pH control. J Biotechnol 231:149–159. https://doi.org/10.1016/j.jbiotec.2016.06.019
92. Beutel S, Henkel S (2011) In situ sensor techniques in modern bioprocess monitoring. Appl Microbiol Biotechnol 91(6):1493–1505. https://doi.org/10.1007/s00253-011-3470-5
93. Busse C, Biechele P, de Vries I, Reardon KF, Solle D, Scheper T (2017) Sensors for disposable bioreactors. Eng Life Sci 17(8):940–952. https://doi.org/10.1002/elsc.201700049
94. Gupta BD, Sharma S (1998) A long-range fiber optic pH sensor prepared by dye doped sol-gel immobilization technique. Opt Commun 154(5):282–284. https://doi.org/10.1016/S0030-4018(98)00321-6
95. Steinegger A, Wolfbeis OS, Borisov SM (2020) Optical sensing and imaging of pH values: spectroscopies, materials, and applications. Chem Rev 120(22):12357–12489. https://doi.org/10.1021/acs.chemrev.0c00451
96. Anane E, Knudsen IM, Wilson GC (2021) Scale-down cultivation in mammalian cell bioreactors – the effect of bioreactor mixing time on the response of CHO cells to dissolved oxygen gradients. Biochem Eng J 166:107870. https://doi.org/10.1016/j.bej.2020.107870
97. Ast T, Mootha VK (2019) Oxygen and mammalian cell culture: are we repeating the experiment of Dr. Ox? Nat Metab 1(9):858–860. https://doi.org/10.1038/s42255-019-0105-0
98. Martin Y, Vermette P (2005) Bioreactors for tissue mass culture: design, characterization, and recent advances. Biomaterials 26(35):7481–7503. https://doi.org/10.1016/j.biomaterials.2005.05.057
99. Goudar CT, Piret JM, Konstantinov KB (2011) Estimating cell specific oxygen uptake and carbon dioxide production rates for mammalian cells in perfusion culture. Biotechnol Prog 27(5):1347–1357. https://doi.org/10.1002/btpr.646
100. Link T, Bäckström M, Graham R, Essers R, Zörner K, Gätgens J, Burchell J, Taylor-Papadimitriou J, Hansson GC, Noll T (2004) Bioprocess development for the production of a recombinant MUC1 fusion protein expressed by CHO-K1 cells in protein-free medium. J Biotechnol 110(1):51–62. https://doi.org/10.1016/j.jbiotec.2003.12.008
101. Al-Ani A, Toms D, Kondro D, Thundathil J, Yu Y, Ungrin M (2018) Oxygenation in cell culture: critical parameters for reproducibility are routinely not reported. PLoS One 13(10): e0204269. https://doi.org/10.1371/journal.pone.0204269
102. Blombach B, Takors R (2015) CO2 – intrinsic product, essential substrate, and regulatory trigger of microbial and mammalian production processes. Front Bioeng Biotechnol 3. https://doi.org/10.3389/fbioe.2015.00108
103. Pattison RN, Swamy J, Mendenhall B, Hwang C, Frohlich BT (2000) Measurement and control of dissolved carbon dioxide in mammalian cell culture processes using an in situ fiber optic chemical sensor. Biotechnol Prog 16(5):769–774. https://doi.org/10.1021/bp000089c
104. Chatterjee M, Ge X, Uplekar S, Kostov Y, Croucher L, Pilli M, Rao G (2015) A unique noninvasive approach to monitoring dissolved O2 and CO2 in cell culture. Biotechnol Bioeng 112(1):104–110. https://doi.org/10.1002/bit.25348
105. Chopda VR, Holzberg T, Ge X, Folio B, Tolosa M, Kostov Y, Tolosa L, Rao G (2020) Real-time dissolved carbon dioxide monitoring I: application of a novel in situ sensor for CO2 monitoring and control. Biotechnol Bioeng 117(4):981–991. https://doi.org/10.1002/bit.27253
106. Matsunaga N, Kano K, Maki Y, Dobashi T (2009) Estimation of dissolved carbon dioxide stripping in a large bioreactor using model medium. J Biosci Bioeng 107(4):419–424. https://doi.org/10.1016/j.jbiosc.2008.11.017
107. Severinghaus JW, Bradley AF (1958) Electrodes for blood pO2 and pCO2 determination. J Appl Physiol 13(3):515–520. https://doi.org/10.1152/jappl.1958.13.3.515
108. Uttamlal M, Walt DR (1995) A fiber-optic carbon dioxide sensor for fermentation monitoring. Bio/Technology 13(6):597–601. https://doi.org/10.1038/nbt0695-597

109. Ali R, Saleh SM, Meier RJ, Azab HA, Abdelgawad II, Wolfbeis OS (2010) Upconverting nanoparticle based optical sensor for carbon dioxide. Sensors Actuators B Chem 150(1): 126–131. https://doi.org/10.1016/j.snb.2010.07.031
110. Schulte A, Brockmann J, Müller N, Anderlei T, Büchs J (2025) A new approach to off-gas analysis for shaken bioreactors showing high CTR and RQ accuracy. J Biol Eng 19(1):11. https://doi.org/10.1186/s13036-025-00480-5
111. Heathman TRJ, Nienow AW, Rafiq QA, Coopman K, Kara B, Hewitt CJ (2018) Agitation and aeration of stirred-bioreactors for the microcarrier culture of human mesenchymal stem cells and potential implications for large-scale bioprocess development. Biochem Eng J 136:9–17. https://doi.org/10.1016/j.bej.2018.04.011
112. Sen A, Kallos MS, Behie LA (2001) Effects of hydrodynamics on cultures of mammalian neural stem cell aggregates in suspension bioreactors. Ind Eng Chem Res 40(23):5350–5357. https://doi.org/10.1021/ie001107y
113. Ebrahimian A, Schalk M, Dürkop M, Maurer M, Bliem R, Kühnel H (2024) Seed train optimization in microcarrier-based cell culture post in situ cell detachment through scale-down hybrid modeling. Bioengineering 11(3):268. https://doi.org/10.3390/bioengineering11030268
114. Abu-Reesh I, Kargi F (1991) Biological responses of hybridoma cells to hydrodynamic shear in an agitated bioreactor. Enzym Microb Technol 13(11):913–919. https://doi.org/10.1016/0141-0229(91)90108-M
115. Kretzmer G, Schügerl K (1991) Response of mammalian cells to shear stress. Appl Microbiol Biotechnol 34:613–616. https://doi.org/10.1007/BF00167909
116. Zhan C, Bidkhori G, Schwarz H, Malm M, Mebrahtu A, Field R, Sellick C, Hatton D, Varley P, Mardinoglu A, Rockberg J, Chotteau V (2020) Low shear stress increases recombinant protein production and high shear stress increases apoptosis in human cells. iScience 23(11). https://doi.org/10.1016/j.isci.2020.101653
117. Chang D, Fox R, Hicks E, Ferguson R, Chang K, Osborne D, Hu W, Velev OD (2017) Investigation of interfacial properties of pure and mixed poloxamers for surfactant-mediated shear protection of mammalian cells. Colloids Surf B Biointerfaces 156:358–365. https://doi.org/10.1016/j.colsurfb.2017.05.040
118. Ahleboot Z, Khorshidtalab M, Motahari P, Mahboudi R, Arjmand R, Mokarizadeh A, Maleknia S (2021) Designing a strategy for pH control to improve CHO cell productivity in bioreactor. Avicenna J Med Biotechnol 13:123–130. https://doi.org/10.18502/ajmb.v13i3.6365
119. Dong J, Gu Y, Li C, Wang C, Feng Z, Qiu R, Chen B, Li J, Zhang S, Wang Z, Zhang J (2009) Response of mesenchymal stem cells to shear stress in tissue-engineered vascular grafts. Acta Pharmacol Sin 30(5):530–536. https://doi.org/10.1038/aps.2009.40
120. Kwon T, Leroux A-C, Zang H, Pollard D, Zehe C, Akbari S (2024) Cell-based shear stress sensor for bioprocessing. J Biotechnol 390:71–79. https://doi.org/10.1016/j.jbiotec.2024.04.016
121. Verma R, Mehan L, Kumar R, Kumar A, Srivastava A (2019) Computational fluid dynamic analysis of hydrodynamic shear stress generated by different impeller combinations in stirred bioreactor. Biochem Eng J 151:107312. https://doi.org/10.1016/j.bej.2019.107312
122. Mulukutla BC, Yongky A, Le T, Mashek DG, Hu W-S (2016) Regulation of glucose metabolism – a perspective from cell bioprocessing. Trends Biotechnol 34(8):638–651. https://doi.org/10.1016/j.tibtech.2016.04.012
123. Lao MS, Toth D (1997) Effects of ammonium and lactate on growth and metabolism of a recombinant Chinese hamster ovary cell culture. Biotechnol Prog 13(5):688–691. https://doi.org/10.1021/bp9602360
124. Ozturk SS, Riley MR, Palsson BO (1992) Effects of ammonia and lactate on hybridoma growth, metabolism, and antibody production. Biotechnol Bioeng 39(4):418–431. https://doi.org/10.1002/bit.260390408
125. Schneider M, Marison IW, von Stockar U (1996) The importance of ammonia in mammalian cell culture. J Biotechnol 46(3):161–185. https://doi.org/10.1016/0168-1656(95)00196-4

126. Synoground BF, McGraw CE, Elliott KS, Leuze C, Roth JR, Harcum SW, Sandoval NR (2021) Transient ammonia stress on Chinese hamster ovary (CHO) cells yield alterations to alanine metabolism and IgG glycosylation profiles. Biotechnol J 16(7):2100098. https://doi.org/10.1002/biot.202100098
127. Yang M, Butler M (2000) Effects of ammonia on CHO cell growth, erythropoietin production, and glycosylation. Biotechnol Bioeng 68(4):370–380. https://doi.org/10.1002/(SICI)1097-0290(20000520)68:4<370::AID-BIT2>3.0.CO;2-K
128. Derfus GE, Abramzon D, Tung M, Chang D, Kiss R, Amanullah A (2010) Cell culture monitoring via an auto-sampler and an integrated multi-functional off-line analyzer. Biotechnol Prog 26(1):284–292. https://doi.org/10.1002/btpr.303
129. Druhmann D, Reinhard S, Schwarz F, Schaaf C, Greisl K, Nötzel T (2011) Utilizing Roche Cedex Bio analyzer for in process monitoring in biotech production. BMC Proc 5(Suppl 8): P106. https://doi.org/10.1186/1753-6561-5-S8-P106
130. Gibbons L, Maslanka F, Le N, Magill A, Singh P, Mclaughlin J, Madden F, Hayes R, McCarthy B, Rode C, O'Mahony J, Rea R, O'Mahony-Hartnett C (2023) An assessment of the impact of Raman based glucose feedback control on CHO cell bioreactor process development. Biotechnol Prog 39(5):e3371. https://doi.org/10.1002/btpr.3371
131. Karnachoriti M, Chatzipetrou M, Touloupakis E, Kontos AG, Zergioti I (2025) Raman spectroscopy as a tool for real-time nutrient monitoring in bioreactor cultivation of microalgae. J Raman Spectrosc. https://doi.org/10.1002/jrs.6841
132. Domján J, Pantea E, Gyürkés M, Madarász L, Kozák D, Farkas A, Horváth B, Benkő Z, Nagy ZK, Marosi G, Hirsch E (2022) Real-time amino acid and glucose monitoring system for the automatic control of nutrient feeding in CHO cell culture using Raman spectroscopy. Biotechnol J 17(5):2100395. https://doi.org/10.1002/biot.202100395
133. Kozma B, Salgó A, Gergely S (2019) On-line glucose monitoring by near infrared spectroscopy during the scale up steps of mammalian cell cultivation process development. Bioprocess Biosyst Eng 42(6):921–932. https://doi.org/10.1007/s00449-019-02091-z
134. Berry BN, Dobrowsky TM, Timson RC, Kshirsagar R, Ryll T, Wiltberger K (2016) Quick generation of Raman spectroscopy based in-process glucose control to influence biopharmaceutical protein product quality during mammalian cell culture. Biotechnol Prog 32(1): 224–234. https://doi.org/10.1002/btpr.2205
135. Hubli GB, Banerjee S, Rathore AS (2023) Near-infrared spectroscopy based monitoring of all 20 amino acids in mammalian cell culture broth. Talanta 254:124187. https://doi.org/10.1016/j.talanta.2022.124187
136. Lee HLT, Boccazzi P, Gorret N, Ram RJ, Sinskey AJ (2004) In situ bioprocess monitoring of *Escherichia coli* bioreactions using Raman spectroscopy. Vib Spectrosc 35(1):131–137. https://doi.org/10.1016/j.vibspec.2003.12.015
137. Rafferty C, Johnson K, O'Mahony J, Burgoyne B, Rea R, Balss KM (2020) Analysis of chemometric models applied to Raman spectroscopy for monitoring key metabolites of cell culture. Biotechnol Prog 36(4):e2977. https://doi.org/10.1002/btpr.2977
138. Whelan J, Craven S, Glennon B (2012) In situ Raman spectroscopy for simultaneous monitoring of multiple process parameters in mammalian cell culture bioreactors. Biotechnol Prog 28(5):1355–1362. https://doi.org/10.1002/btpr.1590
139. Rösner LS, Walter F, Ude C, John GT, Beutel S (2022) Sensors and techniques for on-line determination of cell viability in bioprocess monitoring. Bioengineering 9(12):762. https://doi.org/10.3390/bioengineering9120762
140. Yardley JE, Kell DB, Barrett J, Davey CL (2000) On-line, real-time measurements of cellular biomass using dielectric spectroscopy. Biotechnol Genet Eng Rev 17(1):3–36. https://doi.org/10.1080/02648725.2000.10647986
141. Fernandes J, Currie J, Ramer K, Zhang A (2018) Development of capacitance tools: at-line method for assessing biomass of mammalian cell culture and fixed cell calibration standard. Biotechnol J 14(4):1800283. https://doi.org/10.1002/biot.201800283
142. Ferreira AP, Vieira LM, Cardoso JP, Menezes JC (2005) Evaluation of a new annular capacitance probe for biomass monitoring in industrial pilot-scale fermentations. J Biotechnol 116(4):403–409. https://doi.org/10.1016/j.jbiotec.2004.12.006

143. Konakovsky V, Yagtu AC, Clemens C, Müller M, Berger M, Schlatter S, Herwig C (2015) Universal capacitance model for real-time biomass in cell culture. Sensors (Basel, Switzerland) 15:22128–22150. https://doi.org/10.3390/s150922128
144. Metze S, Ruhl S, Greller G, Grimm C, Scholz J (2019) Monitoring online biomass with a capacitance sensor during scale-up of industrially relevant CHO cell culture fed-batch processes in single-use bioreactors. Bioprocess Biosyst Eng 43:193–205. https://doi.org/10.1007/s00449-019-02216-4
145. Carvell JP, Dowd JE (2006) On-line measurements and control of viable cell density in cell culture manufacturing processes using radio-frequency impedance. Cytotechnology 50(1):35–48. https://doi.org/10.1007/s10616-005-3974-x
146. Tamburini E, Marchetti MG, Pedrini P (2014) Monitoring key parameters in bioprocesses using near-infrared technology. Sensors 14(10):18941. https://doi.org/10.3390/s141018941
147. Fan R, Ebrahimi M, Quitmann H, Aden M, Czermak P (2016) An innovative optical sensor for the online monitoring and control of biomass concentration in a membrane bioreactor system for lactic acid production. Sensors 16(3):411. https://doi.org/10.3390/s16030411
148. Ude C, Schmidt-Hager J, Findeis M, John GT, Scheper T, Beutel S (2014) Application of an online-biomass sensor in an optical multisensory platform prototype for growth monitoring of biotechnical relevant microorganism and cell lines in single-use shake flasks. Sensors 14(9):17390. https://doi.org/10.3390/s140917390
149. Pappenreiter M, Sissolak B, Sommeregger W, Striedner G (2019) Oxygen uptake rate soft-sensing via dynamic kLa computation: cell volume and metabolic transition prediction in mammalian bioprocesses. Front Bioeng Biotechnol 7. https://doi.org/10.3389/fbioe.2019.00195
150. Wechselberger P, Sagmeister P, Herwig C (2012) Real-time estimation of biomass and specific growth rate in physiologically variable recombinant fed-batch processes. Bioprocess Biosyst Eng 36:1205–1218. https://doi.org/10.1007/s00449-012-0848-4
151. Guez J-S, Lacroix P-Y, Château T, Vial C (2023) Deep in situ microscopy for real-time analysis of mammalian cell populations in bioreactors. Sci Rep 13(1):22045. https://doi.org/10.1038/s41598-023-48733-x
152. Kuystermans D, Mohd A, Al-Rubeai M (2012) Automated flow cytometry for monitoring CHO cell cultures. Methods 56(3):358–365. https://doi.org/10.1016/j.ymeth.2012.03.001
153. Flynn J, Breen L, Narayanan S, Butler M (2024) Measurement and control of foam generation in a mammalian cell culture. Biotechnol Prog 40(4):e3450. https://doi.org/10.1002/btpr.3450
154. Routledge SJ (2012) Beyond de-foaming: the effects of antifoams on bioprocess productivity. Comput Struct Biotechnol J 3(4):e201210001. https://doi.org/10.5936/csbj.201210014
155. Delvigne F, Lecomte J (2010) Foam formation and control in bioreactors. In: Encyclopedia of industrial biotechnology. Wiley, pp 1–13. https://doi.org/10.1002/9780470054581.eib326
156. Zhang S, Handa-Corrigan A, Spier RE (1992) Foaming and media surfactant effects on the cultivation of animal cells in stirred and sparged bioreactors. J Biotechnol 25(3):289–306. https://doi.org/10.1016/0168-1656(92)90162-3
157. Junker B (2007) Foam and its mitigation in fermentation systems. Biotechnol Prog 23(4):767–784. https://doi.org/10.1021/bp070032r
158. Tiso T, Demling P, Karmainski T, Oraby A, Eiken J, Liu L, Bongartz P, Wessling M, Desmond P, Schmitz S, Weiser S, Emde F, Czech H, Merz J, Zibek S, Blank LM, Regestein L (2024) Foam control in biotechnological processes—challenges and opportunities. Discov Chem Eng 4(1):2. https://doi.org/10.1007/s43938-023-00039-0
159. Brown AK, Isbell C, Gallagher S, Dodd PW, Varley J (2001) An improved method for controlling foams produced within bioreactors. Food Bioprod Process 79(2):114–121. https://doi.org/10.1205/096030801750286276
160. Austerjost J, Söldner R, Edlund C, Trygg J, Pollard D, Sjögren R (2021) A machine vision approach for bioreactor foam sensing. SLAS Technol Transl Life Sci Innov 26(4):408–414. https://doi.org/10.1177/24726303211008861
161. Lee M, Choi W, Lee JM, Lee ST, Koh W-G, Hong J (2024) Flavor-switchable scaffold for cultured meat with enhanced aromatic properties. Nat Commun 15(1):5450. https://doi.org/10.1038/s41467-024-49521-5

162. Bomkamp C, Skaalure SC, Fernando GF, Ben-Arye T, Swartz EW, Specht EA (2021) Scaffolding biomaterials for 3D cultivated meat: prospects and challenges. Adv Sci 9(3): 2102908. https://doi.org/10.1002/advs.202102908
163. Nguyen T-H, Paprzycki L, Legrand A, Declèves A-E, Heher P, Limpens M, Belayew A, Banerji CRS, Zammit PS, Tassin A (2023) Hypoxia enhances human myoblast differentiation: involvement of HIF1α and impact of DUX4, the FSHD causal gene. Skelet Muscle 13(1):21. https://doi.org/10.1186/s13395-023-00330-2
164. Pierezan MD, Verruck S (2024) Cattle, swine, poultry, and fish cells for meat production: addressing challenges of co-culture and the innovative approaches for muscle and fat integration. Curr Opin Food Sci 60:101230. https://doi.org/10.1016/j.cofs.2024.101230
165. Sugii S, Wong CYQ, Lwin AKO, Chew LJM (2022) Reassessment of adipocyte technology for cellular agriculture of alternative fat. Compr Rev Food Sci Food Saf 21(5):4146–4163. https://doi.org/10.1111/1541-4337.13021
166. Braun T, Gautel M (2011) Transcriptional mechanisms regulating skeletal muscle differentiation, growth and homeostasis. Nat Rev Mol Cell Biol 12(6):349–361. https://doi.org/10.1038/nrm3118
167. Melzener L, Schaeken L, Fros M, Messmer T, Raina D, Kiessling A, van Haaften T, Spaans S, A, Post MJ, Flack JE (2024) Optimisation of cell fate determination for cultivated muscle differentiation. Commun Biol 7(1):1–12. https://doi.org/10.1038/s42003-024-07201-6
168. Kang D-H, Louis F, Liu H, Shimoda H, Nishiyama Y, Nozawa H, Kakitani M, Takagi D, Kasa D, Nagamori E, Irie S, Kitano S, Matsusaki M (2021) Engineered whole cut meat-like tissue by the assembly of cell fibers using tendon-gel integrated bioprinting. Nat Commun 12(1):5059. https://doi.org/10.1038/s41467-021-25236-9
169. Pirkmajer S, Chibalin AV (2011) Serum starvation: caveat emptor. Am J Phys Cell Phys 301(2):C272–C279. https://doi.org/10.1152/ajpcell.00091.2011
170. Messmer T, Klevernic I, Furquim C, Ovchinnikova E, Dogan A, Cruz H, Post MJ, Flack JE (2022) A serum-free media formulation for cultured meat production supports bovine satellite cell differentiation in the absence of serum starvation. Nat Food 3(1):74–85. https://doi.org/10.1038/s43016-021-00419-1
171. Sugii S, Wong CYQ, Lwin AKO, Chew LJM (2023) Alternative fat: redefining adipocytes for biomanufacturing cultivated meat. Trends Biotechnol 41(5):686–700. https://doi.org/10.1016/j.tibtech.2022.08.005
172. Ogawa M, Kermani AS, Huynh MJ, Baar K, Leach JK, Block DE (2024) Edible mycelium as proliferation and differentiation support for anchorage-dependent animal cells in cultivated meat production. npj Sci Food 8(1):23. https://doi.org/10.1038/s41538-024-00263-0
173. VanGenderen CA, Granet JA, Filippelli RL, Liu Y, Chang NC (2022) Modulating myogenesis: an optimized in vitro assay to pharmacologically influence primary myoblast differentiation. Curr Protocols 2(9):e565. https://doi.org/10.1002/cpz1.565
174. Murphy SM, Kiely M, Jakeman PM, Kiely PA, Carson BP (2016) Optimization of an in vitro bioassay to monitor growth and formation of myotubes in real time. Biosci Rep 36(3):e00330. https://doi.org/10.1042/BSR20160036
175. Kinkel AD, Fernyhough ME, Helterline DL, Vierck JL, Oberg KS, Vance TJ, Hausman GJ, Hill RA, Dodson MV (2004) Oil red-O stains non-adipogenic cells: a precautionary note. Cytotechnology 46(1):49–56. https://doi.org/10.1007/s10616-004-3903-4
176. O'Donnell BT, Al-Ghadban S, Ives CJ, L'Ecuyer MP, Monjure TA, Romero-Lopez M, Li Z, Goodman SB, Lin H, Tuan RS, Bunnell BA (2020) Adipose tissue-derived stem cells retain their adipocyte differentiation potential in three-dimensional hydrogels and bioreactors. Biomolecules 10(7):1070. https://doi.org/10.3390/biom10071070
177. Yuan C, Chakraborty S, Chitta KK, Subramanian S, Lim TE, Han W, Bhanu Prakash KN, Sugii S (2019) Fast Adipogenesis Tracking System (FATS) – a robust, high-throughput, automation-ready adipogenesis quantification technique. Stem Cell Res Ther 10(1):38. https://doi.org/10.1186/s13287-019-1141-0
178. Bakhsh A, Kim B, Ishamri I, Choi S, Li X, Li Q, Hur SJ, Park S (2025) Cell-based meat safety and regulatory approaches: a comprehensive review. Food Sci Anim Resour 45(1):145. https://doi.org/10.5851/kosfa.2024.e122

179. de Carvalho JC, Karp SG, Mamani LDG, Biagini G, dos Santos Costa G, Herrmann LW, Soccol CR (2024) Downstream processes for cultivated meat. In: Soccol CR, Molento CFM, Reis GG, Karp SG (eds) Cultivated meat: technologies, commercialization and challenges. Springer Nature, Cham, pp 145–166. https://doi.org/10.1007/978-3-031-55968-6_8
180. Gupta A, Amara JP, Gousseinov E, Cacace B (2020) Chapter 6 – Recent advances in harvest clarification for antibodies and related products. In: Matte A (ed) Approaches to the purification, analysis and characterization of antibody-based therapeutics. Elsevier, pp 117–136. https://doi.org/10.1016/B978-0-08-103019-6.00006-0
181. Bodiou V, Moutsatsou P, Post MJ (2020) Microcarriers for upscaling cultured meat production. Front Nutr 7:10. https://doi.org/10.3389/fnut.2020.00010
182. Hanga MP, Ali J, Moutsatsou P, de la Raga FA, Hewitt CJ, Nienow A, Wall I (2020) Bioprocess development for scalable production of cultivated meat. Biotechnol Bioeng 117(10):3029–3039. https://doi.org/10.1002/bit.27469
183. Nienow AW, Hewitt CJ, Heathman TRJ, Glyn VAM, Fonte GN, Hanga MP, Coopman K, Rafiq QA (2016) Agitation conditions for the culture and detachment of hMSCs from microcarriers in multiple bioreactor platforms. Biochem Eng J 108:24–29. https://doi.org/10.1016/j.bej.2015.08.003
184. Nienow AW, Rafiq QA, Coopman K, Hewitt CJ (2014) A potentially scalable method for the harvesting of hMSCs from microcarriers. Biochem Eng J 85:79–88. https://doi.org/10.1016/j.bej.2014.02.005
185. Schirmaier C, Jossen V, Kaiser S, Jüngerkes F, Brill S, Safavi-Nab A, Siehoff A, van den Bos C, Eibl D, Eibl R (2014) Scale-up of adipose tissue-derived mesenchymal stem cell production in stirred single-use bioreactors under low-serum conditions. Eng Life Sci. https://doi.org/10.1002/elsc.201300134
186. Schnitzler AC, Verma A, Kehoe DE, Jing D, Murrell JR, Der KA, Aysola M, Rapiejko PJ, Punreddy S, Rook MS (2016) Bioprocessing of human mesenchymal stem/stromal cells for therapeutic use: current technologies and challenges. Biochem Eng J 108:3–13. https://doi.org/10.1016/j.bej.2015.08.014
187. Jossen V, van den Bos C, Eibl R, Eibl D (2018) Manufacturing human mesenchymal stem cells at clinical scale: process and regulatory challenges. Appl Microbiol Biotechnol 102(9):3981–3994. https://doi.org/10.1007/s00253-018-8912-x
188. Roush DJ, Lu Y (2008) Advances in primary recovery: centrifugation and membrane technology. Biotechnol Prog 24(3):488–495. https://doi.org/10.1021/bp070414x
189. Dryden WA, Larsen LM, Britt DW, Smith MT (2021) Technical and economic considerations of cell culture harvest and clarification technologies. Biochem Eng J 167:107892. https://doi.org/10.1016/j.bej.2020.107892
190. Kempken R, Preißmann A, Berthold W (1995) Clarification of animal cell cultures on a large scale by continuous centrifugation. J Ind Microbiol 14(1):52–57. https://doi.org/10.1007/BF01570067
191. Fallatah MH, Otaibi AFA, Aljohani RR, Ghabban WJ, Alshehri AA (2024) Evaluation of centrifugation technique and the effect of centrifugation condition on the laboratory samples. J Healthc Sci 4(12):1003–1009. https://doi.org/10.52533/JOHS.2024.41248
192. Sharifian Gh. M, Norouzi F (2023) Guidelines for an optimized differential centrifugation of cells. Biochem Biophys Rep 36:101585. https://doi.org/10.1016/j.bbrep.2023.101585
193. Yang M, Wang Q, Zhu Y, Sheng K, Xiang N, Zhang X (2023) Cell culture medium cycling in cultured meat: key factors and potential strategies. Trends Food Sci Technol 138:564–576. https://doi.org/10.1016/j.tifs.2023.06.031
194. Silva CPD, Semprebon E (2021) How about cultivated meat? The effect of sustainability appeal, environmental awareness and consumption context on consumers' intention to purchase. J Food Prod Mark 27(3):142–156. https://doi.org/10.1080/10454446.2021.1921090
195. Fraeye I, Kratka M, Vandenburgh H, Thorrez L (2020) Sensorial and nutritional aspects of cultured meat in comparison to traditional meat: much to be inferred. Front Nutr 7:35. https://doi.org/10.3389/fnut.2020.00035
196. Louis F, Furuhashi M, Yoshinuma H, Takeuchi S, Matsusaki M (2023) Mimicking Wagyu beef fat in cultured meat: progress in edible bovine adipose tissue production with controllable

fatty acid composition. Mater Today Bio 21:100720. https://doi.org/10.1016/j.mtbio.2023.100720
197. Lee M, Park S, Choi B, Choi W, Lee H, Lee JM, Lee ST, Yoo KH, Han D, Bang G, Hwang H, Koh W-G, Lee S, Hong J (2024) Cultured meat with enriched organoleptic properties by regulating cell differentiation. Nat Commun 15(1):77. https://doi.org/10.1038/s41467-023-44359-9
198. Joo S-T, Choi J-S, Hur S-J, Kim G-D, Kim C-J, Lee E-Y, Bakhsh A, Hwang Y-H (2022) A comparative study on the taste characteristics of satellite cell cultured meat derived from chicken and cattle muscles. Food Sci Anim Resour 42(1):175–185. https://doi.org/10.5851/kosfa.2021.e72
199. Hospodiuk M, Dey M, Sosnoski D, Ozbolat IT (2017) The bioink: a comprehensive review on bioprintable materials. Biotechnol Adv 35(2):217–239. https://doi.org/10.1016/j.biotechadv.2016.12.006
200. Hashizume T, Ozawa Y, Ying B-W (2023) Employing active learning in the optimization of culture medium for mammalian cells. npj Syst Biol Appl 9(1):20. https://doi.org/10.1038/s41540-023-00284-7
201. Todhunter ME, Jubair S, Verma R, Saqe R, Shen K, Duffy B (2024) Artificial intelligence and machine learning applications for cultured meat. Front Artif Intell 7. https://doi.org/10.3389/frai.2024.1424012
202. Duong-Trung N, Born S, Kim JW, Schermeyer M-T, Paulick K, Borisyak M, Cruz-Bournazou MN, Werner T, Scholz R, Schmidt-Thieme L, Neubauer P, Martinez E (2023) When bioprocess engineering meets machine learning: a survey from the perspective of automated bioprocess development. Biochem Eng J 190:108764. https://doi.org/10.1016/j.bej.2022.108764
203. Walsh I, Myint M, Nguyen-Khuong T, Ho YS, Ng SK, Lakshmanan M (2022) Harnessing the potential of machine learning for advancing "Quality by Design" in biomanufacturing. MAbs 14(1):2013593. https://doi.org/10.1080/19420862.2021.2013593
204. Cosenza Z, Block DE, Baar K, Chen X (2023) Multi-objective Bayesian algorithm automatically discovers low-cost high-growth serum-free media for cellular agriculture application. Eng Life Sci 23(8):e2300005. https://doi.org/10.1002/elsc.202300005
205. Nikkhah A, Rohani A, Zarei M, Kulkarni A, Batarseh FA, Blackstone NT, Ovissipour R (2023) Toward sustainable culture media: using artificial intelligence to optimize reduced-serum formulations for cultivated meat. Sci Total Environ 894:164988. https://doi.org/10.1016/j.scitotenv.2023.164988
206. Zavala-Ortiz DA, Denner A, Aguilar-Uscanga MG, Marc A, Ebel B, Guedon E (2022) Comparison of partial least square, artificial neural network, and support vector regressions for real-time monitoring of CHO cell culture processes using in situ near-infrared spectroscopy. Biotechnol Bioeng 119(2):535–549. https://doi.org/10.1002/bit.27997
207. Ng WL, Tan JS (2024) Application of machine learning in 3D bioprinting of cultivated meat. Int J AI Mater Des 1(1):2279. https://doi.org/10.36922/ijamd.2279
208. Bonatti AF, Vozzi G, Chua CK, Maria CD (2022) A deep learning quality control loop of the extrusion-based bioprinting process. Int J Bioprint 8(4):620. https://doi.org/10.18063/ijb.v8i4.620
209. Cui J, Wang Y, Wang Q, Yang L, Zhang Y, Karrar E, Zhang H, Jin Q, Wu G, Wang X (2022) Prediction of flavor of Maillard reaction product of beef tallow residue based on artificial neural network. Food Chem X 15:100447. https://doi.org/10.1016/j.fochx.2022.100447
210. Park S-Y, Park C-H, Choi D-H, Hong JK, Lee D-Y (2021) Bioprocess digital twins of mammalian cell culture for advanced biomanufacturing. Curr Opin Chem Eng 33:100702. https://doi.org/10.1016/j.coche.2021.100702
211. Moser A, Appl C, Pörtner R, Baganz F, Hass VC (2024) A new concept for the rapid development of digital twin core models for bioprocesses in various reactor designs. Fermentation 10(9):463. https://doi.org/10.3390/fermentation10090463
212. Specht EA, Welch DR, Rees Clayton EM, Lagally CD (2018) Opportunities for applying biomedical production and manufacturing methods to the development of the clean meat industry. Biochem Eng J 132:161–168. https://doi.org/10.1016/j.bej.2018.01.015

213. Al-Kharusi G, Dunne NJ, Little S, Levingstone TJ (2022) The role of machine learning and design of experiments in the advancement of biomaterial and tissue engineering research. Bioengineering 9(10):561. https://doi.org/10.3390/bioengineering9100561
214. Cultivated Meat Modeling Consortium (2025). https://thecmmc.org/. Accessed 11 Jun 2025
215. Garrison GL, Biermacher JT, Brorsen BW (2022) How much will large-scale production of cell-cultured meat cost? J Agric Food Res 10:100358. https://doi.org/10.1016/j.jafr.2022.100358
216. Humbird D (2021) Scale-up economics for cultured meat. Biotechnol Bioeng 118(8):3239–3250. https://doi.org/10.1002/bit.27848
217. Goodwin CM, Aimutis WR, Shirwaiker RA (2024) A scoping review of cultivated meat techno-economic analyses to inform future research directions for scaled-up manufacturing. Nat Food. https://doi.org/10.1038/s43016-024-01061-3
218. Lever VC (2025) Lever VC insights: a second generation of cultivated meat companies breaks through projected cost barriers. Lever
219. Vergeer R, Sinke P, Odegard I (2021) TEA of cultivated meat. CE Delft
220. Kobos PH, Drennen TE, Outkin AV, Webb EK, Paap SM, Wiryadinata S (2020) Techno-economic analysis: best practices and assessment tools (No. SAND-2020-13473). Sandia National Lab. (SNL-CA), Sandia National Lab. (SNL-NM), Hobart and William Smith Colleges, Livermore, Albuquerque, Geneva. https://doi.org/10.2172/1738878
221. Coyne A (2025) Protein pioneers: the countries that have approved cultivated meat. *Just Food*, 10 Mar 2025. https://www.just-food.com/features/protein-pioneers-the-countries-which-have-approved-cultivated-meat/

Adv Biochem Eng Biotechnol (2026) 193: 63–80
https://doi.org/10.1007/10_2025_312

Published online: 24 December 2025

Development of Cell Culture Media for Cultivated Meat Production

M. J. Post

Contents

Abstract The culture medium is a central part of the cultivated meat technology, from biological, economical, and safety perspectives. Many cues to drive proliferation or differentiation of cells relevant for cultivated meat are biochemical and are therefore part of the medium. Traditionally, these cues came from fetal bovine serum (FBS), but that has been replaced by a set of components, often of recombinant origin. The culture medium is the costliest input in cultivated meat and therefore subject to intense efforts to reduce cost through simplification, by downgrading components, by replacing components with cheaper alternatives, and by being cognizant of appropriate timing of administration. In addition, reducing the volume of medium per kg of output needed is a subject of research and development. These modifications are all bounded by the need for culture medium to be food-safe and preferably chemically defined. Importantly, the use of medium needs to be as

M. J. Post (✉)
CARIM, Physiology, Maastricht University, Maastricht, The Netherlands

Mosa Meat, B.V, Maastricht, The Netherlands
e-mail: mpost@mosameat.com

efficient as possible through smart feeding and judicious medium regimes to reduce waste to a minimum. Metabolic models and AI tools will further facilitate medium development leading to less experimental waste. It is expected that this progress will continue for many years leading to ever better media at lower cost, enabling commercialization of cultivated meat.

Graphical Abstract

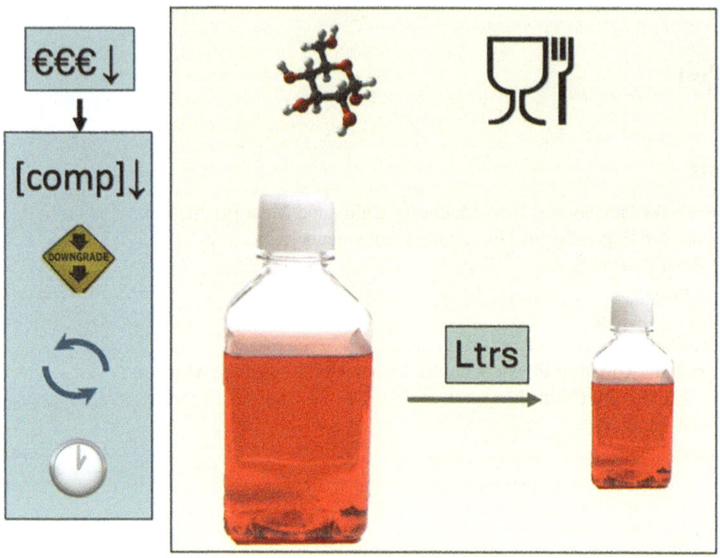

Medium for cultivated meat is preferably chemically defined and is required to be food-safe. Within these boundaries, cost of medium can be reduced by reducing the concentration of components, downgrading components, replacing them with cheaper alternatives, and timing them appropriately. In addition, the volume of medium required for a kg of cultivated meat can be reduced

Keywords Cultivated meat, Medium, Cell culture, Bioprocess

Abbreviations

ATMP	Advanced therapy medicinal product
EGF	Epithelial growth factor
FBS	Fetal bovine serum
FGF-2	Fibroblast growth factor-2
HGF	Hepatocyte growth factor
IGF-1	Insulin-like growth factor-1
IL-6	Interleukin-6
LC-MS	Liquid chromatography-mass spectrometry

LIF-1	Lymphocyte inhibiting factor-1
PDGF-BB	Platelet-derived growth factor BB
STR	Stirred tank bioreactor
TCA	Tricarboxylic acid cycle
TGFβ1	Transforming growth factor beta 1
VEGF-A	Vascular endothelial growth factor-A

1 Rationale for Developing New Media for Cultivated Meat Production

Mammalian cells have been successfully cultured outside of the body for over a century [1]. In addition to conditions like sterility, appropriate temperature, and often a surface that cells can adhere to, the most important and diverse requirement is to provide a fluid containing all the components that cells need to survive or grow. This fluid was named "medium" by Alexis Carrel in 1912 [2] and has carried that name ever since.

To culture cells, one must provide nutrients, minerals, vitamins, and energy supply. This is typically referred to as the basal medium. The medium must have a physiological osmolality and a buffered pH around 7.4. Although this medium can keep the cells alive, the cells typically do not multiply through proliferation. Pro-proliferative cues for the cells traditionally come from serum, most often fetal bovine serum (FBS) [3]. From the very start of the development of cultivated meat, however, FBS is considered an unwanted, unsustainable, unscalable, and uneconomical medium component [4]. The unprecedented scale of cell and tissue culture as well as the techno-economic conditions for cultivated meat emphasize the need for scalable and cheap ingredients. Materials with inherent limited supply such as FBS therefore have no place in cultivated meat production.

For medium components whose production can be scaled up to the extraordinary large scale required – same order of magnitude as animal feed – the traditional small-scale production must be transformed. The low-volume, high-value production of pharma-grade medium must move towards a large-volume, low-value production for cultivated meat to become cost-effective [5, 6]. This is not only true for the basal medium but applies to all medium components, where invariably current supply is far outweighed by future demand. The pressure for components that are used at low concentrations might be less than for abundant components, but the unprecedented scale will affect all components.

Equally important to scalability and cost-effectiveness is the conversion ratio of feedstock in the medium to edible cultivated meat biomass [6]. One of the potential benefits of cultivated meat and indeed an important reason for some pioneers to develop cultivated meat is the promise that the production is more resource efficient than traditional livestock, in particular cows [4]. The benefit can only be achieved if cell and tissue culture is reinvented into an efficient conversion system.

These considerations lead to an approach towards medium development that reduces or eliminates unnecessary components, replaces expensive or unscalable components, or reduces their concentrations. At the same time, the use of feedstock in the medium needs to be optimally used, for instance through selective feeding.

2 Solutions for Expensive or Unscalable Components

2.1 Fetal Bovine Serum

Replacing FBS leads to a so-called serum-free medium, which can be xenofree (i.e., devoid of extracts from another species) and is preferably chemically defined, meaning that each exact chemical description of the component is known and can be analyzed and traced as such. This is already true for components that are specifically synthesized and purified for cell culture, such as amino acids and glucose, under Good Manufacturing Practice principles. However, often used extracts such as serum, transferrin, and albumin are not chemically defined and contain other components that are often unknown and not analyzed. For instance, serum is estimated to contain up to 10,000 proteins [7] which cannot all be replaced by chemically defined alternatives. Fortunately, the required functionalities of serum in cell culture can be achieved by a handful of recombinant proteins.

Serum is added to basal medium to stimulate the cells in culture to proliferate, in addition to some other functions such as adhesion [8] and shear protection [9], with the latter two functions being important for suspension cultures and less relevant for planar cultures.

To provide sufficient proliferative cues to the cells in the absence of serum, a cocktail of growth factors and cytokines is usually sufficient. Commonly tested and used growth factors are fibroblast growth factor-2 (FGF-2), platelet-derived growth factor BB (PDGF-BB), transforming growth factor beta 1 (TGFβ1), hepatocyte growth factor (HGF), insulin-like growth factor-1 (IGF-1), vascular endothelial growth factor-A (VEGF-A), epithelial growth factor (EGF), interleukin-6 (IL-6), lymphocyte inhibiting factor-1 (LIF-1), and activin [10, 11]. In a design of experiment setup, it is relatively easy and fast to establish which cocktail works best at what concentrations, given that the number of growth factors and cytokines to choose from is limited to the abovementioned. Nevertheless, it is extremely useful to have a stable cell line and a robust high-throughput cell-cultured platform with an equally high-throughput readout. The readout is usually number of cells over time in a short-term 96- or 348-well culture, assessed by either an optical cell counting method or a metabolic assay. This system has screening capacity but is not fully predictive of the need for growth factors in a large-scale and high-density suspension culture. For instance, suspension culture using microcarriers might include factors that support adhesion of the cells to the carrier surface in a high agitation environment such as spinner or shaker flask or stirred tank bioreactor (STR). Factors that facilitate adhesion of cells to a polymeric surface are typically extracellular matrix proteins that bind to integrins on the surface of the cell. Examples are vitronectin, fibronectin, collagen, and laminin [12]. Verification of the culture conditions that are established in planar culture screens in longer-term culture and in suspension culture is therefore necessary. Also in higher-density cultures, the concentration of growth factors may need positive or negative adjustment, respectively, depending on the specific consumption or binding of growth factors or increased production and secretion by cells. Mostly, a combination of two or three of these factors suffices, and FGF-2, a universal growth factor that is pro-proliferative on every mammalian cell capable of division, is one of them. Such a combination is cell line and species specific [13], so that for each, a new cocktail must be defined and fine-tuned. Using primary cells with inevitable donor-to-donor variability, robustness of the growth factor cocktail with minimal donor dependency is required. This increases the effort tremendously if growth factor cocktails need to be tested in, for instance, three donors or, in a worst-case scenario, where consistency cannot be achieved, for each new donor.

Another layer of complexity is added when timing of growth factor addition is also varied. The rationale for variation in timing can be that a particular growth factor is extremely unstable, thus requiring frequent spiking in the medium. Alternatively, during a growth or differentiation phase of cell culture, the cells may start to produce and excrete growth factors, partially or totally obviating the need for exogenous administration. These timing considerations apply to all medium components, and an optimal strategy of component supplementation or deprivation can therefore only be designed based on thorough analysis of the spent medium that includes all components.

2.2 Cheaper

As the technologies to develop cultivated meat stem from cell culture and tissue engineering for biomedical purposes, the media were never designed with cost in mind. Rather, they were assembled according to strict quality and uniformity criteria to reduce experimental variation because of changing medium composition or impurities. In addition to standardization, the regulatory requirements for medical use are stringent, forcing producers to purify these ingredients to very high grades. For instance, for advanced therapy medicinal products (ATMPs), the levels of endotoxins need to be carefully minimized and controlled [14], whereas cells in culture are not sensitive to endotoxins [15]. High-grade, highly purified medium components are expensive and not scaled to the prospective needs for consumer products, such as meat, milk, or leather. As an example, at the start of the cultivated meat development, FGF-2 cost €800,000–2,000,000 per g. Some suppliers currently provide less pure, but equally functional, FGF-2 for €1,500 per g, indicating a more than 1,000-fold reduction in price in a matter of a few years. For some recombinant proteins, it is assumed that the cost of production will drop to USD 10/g when produced at large scale [16], and indeed for food-grade enzymes such as lipase, this cost range is within reach [17]. This will likely also be true for most growth factors, cytokines, and essential vitamins [18]. Here, we are facing a practical Catch-22 situation where scaling up is expensive because of current medium prices and therefore limited, thus delaying the demand-driven large-scale production of cheap medium components.

Medium will be made cheaper by:

1. Eliminating unnecessary components
2. Minimizing the concentration of components
3. Replacing them with cheap alternatives

Prime example of eliminating unnecessary medium components is antimicrobial agents, such as antibiotics. Antibiotics are commonly used in cell culture to prevent infections that follow accidental contamination of the culture [19]. The medium and temperature of the culture are highly conducive to microbial overgrowth in case of contamination. However, in addition to being expensive medium components, antimicrobial agents are also unwanted for food safety and public health reasons related to antibiotic-resistant bacteria in the environment [20]. Antibiotics negatively affect cell growth and viability, and one should expect improved cell performance in the absence of antimicrobial agents [21]. With proper cell culture techniques and the design of a culture system that minimizes or avoids altogether exposure to air, it is possible and indeed desirable to refrain from using antimicrobial agents. A notable exception might be the medium in which biopsy samples from livestock are taken up, immediately after the biopsy is taken under aseptic but non-sterile environment [22]. Since this is at the entry stage of the process and cultures are still very small, cost is not affected and any concerns about having antibiotics at this stage are limited. To simplify the medium and eliminate all the components that are not necessary for the cell type at hand, a thorough analysis should be performed by eliminating components one by one and in combination. This is a massive undertaking and a good candidate for artificial intelligence (AI) methodologies.

Standard concentrations of components in medium are not optimized for every cell type but are generally sufficient. It is however quite possible and indeed likely that they are either too high or too low for a particular cell line. The large differences between, for instance, amino acid concentrations in Dulbecco's minimum essential medium and F12 medium suggest that further optimization for cell lines and species is likely needed. Once the optimal concentrations have been established, it is possible of course to trade a slightly reduced cell growth for a substantial cheaper medium.

An interesting way to reduce concentrations of components is by adding stabilizers. Stabilizing growth factors by changing their sequence as is the case for thermally stable FGF-2, or by adding stabilizers to the medium, can reduce the need for frequent spiking as earlier discussed [23].

Currently, growth factors are added to the medium from start to finish of the culture, and it is doubtful that this is necessary as shown recently for HGF [23]. Since cells in culture start to produce and excrete some growth factors, it is likely that in certain culture conditions, growth factor administration can be reduced or omitted.

Table 1 Characteristics of common medium component categories

Component	Concentration	Cost (€/g)	prol/diff	Alternative	Remarks
Basal	17 g/L	0.25	Both	Downgrade	Conservation
Growth factors	100 μg/L	1,500	prol	Lower purity	Timing
Albumin	5 g/L	6.00	prol	Replacer	Conservation/ recycling
Transferrin	10 mg/L	700	Both	Fe chelator	
Insulin	20 mg/L	200	prol/ (diff)	IGF, others	Conservation/ recycling
Glutamax	1 g/L	0.70	Both	Glutamine	Ammonia toxicity
PSA	1% v/v	3.00 (1 mL)	Both	PSA-free	Contamination risk

prol proliferation, *diff* differentiation
Source: Various commercial suppliers for prices and known concentrations of components in E8 [27]

For instance, when cells are grown in aggregates and are in close contact, paracrine stimulation by locally produced growth factors that are captured in the intercellular space might be sufficient to sustain further growth. On the other hand, 3D structures complexed with extracellular matrix that is laid down by the cells likely pose diffusion obstacles to medium components [24], thus requiring perhaps even higher growth factor concentrations than in a planar culture. No data is publicly available yet to support these notions, so until then, empirical evidence needs to be gathered to fully optimize growth factor administration through every phase of the culture.

Replacing expensive components with cheaper equivalents has led to tremendous reductions in the cost of medium as exemplified by the previously mentioned case of FGF-2. A lot of studies and discussion on the cost of medium have focused on growth factors because of the historic high prices. Given that growth factors are used at low concentrations, and we likely have not achieved the optimal regime of administration yet, recombinant growth factors will not be important cost contributors in large-scale cell culture for cultivated meat in the future. Nevertheless, research is ongoing to identify nonrecombinant functional replacements for serum and serum-replacing proteins such as growth factors [25]. Although not animal-free, it was also recently shown that cow milk fractions can partially replace serum in mouse and bovine myoblast cultures [26].

For medium components that are present at high concentrations, such as albumin, transferrin, and insulin, it is likely that the cost of production of recombinant analogues will remain too high for them to be cost-effective (Table 1). This means that alternative solutions need to be sought.

Albumin is a necessary component in cell medium although its precise function or set of functions remains elusive. It is therefore difficult to establish a replacement on rational grounds. Based on perceived functions and protein characteristics, a host of plant-based alternatives have been suggested [10, 28–30]. While some successes have been reported, they rely on trial and error and appear to be cell and context specific. These developments started with Chinese hamster ovary (CHO) cells where canola protein was successfully used to replace albumin [31]. Recently, this concept was adopted for bovine stem cells with the intention to develop a medium for cultivated meat [10]. Most likely because of regulatory concerns about the batch-to-batch variability and impurities of a crop derivative, this discovery has not led to widespread use of canola in CHO bioprocesses, which are predominantly biopharmaceutical oriented. Biological, seasonal, and geographical dependent variability of crop-derived components can be equally problematic for a stable cultivated meat bioprocess. For instance, when relying on plant-protein hydrolysates to provide feed, this variability can become a concern. On the plus side, for cultivated meat, there is no regulatory issue provided that the plant extract is food-safe and free of pest-control chemicals or other contaminations. The search for suitable and cheap albumin alternatives is still ongoing, and solutions are again likely cell and species specific. If for instance adipogenic precursors are grown with the intention to differentiate them into adipocytes, canola might not be a good alternative given its inhibition of adipogenesis [32].

Insulin can most likely be partially replaced by IGF-1. The effective concentration at half of the efficacy (EC50) for IGF-1 is 100-fold lower than for insulin [33] while showing a similar quality of effect on the insulin and IGF-1 receptor [34]. Further in favor of IGF-1 is the development of stable variants such as LR3-IFG-1 [35]. Both insulin and IGF-1 being recombinantly made proteins with similar future cost profiles, IGF-1 will be a suitable insulin alternative. A single-chain, stable, insulin has recently entered the market specifically for cellular agriculture applications [36]. Some plant-based compounds such as aglycin, vglycin, and ginseng oligopeptides can mimic insulin's effect on cell culture [37], although most proposed insulin mimetics of plant origin act on insulin release rather than mimicking its cellular action. Summarizing, solutions for insulin are expected or already around.

Transferrin delivers iron to the cell as its sole function, which suggests that an appropriate iron chelator can replace transferrin. In vivo, an effect on the immune system has been described for transferrin primarily in fish [38], but this is unlikely to play a role in cell culture. Although iron chelation seems to be a relatively simple mechanism, most iron chelators were developed to sequester iron in diseases that lead to accumulation of iron in the blood and not to deliver iron to cells. They therefore were designed to have a high affinity and binding capacity for iron. Characteristics of iron chelators such as their affinity to Fe^{2+} and Fe^{3+} and pH dependency of affinity are important for their functionality in cell culture [39]. Indeed, efficacy of iron chelators as replacement for transferrin has been shown for erythropoietic as well as kidney and (CHO) cells [40–42], and it is likely that among the large array of iron chelators, a suitable transferrin replacer will be found.

Remaining components such as vitamins, minerals, and amino acids and the energy components glucose and glutamine are typically prevalent as well as cheap, especially if they are downgraded from pharmaceutical-grade to food-safe levels. Especially for vitamins, stability and consumption are important considerations. When stable and not consumed, they are obvious targets for savings in a culture, where only components that are being consumed are supplemented (see below).

A thorough metabolic analysis of spent medium will reveal the specific consumption rates of all components and especially nutrients such as glucose and glutamine. Glucose and glutamine are typically depleted the fastest in proliferating cells, and their timely supplementation needs to keep up with consumption. Because of the release of ammonia from glutamine in the absence of cells, but more pronounced because of cellular metabolism, glutamine is held responsible for most of the toxic ammonia buildup in the cell culture. It is for sure not the only source, as amino acid metabolism also leads to release of ammonia. To prevent the ammonia release in cell-free medium, glutamine is commonly incorporated in medium in the form of Glutamax, a dipeptide of glutamine and alanine. The dipeptide is cleaved by the cell, thereby regulating release of glutamine as a function of the cell's metabolic activity. However, Glutamax is expensive and can be replaced by glutamine or other precursors of the TCA cycle substrates. In case of glutamine addition, ammonia buildup needs to be monitored and controlled.

Replacement of glutamine by precursors of the TCA cycle substrates led recently to a non-ammoniogenic medium where glutamine was successfully replaced by α-ketoglutarate (α-KG) [43]. α-KG is relatively cheap and already used in agriculture [44]. Adding pyruvate or galactose is another alternative [45, 46]. Since the accumulation of toxic metabolic waste, in particular ammonia, requires medium refreshment and thereby wasting of precious medium components, non-ammoniogenic strategies are likely to play an important role in efficient medium and feedstock use. The effect of non-ammoniogenic media can be accompanied by glutamine synthase activity in cells that spontaneously reduce ammonia. Both effects are again cell and species specific [47].

In addition to cost-effectiveness, it is important to make efficient use of the feedstock. After all, it is a goal of cellular agriculture to reduce the impact of meat production on natural resources.

3 Resource Efficiency

Traditional cell culture is not only expensive but also inefficient. The inefficiency comes from the general practice to refresh the entire medium when the pH drops below levels compatible with cell growth or even cell survival. When cells grow and metabolize energy components such as glucose and nutrients, the medium acidifies by production and release of protons. In a static planar culture, protons are buffered to some degree until the finite buffer capacity of the medium is reached. Further release of protons leads to a drop in pH of the medium. When the cells are cultured in a bioreactor, additional control of pH can be secured by reducing carbon dioxide gassing making use of the carbo-anhydrase activity in cells and using bicarbonate to buffer protons, thus increasing pH. However, both buffer systems are easily saturated, especially at the high cell densities than can be obtained in suspension culture.

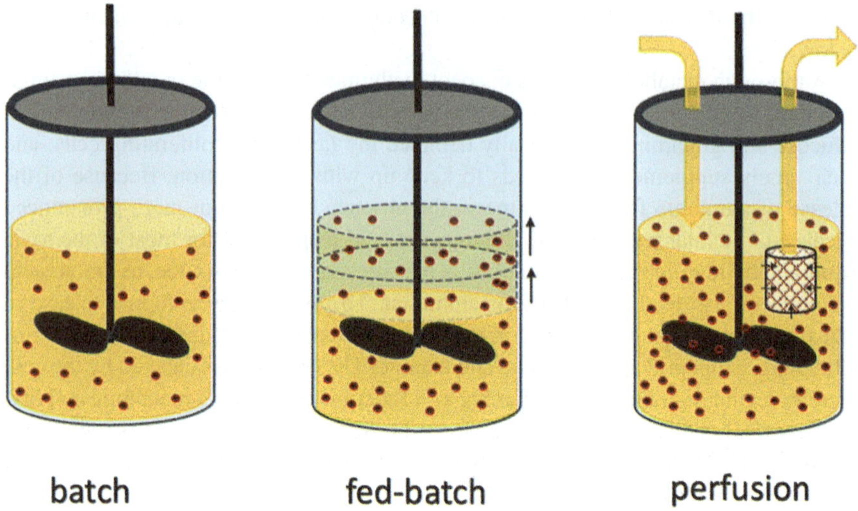

Fig. 1 Three types of bioprocess, a batch, fed-batch, or perfusion process. The depiction is schematic and does not reflect the actual setup. Filtration of medium in the perfusion setup typically occurs outside of the vessel, with a separate back flow of cells and waste flow of spent medium. Fed-batch and perfusion cultures usually allow higher cell densities

Adding alkaline solutions such as bicarbonate is an alternative solution, although limited by osmolality boundaries [48] as it is usually administered as a salt.

There are two considerations when designing an efficient cell culture system, and they both are intricately related to the bioreactor and bioprocess that are developed (Fig. 1). One is to supply those nutrients, vitamins, growth factors, and other components that are being consumed by the cells or are otherwise depleted, for instance, by degradation or irreversible binding [49]. The other consideration is indeed washing out accumulated toxic waste during prolonged cell culture. In a batch culture (Fig. 1), a medium that is enriched for highly consumed components is preferred to achieve a high cell density. In a fed-batch culture, components that are being highly consumed are added to the medium, gradually increasing the volume of the culture. A perfusion culture allows continuous medium refreshment, while the outflow is filtered to keep the cells in the culture. The system is versatile as it allows the use of enriched medium [50] as well as feeding.

Enrichment obviates the need for handling the culture until certain cell densities are reached and additional feeding becomes necessary. The feeding can be done in a fed-batch culture where the volume of the culture gradually increases upon feeding or in perfusion mode, where excess medium is filtered and taken to waste. In perfusion mode, the rate of perfusion and consequently medium refreshment can then be governed by the rate of toxic waste accumulation, such as ammonia and lactate [51]. To obtain the highest resource efficiency possible, the perfusion rate must obviously be kept to a minimum.

To optimize these processes for a particular cell line, a good documentation and understanding of the metabolism are required. Ascertaining depletion of essential medium components is typically done by spent medium analysis, either online, in-line, or off-line. Commercial equipment is available for analysis of commonly used components such as glucose and glutamine. For elaborate analysis of spent medium, LC-MS techniques have been developed [52]. A time course of these analytes together with assessment of cell density provides specific consumption patterns of medium components that allow the design of a feeding strategy. This feeding strategy leads to resource efficiency as only those medium components that are depleted will be supplemented.

Alternatively, one could selectively remove toxic waste or recycle medium after the end of culture. Neither has been practiced at large scale yet. Research into selectively filtering ammonia from the medium or a circular culture system including ammonia-reducing microorganisms [53] is ongoing, and promising technologies have been developed, especially in wastewater management [54], but recently also for cultivated meat [55]. The biggest challenge is to design an efficient and cost-effective method. Ideally, the resulting nitrogen can be reused for valuable side-stream products to help recovering the additional cost of filtration. The other potentially toxic waste product is lactate coming from glycolysis of the glucose present in the medium. The toxic effect of lactate is not very well established as the production of lactate leads to an equivalent number of protons causing a harmful drop in pH. Thus, the assumed negative effect of lactate is difficult to distinguish from the concomitant acidification. Adding sodium lactate to a culture has led to the conclusion that up to a concentration of 20 mM and for some cells even considerably higher, lactate does not affect cell viability [56, 57].

Recycling of medium refers to extracting valuable medium components from the spent medium. From a cost perspective for instance, it would target specific valuable proteins such as stable growth factors, albumin, insulin, and others. From a resource efficiency perspective, an off-line strategy to remove toxic waste is the better strategy making the distinction with selectively removing toxic waste or using a non-ammoniogenic medium unclear.

4 Differentiation

Despite successful undifferentiated cell-based cultivated meat products, it is the goal of cultivated meat developers to replace conventional meat with a product that is similar if not equal. This requires not only large-scale cell production, but also differentiation of the cells into either muscle or fat and preferably both. Medium composition and medium requirement during differentiation are different from proliferation but have the same requirements for being serum-free and food-safe [58, 59]. During differentiation, proliferation-promoting growth factors are typically withdrawn and replaced by differentiation-stimulating triggers. Nutrient composition might also be different to cater to the high protein production in differentiating muscle or fatty acid production in developing adipocytes. It is assumed that metabolism during differentiation is less high than during proliferation, as indicated by reduced glucose and glutamine consumption [60], suggesting that less toxic waste is produced. Consequently, less medium refreshment is needed. On the other hand, differentiation typically takes longer than proliferation but still leading to an overall greatly lower amount of medium than during proliferation. This long residence time in bioreactors does introduce additional bioreactor cost.

The big differences between proliferation and differentiation medium mean that both need to be optimized with respect to initial composition, additional feeding, and change of regime/bioreactor setup.

5 Components Affecting Physicochemical Characteristics of the Medium

Culturing mammalian cells in suspension in a bioreactor comes with specific considerations on foam formation and shear stress. Both can be detrimental to the culture, and therefore measures are taken to prevent foam formation and keep shear stress at low enough levels for mammalian cells. Anti-foaming and shear stress-protecting agents are added to the medium to avoid these hazards.

Foaming is especially prevalent when the culture needs to be sparged, for instance, to provide sufficient oxygen to the cells [61], and it is related to the protein and polar lipid content of the medium [62, 63]. Foaming also occurs in serum-free or chemically defined medium [64]. While mechanical methods are being considered to break down foam, the majority of strategies aim to prevent foaming by adding anti-foam compounds to the medium. With selecting the appropriate anti-foam considerations such as cell compatibility [65], price and food safety in addition to specific functional criteria need to be taken into account. A large array of anti-foams are being used in fermentation, and they all have a surface activity by reducing surface tension. The chemical nature is widely variable. Some are derived from oils, silicon, or polyalkylene glycol. There are many interactions of anti-foaming agents with other medium components and also with functions such as metabolism and oxygen, mass and heat transfer [62]. Establishing a medium with appropriate anti-foam activity is therefore an important subject of optimization.

Shear stress protection for mammalian cells in a highly agitated environment was introduced in serum-free medium and has shown to improve cell performance. Shear stress comes from mechanical stress induced by the impeller in a stirred tank bioreactor or by air bubbles during sparging. Shear stress-protecting agents such as poloxamer or methylcellulose are macromolecules in the medium that improve cell density in suspension cultures in bioreactors [66]. For poloxamer, the mechanism includes incorporation or attachment of the compound to the cell membrane, while the mechanism for methylcellulose is unknown but referred to as either a direct shear stress suppression or an interfacial compound separating air bubbles from cell membranes. Having to add shear stress-protecting agents again adds complexity to establishing a functional medium.

6 Safety and Quality Control

Because medium components might end up in the final consumer product, food safety consideration extends to analysis of the medium composition and potential hazards associated with these components. For most components such as minerals, nutrients, and vitamins, food safety is clear, although the manufacturing method must be documented and monitored [67]. For some more unusual components such as growth factors [68] and antimicrobial agents, specific analysis and perhaps even regulation need to be developed. For components where food safety is unknown, they should be replaced by food-safe alternatives or should be reduced to below detectable levels. Alternatively, the manufacturer must provide toxicity data for the specific component that requires a formal toxicity study. This is true not only for unknown components but also for contaminants of known components. Quality control of incoming materials is therefore highly recommended if not required to exclude contaminants. Another safety concern is allergenicity through the introduction of unfamiliar protein sequences or other known allergenic medium components. A known medium component that is allergenic is phenol red, a pH indicator usually added to medium. This indicator can be easily avoided; it is added to cell culture medium for convenience only. A relatively new topic of allo-allergenicity was brought in the context of using human sequences of recombinant proteins such as growth factors. Allo-allergenicity has been described in transplantation medicine but has not been described for food yet. It is therefore unknown if allo-allergenicity to ingested proteins can occur. Summarizing, medium development has many regulatory implications.

7 Artificial Intelligence

The call for using AI to speed up development or optimization of medium composition is becoming louder and louder. Being a multi-independent-variable problem, this is understandable, and indeed of all the optimization processes related to cultivated meat, medium optimization seems especially eligible for an AI approach [69]. In the absence of a clear hypothesis on what medium components should be present and optimized, the use of AI undoubtedly is the most efficient way to find an optimal composition. Various tools have been developed for this, for instance Bayesian optimization [70, 71]. This is true for growth factors, where there is no preexistent understanding of what cells require which growth factors given a particular culture condition: It is simply trial and error. The simplest form of AI is a design of experiments where full factorial analysis of every potential combination of growth factors is avoided by smartly using the data of a limited set of combinations. To optimize a feeding strategy, existing AI algorithms can be used, although it is likely that supplementing depleted medium components is a more efficient optimization strategy. Here, the theory to base one's intervention on seems clear, and there will be less efficiency benefit from a black-box AI approach [20].

Using more elaborate and powerful generative AI methods typically requires large preexistent datasets on relevant culture systems. Although those datasets are likely present among the many groups working on cultivated meat-relevant cell lines, this data is unfortunately not public and can therefore not be used to teach neural networks. One therefore needs to implement AI judicially and critically.

8 Summary

Medium development is a complex problem but scientifically not a very difficult problem. Finding the optimum basal medium to support proliferation or differentiation is a matter of time and lots of experimentation. A high-throughput assay system is paramount. A medium is first optimized for efficacy with a secondary goal of cost-effectiveness that is especially relevant for cultivated meat. Many suitable serum-free media have been developed in the cultivated meat field, while great advance has been made towards very-low-cost solutions. It is expected that this progress will continue for many years leading to ever better media at lower cost. Large-scale demand and production will further lead to reduced cost. Importantly, the use of medium needs to be as efficient as possible through smart feeding and judicious medium regimes to reduce waste to a minimum. Metabolic models and AI tools will further facilitate medium development leading to less experimental waste.

References

1. Taylor MW (2014) A history of cell culture. In: Viruses and man: a history of interactions. Springer, pp 41–52
2. Carrel A (1912) Pure cultures of cells. J Exp Med 16(2):165–168
3. van der Valk J, Brunner D, De Smet K, Fex Svenningsen A, Honegger P, Knudsen LE et al (2010) Optimization of chemically defined cell culture media – replacing fetal bovine serum in mammalian in vitro methods. Toxicol In Vitro 24(4):1053–1063
4. Post MJ (2012) Cultured meat from stem cells: challenges and prospects. Meat Sci 92(3): 297–301
5. Specht EA, Welch DR, Rees Clayton EM, Lagally CD (2018) Opportunities for applying biomedical production and manufacturing methods to the development of the clean meat industry. Biochem Eng J 132:161–168
6. Hubalek S, Post MJ, Moutsatsou P (2022) Towards resource- and cost-efficient cultured meat. Curr Opin Food Sci 47:100885
7. Adkins JN, Varnum SM, Auberry KJ, Moore RJ, Angell NH, Smith RD et al (2002) Toward a human blood serum proteome: analysis by multidimensional separation coupled with mass spectrometry. Mol Cell Proteomics 1(12):947–955

8. Underwood PA, Bean PA, Mitchell SM, Whitelock JM (2001) Specific affinity depletion of cell adhesion molecules and growth factors from serum. J Immunol Methods 247(1):217–224
9. Bryant JC (1966) Mammalian cells in chemically defined media in suspension cultures. Ann N Y Acad Sci 139(1):143–161
10. Stout AJ, Rittenberg ML, Shub M, Saad MK, Mirliani AB, Dolgin J et al (2023) A Beefy-R culture medium: replacing albumin with rapeseed protein isolates. Biomaterials 296:122092
11. Kolkmann AM, Van Essen A, Post MJ, Moutsatsou P (2022) Development of a chemically defined medium for in vitro expansion of primary bovine satellite cells. Front Bioeng Biotechnol 10:895289
12. Bodiou V, Kumar AA, Massarelli E, van Haaften T, Post MJ, Moutsatsou P (2024) Attachment promoting compounds significantly enhance cell proliferation and purity of bovine satellite cells grown on microcarriers in the absence of serum. Front Bioeng Biotechnol 12:1443914
13. Barnes D, Sato G (1980) Serum-free cell culture: a unifying approach. Cell 22(3):649–655
14. EMA (2014) Guidelines on Good manufacturing practice for advanced therapy medicinal products. European Committee
15. Harlan JM, Harker LA, Striker GE, Jean WL (1983) Effects of lipopolysaccharide on human endothelial cells in culture. Thromb Res 29(1):15–26
16. Schwartz E, Ravi A, Reeber A, Levink J, Huang TB, Smith B (2023) Anticipated growth factor and recombinant protein costs and volumes necessary for cost-competitive cultivated meat, Jan 2023
17. Khootama A, Putri DN, Hermansyah H (2018) Techno-economic analysis of lipase enzyme production from Aspergillus niger using agro-industrial waste by solid state fermentation. Energy Procedia 153:143–148
18. Alibaba (2025) https://www.alibaba.com/product-detail/China-Manufacturer-s-Food-Grade-Vitamin_1601463480281.html?spm=a2700.7724857.0.0.48b2edb7Iur0TX
19. Weiskirchen S, Schröder SK, Buhl EM, Weiskirchen R (2023) A beginner's guide to cell culture: practical advice for preventing needless problems. Cells 12(5):682
20. O'Neill EN, Cosenza ZA, Baar K, Block DE (2021) Considerations for the development of cost-effective cell culture media for cultivated meat production. Compr Rev Food Sci Food Saf 20(1):686–709
21. Kolkmann AM, Post MJ, Rutjens MAM, van Essen ALM, Moutsatsou P (2020) Serum-free media for the growth of primary bovine myoblasts. Cytotechnology 72(1):111–120
22. Dohmen RGJ, Hueber R, Martins B, Gouveia MA, Zschieschang S, de Vries M et al (2025) Optimisation of skeletal muscle sampling for cultured fat production. bioRxiv. https://doi.org/10.1101/2025.01.22.634340
23. Schenzle L, Egger K, Spangl B, Hussein M, Ebrahimian A, Kuehnel H et al (2025) Low-cost food-grade alternatives for serum albumins in FBS-free cell culture media. Sci Rep 15(1):15296
24. McMurtrey RJ (2017) Roles of diffusion dynamics in stem cell signaling and three-dimensional tissue development. Stem Cells Dev 26(18):1293–1303
25. Golshan M, Dortaj H, Rajabi M, Omidi Z, Golshan M, Pourentezari M et al (2024) Animal origins free products in cell culture media: a new frontier. Cytotechnology 77(1):12
26. Shima A, Oda H, Takeuchi S (2025) Milk-derived extract for skeletal muscle cell cultivation. Food Biosci 66:106218
27. Chen G, Gulbranson DR, Hou Z, Bolin JM, Ruotti V, Probasco MD et al (2011) Chemically defined conditions for human iPSC derivation and culture. Nat Methods 8(5):424–429
28. Obaidi I, Mota LM, Quigley A, Butler M (2021) The role of protein hydrolysates in prolonging viability and enhancing antibody production of CHO cells. Appl Microbiol Biotechnol 105(8): 3115–3129
29. Amirvaresi A, Sarkarat R, Jones C, Shahsavari A, Ovissipour R (2025) Sustainable alternatives to fetal bovine serum: evaluating the role of plant and insect protein isolates in serum-free media for bovine satellite cell proliferation in cultivated meat production. ACS Food Sci Technol 5(4): 1614–1624
30. Mi J, Guo Z, Qian H, Birch WR, Srinivas V, Zhou W (2025) Sustainable food-grade serum-reducing biomaterial: plant protein hydrolysates for scalable cultivated meat manufacturing. Chem Eng J 519:164496

31. Farges-Haddani B, Tessier B, Chenu S, Chevalot I, Harscoat C, Marc I et al (2006) Peptide fractions of rapeseed hydrolysates as an alternative to animal proteins in CHO cell culture media. Process Biochem 41(11):2297–2304
32. Alashi AM, Blanchard CL, Mailer RJ, Agboola SO, Mawson AJ, Aluko RE et al (2015) Effects of canola proteins and hydrolysates on adipogenic differentiation of C3H10T/2 mesenchymal stem cells. Food Chem 185:226–232
33. Galal MA, Alouch SS, Alsultan BS, Dahman H, Alyabis NA, Alammar SA et al (2023) Insulin receptor isoforms and insulin growth factor-like receptors: implications in cell signaling, carcinogenesis, and chemoresistance. Int J Mol Sci 24(19):15006
34. Boucher J, Tseng YH, Kahn CR (2010) Insulin and insulin-like growth factor-1 receptors act as ligand-specific amplitude modulators of a common pathway regulating gene transcription. J Biol Chem 285(22):17235–17245
35. Tomas FM, Lemmey AB, Read LC, Ballard FJ (1996) Superior potency of infused IGF-I analogues which bind poorly to IGF-binding proteins is maintained when administered by injection. J Endocrinol 150(1):77–84
36. Biologicals S (2025). https://sticta.webflow.io/, Accessed: 14-9-2025
37. Costa IS, Medeiros AF, Piuvezam G, Medeiros G, Maciel BLL, Morais AHA (2020) Insulin-like proteins in plant sources: a systematic review. Diabetes Metab Syndr Obes 13:3421–3431
38. Kamińska-Gibas T, Szczygieł J, Jurecka P, Irnazarow I (2020) The many faces of transferrin: does genotype modulate immune response? Fish Shellfish Immunol 102:511–518
39. Evans RW, Kong X, Hider RC (2012) Iron mobilization from transferrin by therapeutic iron chelating agents. Biochim Biophys Acta Gen Subj 1820(3):282–290
40. Landschulz W, Thesleff I, Ekblom P (1984) A lipophilic iron chelator can replace transferrin as a stimulator of cell proliferation and differentiation. J Cell Biol 98(2):596–601
41. Gallego-Murillo JS, Yağcı N, Pinho EM, Wahl SA, van den Akker E, von Lindern M (2023) Iron-loaded deferiprone can support full hemoglobinization of cultured red blood cells. Sci Rep 13(1):6960
42. Xu J, Yan F-r, Li Z-h, Wang D, Sheng H-l, Liu Y (2014) Serum-free medium optimization based on trial design and support vector regression. Biomed Res Int 2014(1):269305
43. Hubalek S, Melke J, Pawlica P, Post MJ, Moutsatsou P (2023) Non-ammoniagenic proliferation and differentiation media for cultivated adipose tissue. Front Bioeng Biotechnol 11:1202165
44. Du G, Liu L, Chen J (2015) Chapter 11: White biotechnology for organic acids. In: Pandey A, Höfer R, Taherzadeh M, Nampoothiri KM, Larroche C (eds) Industrial biorefineries and white biotechnology. Elsevier, Amsterdam, pp 409–444
45. Altamirano C, Illanes A, Becerra S, Cairo JJ, Godia F (2006) Considerations on the lactate consumption by CHO cells in the presence of galactose. J Biotechnol 125(4):547–556
46. Genzel Y, Ritter JB, König S, Alt R, Reichl U (2005) Substitution of glutamine by pyruvate to reduce ammonia formation and growth inhibition of mammalian cells. Biotechnol Prog 21(1): 58–69
47. Meister A (1985) Chapter 27: Glutamine synthetase from mammalian tissues. In: Methods in enzymology, vol 113. Academic Press, pp 185–199
48. Zhu MM, Goyal A, Rank DL, Gupta SK, Van den Boom T, Lee SS (2005) Effects of elevated pCO2 and osmolality on growth of CHO cells and production of antibody-fusion protein B1: a case study. Biotechnol Prog 21(1):70–77
49. van der Wielen LAM, Mussatto SI, van Breugel J (2021) Bioprocess intensification: cases that (don't) work. New Biotechnol 61:108–115
50. Xu J, Xu X, Huang C, Angelo J, Oliveira CL, Xu M et al (2020) Biomanufacturing evolution from conventional to intensified processes for productivity improvement: a case study. MAbs 12(1):1770669
51. Mercier SM, Rouel PM, Lebrun P, Diepenbroek B, Wijffels RH, Streefland M (2016) Process analytical technology tools for perfusion cell culture. Eng Life Sci 16(1):25–35
52. O'Neill EN, Ansel JC, Kwong GA, Plastino ME, Nelson J, Baar K et al (2022) Spent media analysis suggests cultivated meat media will require species and cell type optimization. npj Sci Food 6(1):46

53. Yamanaka K, Haraguchi Y, Takahashi H, Kawashima I, Shimizu T (2023) Development of serum-free and grain-derived-nutrient-free medium using microalga-derived nutrients and mammalian cell-secreted growth factors for sustainable cultured meat production. Sci Rep 13(1):498
54. Radu G, Racoviteanu G. Removing ammonium from water intended for human consumption. A review of existing technologies. IOP Conference Series: Earth and Environmental Science. 2021;664:012029
55. Yang M, Xiang N, Tang Y, Li Y, Zhang X (2025) Layered sodium zirconium phosphate adsorbent for ammonia removal in cell culture media for cultivated meat application. Sep Purif Technol 364:132415
56. Bethune Q, Harrington C, Mulukutla B (2021) Lactate control enhances growth advantage in fed-batch cultures of metabolically engineered CHO cells with reduced novel growth-inhibitory compound formation
57. Möller J, Bhat K, Guhl L, Pörtner R, Jandt U, Zeng AP (2021) Regulation of pyruvate dehydrogenase complex related to lactate switch in CHO cells. Eng Life Sci 21(3–4):100–114
58. Mitić R, Cantoni F, Börlin CS, Post MJ, Jackisch L (2022) A simplified and defined serum-free medium for cultivating fat across species. iScience 26:105822
59. Messmer T, Klevernic I, Furquim C, Ovchinnikova E, Dogan A, Cruz H et al (2022) A serum-free media formulation for cultured meat production supports bovine satellite cell differentiation in the absence of serum starvation. Nat Food:1–12
60. Jackson BT, Finley LWS (2024) Metabolic regulation of the hallmarks of stem cell biology. Cell Stem Cell 31(2):161–180
61. Delvigne F, Lecomte J (2010) Encyclopedia on industrial biotechnology: bioprocess, bioseparation and cell technologies. Foam formation and control in bioreactors. Wiley, New York
62. Junker B (2007) Foam and its mitigation in fermentation systems. Biotechnol Prog 23(4): 767–784
63. Deotale SM, Dutta S, Moses JA, Anandharamakrishnan C (2023) Foaming and defoaming–concepts and their significance in food and allied industries: a review. Discov Chem Eng 3(1):9
64. Flynn J, Breen L, Narayanan S, Butler M (2024) Measurement and control of foam generation in a mammalian cell culture. Biotechnol Prog 40(4):e3450
65. Pelton R (2002) A review of antifoam mechanisms in fermentation. J Ind Microbiol Biotechnol 29(4):149–154
66. Katz JS, Jordan S, Flannery H, Rigdon C, Serrano S, Burgett K et al (2025) Methylcellulose has synergistic growth benefits with poloxamer in suspension CHO culture. Biotechnol Prog: e70064. https://doi.org/10.1002/btpr.70064
67. Ong KJ, Johnston J, Datar I, Sewalt V, Holmes D, Shatkin JA (2021) Food safety considerations and research priorities for the cultured meat and seafood industry. Compr Rev Food Sci Food Saf 20(6):5421–5448
68. Huang T, Ko C, Paes D, Smeets E, Post M, Smith B (2024) A review on the safety of growth factors commonly used in cultivated meat production. Compr Rev Food Sci Food Saf 23(3): e13350
69. Todhunter ME, Jubair S, Verma R, Saqe R, Shen K, Duffy B (2024) Artificial intelligence and machine learning applications for cultured meat. Front Artif Intell 7. https://doi.org/10.3389/frai.2024.1424012
70. Lapierre FM, Mattaliano P, Raith D, Castillo-Cota M, Bermeitinger J, Huber R (2025) Multi-cycle high-throughput growth media optimization using batch Bayesian optimization. J Chem Technol Biotechnol 100(8):1571–1583
71. Narayanan H, Hinckley JA, Barry R, Dang B, Wolffe LA, Atari A et al (2025) Accelerating cell culture media development using Bayesian optimization-based iterative experimental design. Nat Commun 16(1):6055

Adv Biochem Eng Biotechnol (2026) 193: 81–154
https://doi.org/10.1007/10_2026_314

Published online: 18 March 2026

Bioreactor Design and Engineering for Cultivated Meat Manufacturing

Marie-Luise Schlieker, David Vorländer, Laura Pasitka, Sebastian Bode, Tatjana Krampitz, Chantal Treinen, and Marius Henkel

Contents

M.-L. Schlieker and M. Henkel
Cellular Agriculture, TUM School of Life Sciences, Technical University of Munich, Freising, Germany

Munich Institute of Integrated Materials, Energy and Process Engineering, Technical University of Munich, Garching, Germany
e-mail: marie.schlieker@tum.de; marius.henkel@tum.de

D. Vorländer (✉), L. Pasitka, S. Bode, and T. Krampitz
GEA Liquid Technologies Germany GmbH, Sarstedt, Germany
e-mail: david.vorlaender@gea.com; laura.pasitka@gea.com; sebastian.bode@gea.com; tatjana.krampitz@gea.com

C. Treinen
Cellular Agriculture, TUM School of Life Sciences, Technical University of Munich, Freising, Germany
e-mail: chantal.treinen@tum.de

Abstract Bioreactors encompass a wide spectrum of engineering concepts, yet for decades, industrial biotechnology has seen little fundamental innovation, with large-scale production mainly confined to stirred-tank and bubble-column reactors for microbial fermentations. In pharmaceutical biotechnology, various cultivation systems for eukaryotic cells are well established, including microcarrier processes, suspension in stirred vessels, and disposable bags or wave (rocking) bioreactors. However, these systems have typically been optimized to produce specific biomolecules, such as recombinant proteins, vaccines, or monoclonal antibodies, rather than for generating cellular biomass itself. The emergence of tissue engineering for medical applications brought new technological impulses, particularly in the development of scaffold materials and the spatial organization of cells in three-dimensional constructs. This was accompanied by advances in perfusion bioreactors, enabling sustained nutrient delivery and waste removal within thick tissue cultures. Such systems were designed for relatively small, high-value applications in regenerative medicine, but they provide valuable conceptual and technical groundwork for structured cultivated meat manufacturing. In cellular agriculture, the goal is to apply and adapt established bioprocessing knowledge to a new challenge: producing animal cells at food-relevant scales and costs. Rather than high-value, low-volume pharmaceutical production, the focus shifts to high-volume, cost-efficient food production. Achieving this will require interdisciplinary collaboration between bioprocess engineers, automation specialists, tissue engineers, cell biologists, and food technologists.

Graphical Abstract

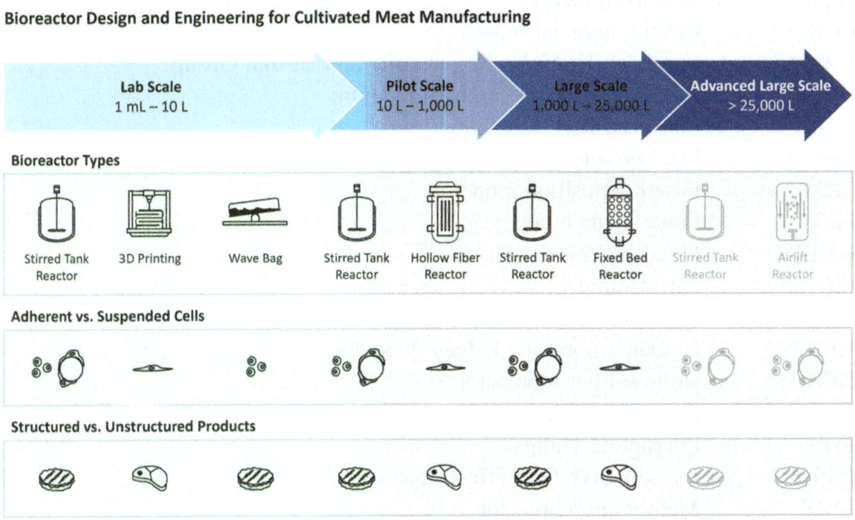

Keywords Bioreactor prototyping, Bioreactor upscaling, Bioscaffold, Cultivated meat, Perfusion bioreactor, Stirred-tank reactor

Abbreviations

2PP	Two-photon polymerization
3DP	Binder jetting
AI	Artificial intelligence
ALB	Air-lift bioreactor
AM	Additive manufacturing
ASME BPE	American Society of Mechanical Engineers: Bioprocessing Equipment
ATF	Alternating tangential flow
AU	Absorbance unit
BCBs	Bubble-column bioreactors
BHK	Baby hamster kidney
CAD	Computer-aided design
CAPEX	Capital expenditures
CDMO	Contract development and manufacturing organization
CFD	Computational fluid dynamics
CHO	Chinese hamster ovary cells
CIP	Cleaning in place
CNC	Computer numerical control

DAMS	Double-acting mechanical sealing
DCW	Dry cell weight
DO	Dissolved oxygen
ECB	Electrochemical biosensors
EHEDG	European Hygienic Engineering and Design Group
EMEM	Eagle's minimum essential medium
FBBR	Fluidized bed bioreactor
FBS	Fetal bovine serum
FDM	Fused deposition modeling
FEA	Finite element analysis
FiBBR	Fixed-bed bioreactor
FIR	Far-infrared
GMP	Good manufacturing practice
HEK293	Human embryonic kidney 293 cells
HFB	Hollow-fiber bioreactor
HMI	Human-machine interface
IoT	Internet of Things
ISFET	Ion-sensitive field-effect transistor
MBR	Membrane bioreactor
MES	Manufacturing execution system
MIR	Mid-infrared
MPC	Model predictive control
MTP	Module type package
NIR	Near-infrared
OD	Optical density
OPEX	Operational expenditures
OTR	Oxygen transfer rate
OUR	Oxygen uptake rate
PAT	Process analytical technologies
PEEK	Polyetheretherketone
PLC	Programmable logic controllers
POL	Process orchestration layer
PVC	Packed cell volume
SAL	Sterility assurance level
SCADA	Supervisory control and data acquisition
sCO_2	Supercritical carbon dioxide
Sf21	*Spodoptera frugiperda* Sf21 cells
Sf9	Clonal isolate of *Spodoptera frugiperda* Sf21 cells
SIP	Sterilization in place
SLA	Stereolithography
SLM/EBM	Selective laser or electron beam melting
SLS	Selective laser sintering
STR	Stirred-tank reactor
SUB	Single-use bioreactor

TCU	Temperature control unit
TFF	Tangential flow filtration
UV	Ultraviolet
VVD	Vessel volume per day
WCW	Wet cell weight

1 History and Revolution of Bioreactor Systems

Bioreactors have long been central to industrial biotechnology, yet large-scale processes remain dominated by stirred-tank and bubble-column bioreactors (STR and BCB) for microbial fermentations [1, 2]. In pharmaceutical applications, diverse eukaryotic cultivation systems, such as suspension cultures or microcarrier processes, are established; however, they are typically optimized for the production of recombinant proteins or antibodies rather than cellular biomass [3, 4]. Tissue engineering introduced new strategies, including scaffolds, 3D cell organization, and perfusion systems for nutrient delivery and waste removal, originally established for regenerative medicine [5]. These innovations now inform cellular agriculture, and, most notably, the research field of cultivated meat, where the challenge is producing animal cells at scale for food. Unlike animal cell culture used in pharmaceutical biotechnology, this requires shifting from low-volume, high-value products to cost-efficient biomass production on a large scale, demanding close collaboration across bioprocess engineering, tissue engineering, and food science [6].

Currently, no proven industrial process for cultivated meat production exists, making direct efficiency comparisons between reactor types difficult [7]. One of the most demanding goals, particularly for structured meat products, is the creation of complex tissues such as muscle fibers or adipose tissue, architectures for which no established large-scale bioreactor designs or process strategies are available [3]. Bioreactors for cultivated meat will likely need to accommodate two primary process phases: proliferation of undifferentiated cells and their subsequent differentiation into specialized tissue, preferably within the same operation unit [6]. The field remains in its early stages, and it is still unclear which types of bioreactors will prove most effective for industrial-scale production.

What is evident is that the unique demands of cultivated meat production will drive innovation in bioreactor technology. The diversity of target products, ranging from unstructured cell aggregates to highly organized and scaffold-based tissues, will likely result in product-specific bioreactor designs tailored to the desired structure, texture, and sensory profile of the final meat product [6].

This chapter highlights the bioreactor as the technological heart of the cultivated meat production process. After a brief history of bioreactor developments and the advances in the field, the chapter covers the fundamentals of the bioreactor itself, focusing on the specific requirements for cultivated meat production. The advantages and disadvantages of different systems are discussed in terms of their impact on product quality and process economics. Emphasis is also placed on the upscaling and downscaling of the process, with prototyping of downsized bioreactor systems being discussed to analyze and test the production of cultivated meat on a lab scale, as contrasted to the upscaling efforts to enable large-scale production. Another important factor that is considered is the integration of the scaffold material for structured products. While the scaffolds themselves are discussed elsewhere in this book, their connection to the bioreactor system is given attention here to complement the evaluation of bioreactor systems suitable for producing structured cultivated meat. The following sections provide an overview of the development of bioreactors, which date back almost 100 years, beginning with the large-scale production of penicillin and continuing through to the latest advances in cultivated meat production.

1.1 The First Bioreactors on an Industrial Scale

A pivotal milestone in bioreactor technology was the industrial-scale production of penicillin during the 1940s, driven by urgent medical needs in World War II. Penicillin, discovered by Alexander Fleming in 1928, proved highly effective in preventing and treating wound infections. Initial production efforts relied on surface culture in shallow trays or bottles, where *Penicillium chrysogenum* grew on nutrient-rich substrates, but yields were limited. The transition to submerged (deep-tank) fermentation, in which microorganisms grow suspended in an agitated liquid medium, marked a turning point, enabling much higher productivity via enhanced oxygen transfer, improved nutrient supply, and better process control [8].

In the post-war period, the expanding demands of the pharmaceutical, biotechnological, and food industries catalyzed the development of larger and more efficient systems. By the 1950s and 1960s, closed, controlled stainless-steel reactors had become the standard. For some commodity fermentations, working volumes reached hundreds of cubic meters. This era established foundational engineering knowledge, including scale-up methodologies, mixing and hydrodynamics, impeller and vessel design, and increasingly sophisticated control strategies. The concepts created are still employed today [9].

From this technological evolution, two reactor types emerged as dominant in large-scale industrial biotechnology: stirred-tank reactors (STRs) and bubble-column bioreactors (BCBs). Over decades, STRs became the "gold standard" owing to flexibility and robust performance across mixing efficiency, power input, shear characteristics, gas-liquid dispersion, and mixing times. They can be adapted to microbial, fungal, or cell culture processes and remain the most widely used design in industrial and pharmaceutical biotechnology [2]. BCBs are tall, vertical vessels where gas sparging provides both mixing and aeration, eliminating the need for mechanical agitation. This configuration offers low energy consumption and simple design, making BCBs especially suitable for high-volume, cost-sensitive products such as organic acids or ethanol [1].

While STRs dominate applications requiring tight process control and flexibility, and BCBs excel in large-scale, cost-sensitive operations, in the past, both were optimized primarily for microbial systems. This distinction is critical for cultivated meat, which presents different engineering constraints, from gentle hydrodynamics and control over differentiation cues to scaffold integration. Whether classical large-scale designs can be adapted or entirely new configurations will be required remains an open question in the development of industrial cultivated meat processes [6].

1.2 Development of Bioreactors for Cell Culture

The emergence of modern biopharmaceuticals created new demands on bioreactor design, leading to systems optimized for mammalian cell culture. Compared to microbes, mammalian cells are more shear-sensitive, grow more slowly, and many require surface attachment. These features drove innovations in reactor configuration, process control, and auxiliary materials (e.g., microcarriers) tailored for large-scale cell expansion [10]. Early scientific foundations date back to Ross Harrison's demonstration of nerve outgrowth from embryonic frog tissue in 1907, the first animal tissue culture. This was followed by the formulation of Eagle's minimum essential medium (EMEM) in 1955, which standardized nutrient supply for mammalian cells [11, 12]. By 1949, Enders, Weller, and Robbins established robust techniques for virus propagation in human embryonic tissues, accelerating vaccine development and further driving the development of larger scale mammalian cell cultivation [13].

From the 1960s onward, STRs were adapted for animal cells using low-shear impellers, careful aeration, and precise environmental control, which enabled industrial production of monoclonal antibodies, recombinant proteins, and vaccines [3]. A key breakthrough for adherent cell lines was microcarrier technology, introduced by van Wezel in 1967, which provided microscale attachment surfaces to cultivate adherent cells in suspension, dramatically increasing volumetric productivity [14].

Hollow-fiber bioreactors, developed in the 1970s and refined thereafter, separated the cell compartment from medium circulation via semipermeable fibers, allowing high-density, long-term cultures and enabling continuous production of therapeutic proteins [15]. In parallel, perfusion strategies provided a continuous nutrient supply and facilitated metabolic waste removal, thereby improving cell longevity and productivity, particularly for unstable or cytotoxic products [16].

By the late 1990s and early 2000s, single-use (disposable) systems, such as wave (rocking) bioreactors and single-use STRs, gained traction in multiproduct facilities due to hygiene, validation, and changeover advantages [17]. Although developed for high-value pharmaceuticals, the engineering principles established across scale-up, shear management, nutrient delivery, and cell-scaffold interactions form a crucial foundation for cultivated meat. Nonetheless, shifting from product-centric (protein yield) to biomass-centric goals introduces new technical and economic challenges that demand further adaptations.

1.3 New Bioreactor Concepts for Cultivated Meat

The development of suitable bioreactor systems is a central engineering challenge in the field of cultivated meat, but particularly important for structured meat products. Unlike traditional bioprocesses that often rely on well-established suspension cultures, the cultivation of muscle, fat, and connective tissue requires highly specialized conditions tailored to adherent cells, scaffold materials, and tissue-like constructs [18].

Between 2000 and 2010, advances in tissue engineering and cell therapies laid essential groundwork for bioreactor innovation in cultivated meat. However, application scenarios and research prototypes were restricted to low-volume, high-value systems for clinical use. Core contributions included scaffold design, three-dimensional culture, and perfusion strategies that maintain viability in thicker constructs [5]. In parallel, 3D cell culture and organ-on-a-chip approaches refined control over microenvironmental cues, enabling structured cell assemblies relevant to meat textures [19].

More recently, groups have begun translating pharmaceutical bioreactor know-how into food-grade platforms. The central challenge of cultivated meat production is to produce far more biomass at far lower cost than in the pharmaceutical sector [10]. Even though single-use bioreactors offer hygiene and turnaround advantages, their economic suitability will depend on material costs, reuse strategies for food-grade operations, and supply chain maturation [6].

Process analytical technologies, automation, and artificial intelligence (AI)-guided optimization are gaining attention to accelerate design-build–test cycles, strengthen online monitoring, and tune proliferation/differentiation kinetics [20]. Concurrently, food-grade design requirements, including materials of construction, cleanability, and regulatory alignment, are being integrated with classic bioprocess criteria to deliver cost-efficient reactors for safe and scalable production [6].

Prominent examples of recent progress on bioreactors for structured cultivated meat are outlined in Table 1, summarizing bioreactor concepts from the last 5 years, along with operating modes, distinctive features, advantages, limitations, and scale-up prospects. Notably, many new systems aim to combine proliferation and differentiation within a single unit to save time, reduce equipment, and simplify logistics, an approach with promise but also with trade-offs in control complexity and harvesting strategies. Peer-reviewed designs sit alongside patent disclosures, reflecting both scientific and competitive dynamics in the field.

For example, Nie et al. describe a hollow-fiber perfusion bioreactor for structured cultivated meat, using closely arranged fibers to achieve microscale uniform perfusion, even cell distribution, alignment, and centimeter-scale tissue formation, which can be scaled to large fiber arrays [21]. Advantages include fine control over nutrient/oxygen transport, as well as tissue architecture. Potential drawbacks include module complexity, the cost of large arrays, and the need for cleaning/validation at the food-scale level.

Table 1 Summary of bioreactor concepts and configurations for adherent cell growth explored for structured cultivated meat production (2020–2025), including laboratory-scale proof-of-concept systems, industrial conceptual designs, and patented approaches

Bioreactor type(s)	Utilization	Investigated working volume/dimensions	Special characteristics	Scaffolds	Original source
Hollow-fiber perfusion bioreactor (HFB)	Proliferation and differentiation	Dense array of perfusable hollow fibers (hundreds to thousands of fibers)	Semipermeable hollow fibers deliver O_2 and nutrients, microfabricated anchors induce alignment	Hydrogel between fibers, edible variants proposed	Nie et al. (2025) [21]
Hollow-fiber perfusion bioreactor (HFB)	Differentiation	Cartridge/module-scale	Edible hollow fibers with perfusable lumens, tuned spacing/porosity	Edible hollow fibers (hydrocolloid/protein)	Merck Patent GmbH (2022) [22]
Stirred-tank reactor (STR)	Proliferation on microcarriers	700 mL	Low-serum (2% FBS) expansion, bead-to-bead microcarrier addition	Non-edible microcarriers	Tzimorotas et al. (2023) [23]
Stirred-tank reactor (STR)	Proliferation on microcarriers	Liter-scale	Bead-to-bead transfer, media and MC addition strategy, spinner-to-STR translation	Microcarriers (various)	Hanga et al. (2021) [24]
Spinner and stirred-tank reactor (STR)	Proliferation and differentiation on microcarriers	Spinner and benchtop STR	Microcarriers with tunable stiffness/topology (grooved vs smooth); formation of cell–MC microtissues	Edible gelatin microcarriers	Norris et al. (2022) [25]
Rotating-wall vessel (HARV) and stirred-tank reactor (STR)	Proliferation and differentiation on microcarriers	HARV and benchtop STR	Microgravity (HARV) vs. 1 g (STR) comparison, target: platform for cultivated meat in space	Edible gelatin microcarriers	Sardroud et al. (2025) [26]
Rocking (wave) bioreactor	Proliferation	n.a.	Simulation and modeling framework with computational fluid dynamics (CFD) for rocking bioreactors	Compatible with microcarriers, but typically omitted in suspension culture	Kim et al. (2025) [27]
Air-lift reactor (internal loop)	Proliferation	Conceptual industrial scale >1 m^3	Gentle mixing with lower shear/power, CM-focused design methodology	Microcarriers (various) and suspension culture	Li et al. (2020) [28]

(continued)

Table 1 (continued)

Bioreactor type(s)	Utilization	Investigated working volume/dimensions	Special characteristics	Scaffolds	Original source
Stirred-tank reactor (STR)	Proliferation on microcarriers followed by downstream assembly	Spinner/benchtop STR	Edible porous microcarriers as building blocks for 3D printing structured fillets	Edible porous microcarriers	Zhou et al. (2025) [29]
Elongated-fiber bioreactor (Rocket)	Proliferation and differentiation of aligned fibrous bundles	n.a.	Extrusion-draw through multi-nozzle, bundles assembled by enzymatic joining	Gel/fiber bundles formed in the reactor	Mirai International AG (2024) [30]
Stirred-tank reactor (STR) and seeding chamber system	Proliferation and differentiation on scaffolds	n.a.	Bioreactor with movable grids maintaining scaffold positions for tissue maturation	Porous textured scaffolds (protein-based)	BioTech Foods S.L. (2021) [31]

A contrasting approach is the elongated-fiber "Rocket" reactor disclosed by Mirai Foods [30]. The patent outlines methods and a corresponding reactor for the parallel formation of long, mature muscle fibers bundled and integrated with cultivated fat to form tenderloin-like structures, with claims of short reactor residency times on the order of days. Advantages include direct whole-cut structuring within the bioreactor, while challenges include translating protected claims into reproducible, cost-effective, and food-grade manufacturing processes, as well as independently validating throughput and quality at scale [30].

Overall, the convergence of tissue engineering perfusion strategies, scalable hollow-fiber architectures, and product-oriented structuring reactors suggests a diversified landscape in which multiple, product-specific bioreactors may coexist, each tuned to the targeted cut, texture, and process economics [6, 10].

2 Fundamentals of Bioreactor Design

This section will highlight the general requirements of bioreactors for cultivated meat, with a focus on how these are addressed in STRs as a dominant and well-established production platform for both small- and large-scale mammalian cell culture. As given by its name, an STR for cell cultivation is characterized by the reactor vessel and the agitator installed inside, creating the stirring. Typically, the vessel is designed as a pressure vessel made of stainlesssteel to withstand pressure and temperature during sterilization or as a pre-sterilized single-use bag. In both cases, the vessel geometry has the highest influence on the process: vessels are usually cylindrical with flat or torispherical bottoms and height-to-diameter ratios of 2:1 or smaller. Agitators need to be designed aseptically and need to deliver the necessary mixing capabilities at low-shear rates. More details are discussed in Sect. 2.2.

In comparison to alternative systems discussed in Sect. 2.4, STRs are widely used in various industrial applications from basic chemicals to biotechnology. Due to their extensive use, there is a lot of data available regarding the design, operation, and scaling of this type of reactor, which simplifies the adaptation to new processes, such as the cultivation of cells for cultivated meat.

Although the term "bioreactor" can be defined in different ways and may encompass various functions, in the context of cultivated meat, it primarily refers to providing a controlled environment that supports proliferation and, in some cases, differentiation of animal-derived cells. Within this chapter, the term bioreactor not only describes the cultivation vessel itself but also includes the associated piping, valves, and instruments required for proper operation. To meet these requirements, several parameters must be controlled, including sterility, homogeneity of culture conditions, and regulation of media components, which are discussed in more detail below.

2.1 Cleaning and Sterilization

The primary risk of biological contamination in cultivated meat cultures stems from ubiquitous microorganisms, such as bacteria, molds, and yeasts. Since production media are typically antibiotic-free, they provide suitable growth conditions for contaminants, whose growth rates can exceed those of cultivated cells by 10–100 times [32]. Therefore, achieving sterility before inoculation and maintaining axenic cultures is essential. Any contamination detected during a batch or continuous process usually requires termination. While the product may still be good to use,[1] further cultivation becomes unprofitable as contaminants outcompete the cells, deplete nutrients, alter pH, and cause oxygen deficiencies.

Effective cleaning and sterilization rely on both equipment design and process execution. Hygienic design standards (e.g., ASME BPE, 3-A, EHEDG) include main criteria for good cleanability, e.g., avoiding sharp edges, dead zones, and dead legs, ensuring drainability, and controlling surface roughness (Ra ≤ 0.8 μm). Cleaning should follow Sinner's circle principles (time, chemicals, mechanics, temperature), with cleaning-in-place (CIP) preferred and minimum flow rates observed for piping and spray devices.

For sterilization, sterile boundaries need to be defined and typically include the bioreactor's inner vessel, sterile filters, and shut-off valves to steam and condensate lines. Depending on scale, stainless steel and glass vessels provide higher durability compared to single-use bags, while the latter eliminate the need for SIP by being provided pre-sterilized. For all bioreactors, sterile filtration is used for liquids and gases that cannot be pre-sterilized or as a safety feature to eliminate aerosols. Inside these boundaries, sterilization media must reach all product-contact surfaces, preferably via the same pathways as CIP. Sterilization-in-place (SIP), as moist heat sterilization by saturated steam, remains the most reliable and easiest method to validate. Sterility assurance is achieved using overkill cycles, typically targeting a 6-log reduction of a challenge organism (e.g., *Geobacillus stearothermophilus*) and doubling the exposure time to achieve sterility assurance levels of 10^{-6} for typical medical applications. Alternative approaches include bioburden-based sterilization, which adjusts treatment based on known contaminant levels, and the industrial F_0 method, which accounts for temperature profiles to optimize sterilization time and efficiency.

2.2 Homogenous Conditions: The Agitator

The key element providing the mixing in the STR is the agitator. While the technical term "agitator" refers to the complete machine, there are two major parts, the

---

[1] Currently, there are no microbiological criteria in place for cultivated meat. Assuming it needs to fulfil the same as traditional minced meat based on EC 2073/2005 a total of 5×10^5 to 5×10^6 cfu g^{-1} would be acceptable at the end of the manufacturing process.

impeller(s) and the shaft sealing, that require further explanation regarding their functionality.

2.2.1 Impellers

The tasks of impellers in STRs for cultivated meat differ from those in microbial fermentations, where gas dispersion dominates. In cultivated meat, the primary task is keeping cells in suspension, favoring low-mounted impellers with (partial) axial flow. The second task is to achieve homogeneous mixing, which includes even distribution of cells, nutrients, and metabolites, as well as minimizing temperature and pH gradients. Since animal cells are highly shear-sensitive, impellers must operate at low shear and minimal rotation speed.

Five basic impeller types are shown in Fig. 1 with low-shear profiles in low-viscosity media. They are commonly used in existing cell cultivation processes and are likely suitable for cultivated meat as well. Typically, one impeller located near the bottom above the sparger suffices. In larger reactors, a second impeller located near the liquid surface improves mixing of additions, such as base or nutrients. A larger impeller diameter relative to the vessel diameter allows the same power input and mixing efficiency at a lower speed, reducing shear stress.

2.2.2 Shaft Sealing

Two shaft sealing options provide sufficient protection against contamination: double-acting mechanical seals (DAMS) and magnetic drives. DAMS use two

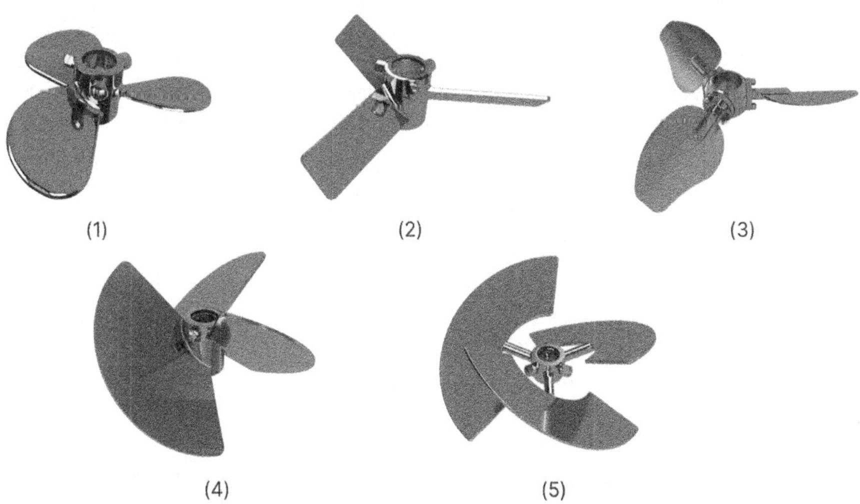

Fig. 1 Typical impellers for cultivated meat application: (1) marine, (2) pitched-blade, (3) butterfly, (4) elephant ear, and (5) segment impeller (Images courtesy of PRG Präzisions-Rührer GmbH)

mechanical sealing surfaces, with cooled steam condensate as a sterile barrier between the inner and outer seal. An overpressure of 1 bar above process pressure and a steam-sterilizable barrier fluid system ensure secure sealing though minor leakage of sterile condensate into the reactor may occur.

Magnetic drives transmit motor torque via magnets, avoiding shaft penetration and eliminating the need for sealing. This maintains sterility even in the case of drive failure and allows temporary dry running during CIP. Torque limitations are negligible for cultivated meat bioreactors, with capacities up to 30 m^3 and potential for even larger volumes [33, 34], given recent developments. While magnetic drives are comparable in cost to DAMS at a small scale, they are initially more expensive for large-scale setups but offer minimal maintenance and lower operational costs.

2.3 Control of Cultivation Conditions

Precise control of cultivation conditions in bioreactors is essential, as cellular metabolism, enzymatic activity, and overall product quality are highly sensitive to changes in temperature, pH, dissolved oxygen, and nutrient availability. Any deviations from the optimal range can reduce cell viability, alter metabolic pathways, or even lead to process failure.

Temperature control represents a key parameter in maintaining stable cultivation environments. The main source of heat in bioreactors is cellular metabolism, which can vary significantly depending on cell type and density. While microbial fermentations may generate up to 15–20 W L^{-1} of metabolic heat, mammalian cell cultures typically produce less than 1 W L^{-1} [35]. Additional heat input arises from agitation, where the energy imparted by impellers is dissipated through viscous shear and friction; however, this contribution is relatively small in cultivated meat bioreactors due to low power input. Other, more minor factors include heat transfer through aeration (via compressed gas), exothermic reactions such as oxygen dissolution, and environmental conditions. For most mammalian cells, temperatures outside the physiological range rapidly reduce viability, making precise and stable control crucial. Bioreactors, therefore, require systems that can both heat the medium during early cultivation stages and provide cooling at higher cell densities. This is typically achieved through jacketed vessels equipped with dimple jackets, pillow plates, or half-pipe coils, combined with thermal insulation for efficiency. Closed-loop temperature control systems featuring integrated pumps, heat exchangers, and sensors allow for precise and sustainable regulation while minimizing thermal stress on the vessel.

Equally important is the regulation of extracellular pH, which directly influences intracellular conditions and enzyme activity. Most mammalian cell lines operate optimally at pH 7.2–7.4, with deviations leading to stress, accumulation of by-products, and impaired growth. Metabolic processes produce acidic (CO_2, lactic acid) and basic (ammonia) by-products that can shift the pH if not controlled. Culture media, therefore, incorporate buffering systems such as sodium bicarbonate or HEPES, yet active regulation via acid and base addition remains necessary. While most buffering systems will require liquid addition for base and acid, bicarbonate systems will use CO_2 as acid, typically introduced into the aeration gas mix, while base is supplied via concentrated carbonate solutions. Effective mixing is critical, as pH sensors measure only locally, and control systems must account for spatial gradients and delays between measurement and corrective input.

Another crucial parameter is dissolved oxygen (DO), as all cultivated meat production processes rely on aerobic metabolism. The balance between the cell-specific oxygen uptake rate (OUR) and the maximum oxygen transfer rate (OTR_{max}) defines whether adequate supply can be maintained. OTR depends on both the solubility of oxygen under given conditions and the volumetric oxygen mass transfer coefficient (k_La), which in turn is influenced by bubble size, gas composition, aeration rate, agitation, and vessel geometry. For example, smaller bubbles increase the transfer area but raise the risk of foaming and shear damage, while oxygen-enriched aeration improves the transfer but increases cost and technical complexity. Agitation enhances dispersion and prevents bubble coalescence, yet it must be carefully balanced to avoid excessive shear stress on fragile cells. Sensors based on either electrochemical (Clark-type) or optical (fluorescence/luminescence) principles are used to monitor DO and provide feedback for aeration and agitation control.

Finally, nutrient composition and accumulation of metabolic by-products must be tightly controlled to ensure long-term culture stability. While online sensors for glucose, glutamine, lactate, and ammonia remain limited and often require frequent replacement, offline sampling is still the most common approach in production. Nutrient supplementation is relatively straightforward, relying on the sterile addition of concentrated feed solutions. The removal of inhibitory by-products, however, presents a greater challenge. Partial medium exchanges dilute waste metabolites but simultaneously reduce cell density, whereas perfusion systems selectively remove cell-free supernatant, thereby maintaining high cell concentrations while limiting by-product accumulation. The choice of feeding and waste management strategy strongly influences productivity and process economics.

In summary, robust cultivation of mammalian cells in bioreactors requires an integrated control of temperature, pH, dissolved oxygen, and nutrient composition. Each parameter is interdependent, and only by maintaining them within tight physiological ranges can cell viability, growth, and product quality be ensured across scales.

2.4 Alternative Design Prospects

2.4.1 Bubble-Column and Air-Lift Bioreactor

Bubble-column bioreactors (BCBs) and air-lift bioreactors (ALBs) are pneumatically driven systems, where gas sparging provides both aeration and mixing, eliminating the need for mechanical agitation. This improves cleanability, reduces contamination risk, and can lower capital and operational costs. The gas flow rate is usually the only independent control parameter. Both types are suitable for media viscosities up to 0.3 Pa s, typically with liquid height-to-diameter ratios between 5:1 and 20:1. For pilot and production-scale systems, a ratio of approximately 10:1 has been established. These reactors are widely used in aerobic wastewater treatment, yeast production, citric acid production, and monoclonal antibody manufacturing [36–40].

BCBs are simple cylindrical vessels with a high height-to-diameter ratio but without internal structures. Gas is sparged at the vessel bottom, and random bubble movement drives non-homogeneous mixing. While this promotes mass and heat transfer, the overall efficiency is lower than in STRs. Additionally, high shear at spargers and bubble rupture can damage suspension cells, limiting use for shear-sensitive animal cells. Small-scale cultivation of hybridoma, lymphoma, BHK-21, and Sf-21 cells has been reported [36, 38, 39].

ALBs are equipped with a guided gas sparging system, which improves circulation and mixing through a riser (aerated upward flow) and downcomer (unaerated return flow). Density differences drive liquid circulation, enhancing mass and heat transfer. ALBs can be internal loops (with a draft tube or baffles separating riser and downcomer compartments) or external loops (two connected but separated columns). Scaling up to cubic-meter volumes for mammalian cells is possible but challenging, e.g., due to cell shear sensitivity, foaming, and back-mixing.

Hydrodynamics and mass transfer in ALBs depend on circulation velocity, influenced by gas flow, sparger design, vessel geometry, and medium properties. Compared to BCBs, bubble behavior is more controlled, which improves mass and heat transfer as well as mixing uniformity. Scale-up strategies maintain geometric similarity, constant $k_L a$, and either constant mixing time, superficial gas velocity, or gas flow rate per unit volume [36, 38, 39, 41].

2.4.2 Fixed-Bed and Fluidized-Bed Bioreactors

Fixed-bed and fluidized-bed bioreactors are widely used in biopharmaceutical and tissue engineering applications, including monoclonal antibody production, vaccines, or tissue propagation. Both systems employ carrier-immobilized animal cells, enabling high volumetric cell densities and productivities while reducing the need for extensive cell line or media optimization.

Fixed-bed bioreactors (FiBBRs) consist of a column packed with carrier material (e.g., microcarriers, macroporous beads, or fibrous beds) supporting cell growth. Medium is circulated through the bed, supplying nutrients and oxygen while removing inhibitory by-products. Cells are protected from shear, allowing bubble aeration or waterfall-like oxygenation. High-volume-specific cell densities (up to 10^8 mL^{-1}) can be achieved over long cultivation periods. Limitations include poor mass and heat transfer as well as nutrient and oxygen depletion along the bed height; therefore, axial-flow beds are usually kept below 15 cm, while new radial-flow designs and systems such as iCELLis™ enable scale-up [36, 38, 42, 43].

Fluidized-bed bioreactors (FBBRs) use porous carriers or beads with immobilized cells. They are kept in suspension by upward medium flow via internal or external loop systems, enhancing mass and heat transfer and providing uniform nutrient and oxygen supply. Internal loop systems simplify cleaning and sterilization. Challenges include axial oxygen gradients in the bioreactor and carrier flotation, which limit bed height and scalability. Solutions, such as membrane modules, are being explored to overcome these issues.

In summary, fixed beds offer simple operation and reliable shear protection but are limited by mass transfer and scale-up constraints. Fluidized beds improve oxygen and nutrient distribution and volumetric productivity but require careful flow and hydrodynamic control. The choice depends on cell type, process requirements, and scale, with ongoing innovations addressing the limitations of both systems [36, 38, 42].

2.4.3 Hollow Fiber Bioreactor

Membrane bioreactors (MBRs) use semipermeable ultrafiltration or microfiltration membranes to retain cells while allowing nutrients and metabolic by-products to pass by diffusion. A common type is the hollow-fiber bioreactor (HFB), where cells grow adherently on the membrane or non-adherently in the extracapillary space. Oxygen- and nutrient-rich medium flows through the intracapillary space, enabling diffusion-driven exchange of media components, gases, and inhibitors. Spent medium can be enriched and recirculated. This supports high-density cultures (up to 10^8 mL^{-1}) with efficient mass transfer and cell retention [36, 39, 44].

Hollow-fiber systems protect cells from shear stress, allow for the targeted delivery of expensive supplements, maintain high local nutrient concentrations, and achieve compact, parallelizable setups suitable for both laboratory and production scales. They also offer high packing density and lower investment compared to other membrane systems.

Scale-up is limited by the length and diameter of the hollow-fiber membrane bundle, which can cause concentration gradients and uneven cell distribution. Flow rates must be carefully controlled to balance nutrient exchange without impairing diffusion, and the feed medium must be clean to prevent fouling. Differences in hydrodynamics and mass transfer between lab and larger scales complicate extrapolation.

2.4.4 Comparison of Alternative Bioreactor Systems

BCBs are unsuitable for large-scale mammalian cell cultivation because of poor mixing and the risk of cell damage. In contrast, air-lift bioreactors offer a promising alternative to STRs if shear sensitivity and mixing time are carefully considered during scale-up.

FiBBRs, FBBRs, and HFBs are applicable to a wide range of biopharmaceutical processes. However, for their implementation in food applications, these systems must be scalable to industrial volumes without incurring higher capital costs than STRs. In addition, several other challenges must be addressed that are negligible in biopharmaceutical production, but are essential for food-related applications, such as cultivated meat, where cells should remain intact. Therefore, if the carrier or scaffold system is not edible, detaching the cells becomes difficult, if not impossible, without causing lasting damage to them, which in turn could affect the final product or product yield. Here, many companies are currently seeking innovative bioreactor vessel designs to achieve higher cell yields in grams per square meter. One significant advantage of FiBBRs, FBBRs, and HFBs is their compatibility with adherent cell cultures. This allows for the use of a broader range of cell types without the substantial investment of time and resources required to adapt them to suspension culture, thereby increasing product diversity. However, it remains necessary to adapt cells to "clean" media to prevent membrane fouling and blockage in HFBs. High protein and particle loads, such as those present in fetal bovine serum (FBS), can rapidly lead to membrane obstruction. Ideally, animal-component-free and protein-free media should be employed to mitigate these issues.

2.5 *Bioreactor Operation Modes*

Bioreactors may account for a significant proportion of the process equipment costs of cultivated meat production facilities [45–48]. The bioreactor costs will primarily depend on the number and size of bioreactors, which in turn are impacted by the volumetric productivity achieved in a bioreactor in a particular operation mode [46]. Cultivation in a bioreactor can be performed in various operation modes (also referred to as cultivation types or process strategies), from simple batch approaches to fed-batch, repeated batch and fed-batch, fully continuous and perfusion processes [49]. The different process strategies will be described and discussed below, with a focus on the proliferation step and assuming carrier-free suspension culture in an ideally homogenous mixed bioreactor. Figure 1 gives a brief comparison of the fluid transfer during the cultivation phase, as well as the vessel volume and liquid flows during the preparation, cultivation, and harvest periods of various operation modes. As will be discussed in the following sections, all the different operation modes have their advantages and disadvantages. From a yield, productivity or economic perspective, one or the other might be the most efficient solution for a specific cultivation process or step within a production process.

2.5.1 Target Values of Bioprocess Optimization

Bioprocess optimization can focus on variables such as product concentration (titer), medium conversion rate, or productivity. In cultivated meat, product titer typically refers to biomass concentration, expressed as cell density, dry cell weight (DCW), wet cell weight (WCW), or packed cell volume (PCV), but may also include protein concentration or other target values. The medium conversion rate, also known as the substrate-specific yield $Y_{X/S}$, when looking for individual substrates, describes the amount of produced biomass ΔX related to the amount of consumed substrate ΔS or medium supplied.[2] Productivity P describes the amount of formed product $\Delta m_{\text{Product}}$ (e.g., biomass) per liter reactor volume $V_{\text{Bioreactor}}$ and time t as defined and is also referred to as space-time yield. Time can be considered incrementally to assess productivity over the cultivation course or as cultivation time; however, the total production time should be considered, including non-productive periods such as preparation, cleaning, sterilization, harvest, and maintenance [32]. Similarly, the volume should account for the total bioreactor or maximum working volume to enable comparisons across different operating modes [50].

2.5.2 Batch Cultivation

Many biotechnological processes are based on batch processes as the simplest operation mode of classical bioreactors, such as the STR. Batch cultivations are "closed" and discontinuous processes, meaning that the reactor vessel is filled with the entire volume of culture medium before the cultivation is started by inoculating the bioreactor with the respective pre-culture (Fig. 2A) [32]. During the actual cultivation, no further medium is added, and the addition and removal of liquids are limited to the supply of concentrated liquids, e.g., acid and base for pH adjustment or antifoam agents, and the removal of small sample quantities [49]. All of them have only minor implications for the liquid volume and concentrations, apart from the additive itself. Due to the low solubility of oxygen, it must be constantly supplied, e.g., through the gassing of air, oxygen-enriched air, or pure oxygen via the sparger. Furthermore, gassing is required to remove the metabolism-derived carbon dioxide from the media [51]. The batch process usually ends with the harvest of the cell suspension, e.g., by transferring them into another differentiation bioreactor or by starting the downstream process. Differentiation may also be performed in the same bioreactor as another (batch) cultivation step.

In batch mode, cultivation conditions constantly change as cell density increases and nutrients are consumed while metabolites accumulate. After inoculation, cells

[2]The calculation of $Y_{X/S}$ is often based on the proportion of the concentration differences ΔX and ΔS or the rates r_X and r_S comparing the actual biomass increase and substrate consumption. However, since different cultivation modes vary in their overall substrate conversion rate, the total amount of supplied substrate must be considered when making economical comparisons.

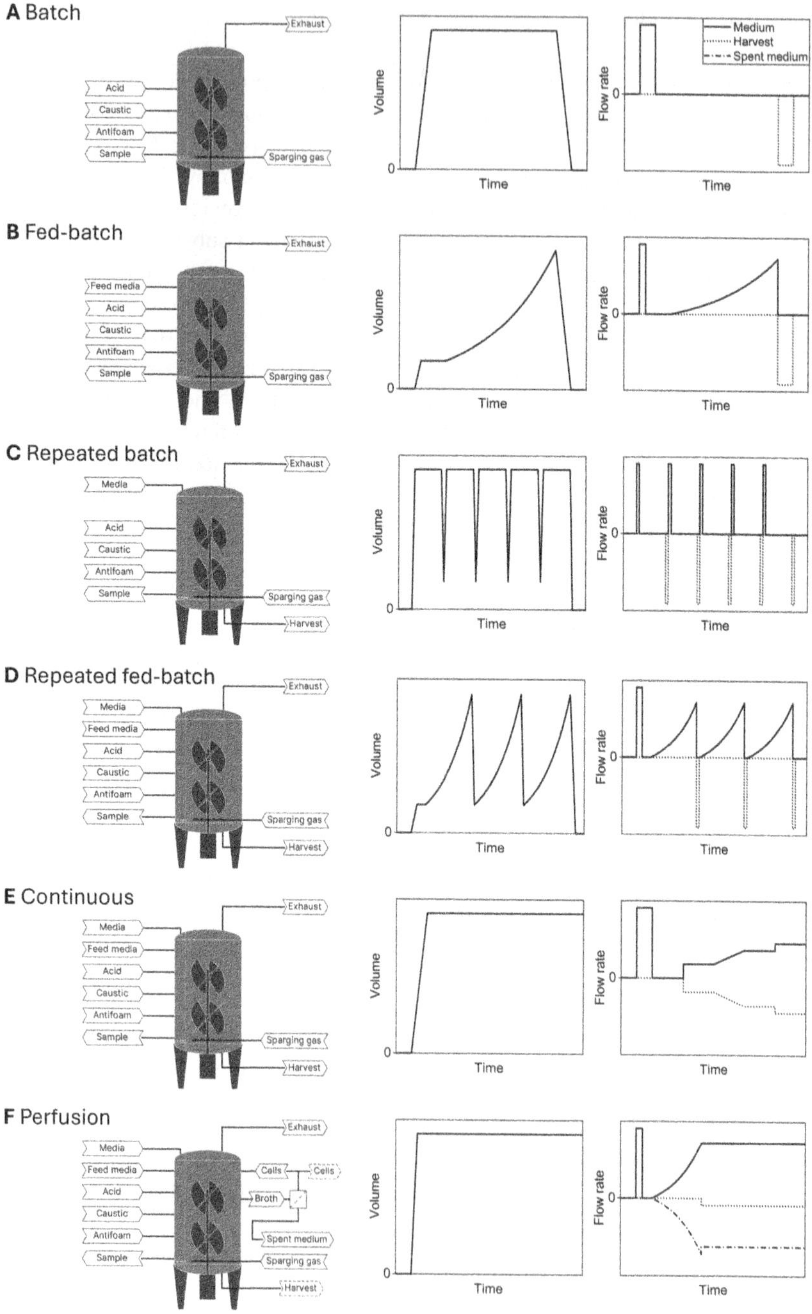

Fig. 2 Comparison of bioreactor operation modes. (**A**) Batch, (**B**) Fed-batch, (**C**) Repeated batch, (**D**) Repeated fed-batch, (**E**) Continuous, (**F**) Perfusion with external system and continuous cell bleeding. The first row shows input/output mass streams during growth; the second and third rows illustrate reactor volume, medium flow (solid), harvest (dotted), and spent medium (dash-dotted, only in **F**). Positive values indicate inflow, negative outflow

undergo a lag and acceleration phase before reaching exponential growth at their maximum growth rate. Once nutrients or oxygen become limiting, or inhibitory metabolites accumulate, growth slows and eventually reaches a stationary state. While an extended stationary phase may be relevant for microbial secondary metabolite formation, later phases are typically not relevant for cultivated meat because the cells themselves represent the product.

Batch processes offer reduced process complexity, lower investment costs, shorter process times, and lower contamination risks compared to fed-batch or continuous modes. However, productivity is limited because cell density is low at the beginning, and substrate availability and metabolite accumulation restrict growth over time [49]. Moreover, repeated cleaning and sterilization reduce effective reactor uptime and increase operational costs. Alternative operation modes and harvest strategies aim to overcome these limitations, yet batch processes remain widely used in seed train steps due to their robustness and simplicity [49].

2.5.3 Fed-Batch Cultivation

Fed-batch cultivations differ from batch cultivations in terms of the addition of fresh nutrients, the feed, during the cultivation process (Fig. 2B). As the removal of liquids is still limited to small sampling quantities, the starting volume must be lower than the maximum working volume, considering the volume increase due to the addition of the feed medium. Fed-batch processes can be distinguished into two categories: those with a significant volume increase due to the feeding of large amounts of media with comparably low substrate concentrations, and those with minor volume increase due to the use of highly concentrated feed media. Such feeds are highly concentrated in terms of nutrients, but not in non-nutrient salts, to minimize dilution effects while keeping appropriate osmolality [49]. Feeding typically starts when nutrient concentrations fall below a defined threshold and can be performed as single or repeated additions (also called pulsed or bolus addition) or in a continuous manner [32]. Continuous feeding can be performed with a constant feed flow, as well as with stepwise, linear, exponential, or otherwise increasing flow. Controlled feeding can also depend on online measurements such as actual substrate concentration, cell density, oxygen consumption, or pH control [49].

2.5.4 Repeated Batch and Fed-Batch Cultivation

Repeated batch (Fig. 2C) or repeated fed-batch cultivation (Fig. 2D) is an approach to further increase the productivity of (fed-) batch processes by reducing non-productive times in the bioreactor [32]. Cultivation is carried out as previously described for the (fed-) batch process. The harvest, however, is only a partial harvest. A (small) part of the culture is kept in the bioreactor, serving as the inoculum for the proceeding (fed) batch. Filling the vessel again can be done either quickly to the actual working volume in case of repeated batch cultivation or slowly and according to different feeding strategies as previously discussed in case of repeated fed-batch cultivation [51]. Repeated fed-batch may combine dilution of the cells with standard batch medium and feeding with fed-batch medium, or only use one feed medium for the refilling. Partial harvest and refilling can be repeated several times, leading to a cyclic pattern as indicated in (Fig. 2C, D). A repeated (fed-) batch is therefore also referred to as a cyclic (fed-) batch cultivation [32]. The duration of this cyclic pattern depends on the quality of process control and the robustness of the cell line. Additionally, dead cells and waste metabolites may accumulate over time, as only parts of the cell suspension are removed.

2.5.5 Continuous Cultivation and Perfusion

Continuous cultivation (Fig. 2E) is characterized by a continuous feed of fresh medium and a constant removal of cell suspension, including the spent media containing inhibiting metabolites such as lactate and ammonia [49]. The continuous operation is usually started after an initial batch or fed-batch phase to reach high cell densities. The feed medium can be the same medium as in the initial batch phase, but it can also be higher concentrated or supported by a second feed, as in a fed-batch. The feed or dilution rate in a continuous cultivation is often constant, but may also vary. In the so-called chemostat, continuous cultivation is carried out with a constant dilution rate, meaning a constant feed of nutrients and an equal removal rate of cell suspension to maintain a constant volume [32]. In the ideal system and within the given boundaries (e.g., dilution rate below max. specific growth rate, applicability of Monod kinetics), the constant dilution rate leads to a self-regulating system based on a growth-determining substrate. Continuous cultivation can also be performed with controlled feed and harvest rates in order to keep a specific parameter constant, such as the cell density, dissolved oxygen concentration, or specific nutrients such as glucose [52].

Maximum cell densities in continuous cultivation are still limited by the accumulation of waste metabolites. To further increase bioreactor cell density and productivity, spent medium and waste metabolites must be removed while retaining or returning the cells to the bioreactor (Fig. 2E) [49, 53]. Such perfusion processes can achieve several-fold higher cell densities compared with fed-batch although doubling times may increase at very high densities. Cell separation can be achieved via filtration- or centrifugation-based technologies, using devices integrated into the bioreactor or external systems that return the cells. Partial or continuous cell bleeds may complement perfused systems, enabling continuous cultivation for maximum productivity over extended periods.

2.5.6 Comparison of the Operation Modes and Their Impact on Bioreactor Design

The introduced operation modes are visualized and compared in Fig. 2, considering that idealized cultivation types and real production processes may vary and possibly be a combination of different types. For example, at least an initial batch phase is typically performed in all cultivation processes, but can be in a fully or only partially filled vessel. Additionally, different cultivation strategies can be applied to the same cells during different steps of the full production process, e.g., in the seed train, proliferation, and differentiation. A comparison of some key characteristics of the different cultivation modes is furthermore given in Table 2.

All runs begin with preparation and initial filling. With the exception of continuous (E) and perfusion (F) modes, they end with complete harvest and cleaning. Feeding in B and D can follow different profiles, such as exponential, linear, or stepwise. Continuous operation (E) may involve gradual or stepwise flow adjustments, while modes B–F typically start with an initial batch phase. In perfusion (F), cultivation begins with full cell retention, after which controlled bleeding is introduced. The diagrams are schematic and illustrate key principles; actual processes may vary or combine elements.

Compared to fed-batch or continuous cultivation, batch processes offer reduced complexity, shorter process times, lower investment costs, and a lower risk of contamination. While they are generally reproducible, they are more suitable for small-scale production due to limited productivity, particularly at low initial cell densities. Constraints arise from limitations of initial substrate concentrations (e.g., osmolality, solubility, or overflow metabolism) and hence proceeding substrate depletion as well as the accumulation of inhibitory metabolites. Repeated cleaning and sterilization also reduce operational time, decrease productivity, and increase costs. Despite these limitations, batch processes remain widely used in seed trains due to their simplicity and reliability [49].

Fed-batch cultivation overcomes some of these limitations by adding fresh, concentrated nutrients during the process, enabling higher cell densities. This is particularly important for growth-inhibiting, unstable, or poorly soluble substrates, as well as for achieving better process control, such as preventing overflow metabolism and the formation of toxic by-products [49, 54]. Feed media typically contain glucose and may include amino acids, vitamins, trace elements, lipids, growth factors, and hormones. The exact demand depends on the species and cell type [55].

Table 2 Comparison of key characteristics of different cultivation modes

Characteristics	Batch	Fed-batch	Repeated (fed-) batch	Continuous	Perfusion[a]
Complexity	–	o	o	+	+++
Investment costs	–	–	+	++	+++
Operation time	–	o	++	+++	+++
Risk of contamination	–	–	+	+	++
Medium conversion rate	+++	+++	++	+	+
Cell density	–	+	–/+	o	+++
Process control efforts	–	+	+	+	+++
Productivity	–	o	+	+	+++

[a] Perfusion means perfusion in addition to continuous operation with continuous or repeated partial cell bleeding after reaching the target density

The supply of nutrients enables extended cultivation at higher cell densities, thereby increasing productivity and mitigating the impact of low initial growth and lag phases. It also reduces cleaning and sterilization requirements compared to batch processes [32]. However, this comes with increased system complexity, requiring sophisticated bioreactor design, feeding control, online monitoring, and careful sampling. Continuous or repeated nutrient additions can also increase the risk of contamination, and increasing the liquid level may lead to foaming and shear stress in the case of partially covered impellers, which must be considered in the design of the process and vessel.

Although fed-batch cultivation improves productivity compared to simple batch processes, it cannot completely overcome the limitations imposed by inhibitory waste metabolites, such as lactate and ammonia, which restrict cultivation time. Extending cultivation through partial harvests, continuous operation, perfusion, or combinations of these strategies can further increase productivity. Typically performed at the end of the exponential growth phase, partial harvests maintain healthy cells but result in slightly lower medium conversion and cell densities compared to full batch or fed-batch cultivation. However, repeated batch or fed-batch cycles increase the productive time by reducing cleaning and sterilization cycles. Using part of the previous culture as inoculum enables staggered use of seed trains to optimize line-specific costs. However, these strategies increase complexity and sterility demands, requiring aseptic intermediate vessels, precise process control, and management of fluctuating liquid levels to prevent foaming, shear stress, or insufficient mixing and oxygen supply.

Compared to batch or fed-batch processes, a continuous media supply, combined with the removal of cells, spent medium, and/or waste metabolites, enables higher cell densities and productivity to be maintained for longer periods, minimizing downtime for cleaning and sterilization. At the same time, the complexity of the bioreactor and process control system increases in order to maintain a constant bioreactor level and ensure stable and contamination-free cultivation. Continuous systems are particularly complex due to the need for precise control of inflow and outflow rates, as well as real-time monitoring of culture conditions. Repeated batch or fed-batch operations offer an alternative to fully continuous cultivation, reducing system complexity and simplifying downstream processing, as would be the case for fully continuous processes. Nevertheless, cell density and productivity are still limited, e.g., by the accumulation of waste metabolites. The latter can be overcome using perfusion technology, enabling increased cell densities and productivity. However, system complexity is high, both in terms of equipment and process control. Compared with batch or fed-batch, medium conversion rates are lower in perfusion, as spent medium contains both waste metabolites and substrates. The complexity increases from batch to fed-batch, continuous, and perfusion processes, affecting bioreactor design, process control, as well as upstream and downstream operations. The need for contamination-free operation for extended cultivation times and with continuous liquid flows necessitates sophisticated designs, including fully aseptic design, redundancy, cleanable and re-sterilizable lines, constant positive pressure, and sophisticated valves. In particular, for perfusion, bioreactors must support high cell densities with adequate mixing, pH and temperature control, and sufficient oxygen transfer. Continuous and perfusion systems often require precise level control via mass flow controllers, weighing systems, or liquid level sensors.

Ultimately, bioreactor-specific productivity determines the number of units required to meet annual production targets. Advanced cultivation strategies can reduce the number of bioreactors needed, but the increased complexity leads to higher costs for the same bioreactor size, e.g., for sensors, automation, flexibility, and sterility. Perfusion systems can increase equipment costs further due to their own direct costs as well as through limited working volumes, leading to higher volume-specific bioreactor costs [48]. However, new, scalable, centrifugation-based systems aim to enable cost-efficient operation in larger bioreactors.

The choice of cultivation type influences bioreactor design, as well as the upstream and downstream processes. Batch and fed-batch operations typically involve high peak labor demands, whereas continuous processes may distribute labor more evenly [32]. Batch media can be prepared and sterilized in one step, whereas fed-batch, repeated batch and fed-batch, continuous, and perfusion processes require repeated or continuous sterile medium preparation or intermediate aseptic medium storage. Whereas downstream processes of batch cultivations are often associated with long idle times and low equipment utilization, continuous cultivation enables fully continuous downstream processing with smaller equipment, optimized utilization, and shorter processing times. However, intermediate harvest vessels are often required to bridge cleaning periods of non-aseptic equipment.

Finally, different cultivation types vary in terms of productivity, cell density, and medium conversion, and the commercial optimum may differ from the highest titer, conversion rate, or productivity [46, 48]. Techno-economic analyses should compare operation modes and bioreactor sizes to identify the most cost-effective approach. Optimal design must strike a balance between costs, throughput, and revenue to ensure economic viability.

2.6 *Cell Retention Devices in Perfused Systems*

Perfusion is a process strategy to increase cell densities and productivity by (continuously) removing spent medium and waste metabolites from the bioreactor. The core of any perfusion system is the cell retention device, chosen based on perfusate flux, retention efficiency, hydrodynamic shear effects, consumable costs, and scalability. Scalability is particularly important for cultivated meat production, as achieving industrial-scale production while maintaining high yields is key to reaching price parity with conventional meat [46, 48].

Perfusion technologies can be classified by suitability for adherent or suspension cultures, internal or external mounting, and their underlying physical principles. The following section focuses on scalable retention technologies, rather than lab-scale or exploratory devices.

2.6.1 Filtration-Based Technologies: Separation by Size

Cell retention by filtration is based on the separation of single cells, aggregates, or microcarriers from components of the cell culture medium that are smaller than the pore size of the respective filter membrane in use. Historically, filtration techniques have been dominating the perfusion field, fostered by the development of synthetic membrane materials and low-shear pumps, and ultimately allowing for high cell densities and scalability to pilot- and industrial-scale. Filtration devices typically include the filter membrane as a consumable part with limited potential for reusability. Filtration area and pore size are the pivotal characteristics that determine membrane durability and the maximum attainable cell density. A trade-off in filtration area comes at the cost of larger pore sizes if filtration efficiency is to remain constant. Increasing pore sizes above 0.22 μm should be considered carefully when the filter membrane functions as a sterile barrier.

Membrane clogging and fouling remain critical challenges in any filtration technology, affecting the lifetime and processing capacity of filter membranes. As a result, it is common practice to change filter membranes during the production process when necessary, due to clogging. Clogging and fouling may be successfully managed through the operating regime, careful selection of membrane material and pore size, application of coating materials, as well as the composition of cell culture medium tailored to a specific bioprocess.

Cross-Flow Filtration: TFF and ATF

Tangential flow filtration (TFF) is a cross-flow method in which cell suspension from the bioreactor is circulated unidirectionally through an external filter cartridge via a circulation pump. Cell-free medium (permeate) passes through the membrane into a collection vessel, while the concentrated cell suspension (retentate) is returned to the bioreactor. The perfusion rate, e.g., the volume of medium exchanged over time, is controlled by the pressure difference across the membrane, which depends on the circulation pump and, optionally, an additional permeate pump. Increasing flow rates raises the pressure difference but also the shear stress, which can damage cells or clog the membrane. Shear tolerance varies between cell types; for example, hybridomas are affected at $\geq$1,266 s^{-1} [56], whereas no adverse effects were observed in chicken fibroblasts at 2,500 s^{-1} [48], Sf-9 cells at 3,000 s^{-1} [57], and CHO cells at 3,400 s^{-1} [58]. Proper optimization of flow and pressure is therefore critical to maintain high cell viability and stable filtration.

Filter cartridges typically consist of hollow fibers or stacked membrane filters, available with pore sizes for ultrafiltration or microfiltration. In biopharmaceutical applications, ultrafiltration can retain high-molecular-weight components, as demonstrated by Clincke et al. reaching a cell density of 1.0×10^8 mL^{-1} CHO cells using a 50 kDa hollow fiber [59]. For cultivated meat, animal cells (>10 μm) do not require pores below 0.2 μm. Removing cell debris can support culture rejuvenation.

Microfiltration TFF has also enabled very high cell densities above 1.0×10^8 mL^{-1}, as shown for CHO [58] and stacked membrane-based chicken fibroblast cultures [48].

Various pumps can be used in the TFF circulation loop. Robust cell lines, like CHO, tolerate shear from peristaltic pumps [58, 59], while more sensitive cells may require low-shear pumps, such as four-piston, rotary lobe, or magnetically levitating types. Pump selection depends on factors such as shear stress, pulsation, pressure control, self-priming, handling of viscous fluids, flow rate range, material abrasion, and compatibility with CIP/SIP.

Alternating tangential flow (ATF) is a cross-flow filtration method using a hollow-fiber module connected to the bioreactor and a diaphragm pump. The pump's oscillation creates tangential flow, drawing cell suspension into the fiber lumen, while culture medium passes through the membrane and is removed via a permeate line with negative pressure. The alternating diaphragm motion pushes cells back into the bioreactor, creating a back-flush effect that reduces filter fouling. Compared with unidirectional TFF, ATF exerts lower shear stress. High cell densities $>1.0 \times 10^8$ mL^{-1} have been achieved with CHO [58] cells and chicken fibroblasts [48, 60].

ATF systems are available from lab scale (V_d = 100 mL; 20 cm^2) to pilot and commercial scale (V_d = 6 L; 11 m^2). While TFF can be operated at similar and potentially higher scales, pump sizing is more challenging. ATF systems until 2012 supported perfusion rates up to 1,200 L d^{-1} or 2,000 L bioreactors [43, 61]. Currently, ATF and TFF can be scaled to ~5,000 L, reducing the number of bioreactors and associated capital and operating costs for large-scale production.

Floating Filters

Floating filters consist of a freely floating membrane inside a single-use perfusion wave bag. The filter membrane is attached to a tube penetrating the wave bag wall and leading into a collection vessel external to the bag. A peristaltic pump may be added to create a negative pressure on the permeate line, thereby increasing perfusion rates through the filter. Despite the high tendency of the floating filter for clogging, it was reported to support high cell densities reaching above 5.0×10^7 mL^{-1} in CHO culture [62] and 1.5×10^8 mL^{-1} in Per.C6 culture [63]. Wave bags prove particularly suitable for the cultivation of highly shear-sensitive cells, such as insect cells. To date, the largest available wave bags scale-up to 500 L [64].

Spin Filters

Spin filters are rotating porous cylinders mounted internally or externally to the bioreactor. Permeate is removed from the center of the cylinder via a permeate pump, while cells are retained outside. Filtration screens are typically made from stainless steel with pore sizes ranging from 10–25 μm for the retention of single suspension cells [65, 66] up to 120 μm for the retention of cell aggregates or microcarriers [67]. Cell densities of 1.58×10^7 mL^{-1} were reported in hybridoma bcl-2 culture [68], and up to 4.1×10^7 mL^{-1} in microcarrier-adherent CHO culture [67] with both systems reaching ≥2 reactor volumes per day (VVD). The primary challenge is filter fouling caused by adhesion of cell debris and nucleic acids to the filter screen. The fouling effect is further promoted by the increase in perfusion rates. Different measures have been described in delaying fouling, such as the use of hydrophobic plastic materials, larger pore sizes, and increasing the rotational speed of the cylinder [65, 67, 69]. However, increased speeds are associated with higher cell damage. Filter longevity may be improved in externally mounted devices, as these can be cleaned or replaced during the process. The scalability of spin filters has been reported to support bioreactors up to 100 L and a filtration area of 2,700 cm^2 [66].

2.6.2 Centrifugation- and Sedimentation-Based Technologies: Separation by Density

Cell retention by centrifugation or sedimentation is based on the density difference between single cells, aggregates, or microcarriers and the liquid culture medium. The particle (e.g., cell) velocity in a gravitational field is further affected by the dynamic viscosity of the fluid (e.g., culture medium) as well as the particle radius. The pivotal characteristic of these techniques is the clarification area required for sufficient cell retention and the flow rate required to reach sufficient perfusion rates. In accelerated systems in particular, high flow rates or *g*-forces may lead to increased shear stress with a risk of causing cell damage. Shear sensitivity, small density differences to the medium (~5%), and small cell sizes (10–30 μm) in single suspension cells are the main challenges for density-based separation technologies [70] and extended separation times. Microcarriers hold some advantages for gravitational methods due to their typically larger sizes (100–300 μm without cells) [71], despite their densities being similar to that of culture medium [72].

Centrifugal Separators

Centrifugal separators apply centripetal acceleration to separate cells from culture media, achieving higher efficiency and faster sedimentation than gravitational settling, as described by Stokes' law. At high cell densities, hindered settling can reduce separation efficiency, making operational parameters crucial [70]. Two main types are used in animal cell culture for perfusion and harvest: tubular bowl and disc stack separators. Early systems, such as the Centritech technology, used disposable rotor bags to separate cells and media in intermittent cycles, returning concentrated cells to the bioreactor. These systems, when connected to a 50 L perfusion bioreactor, supported CHO cell densities of 1.4×10^7 mL^{-1} [73]. Modern single-use systems, including the UniFuge® series, now cover lab (0.5–10 L), pilot (25–500 L), and

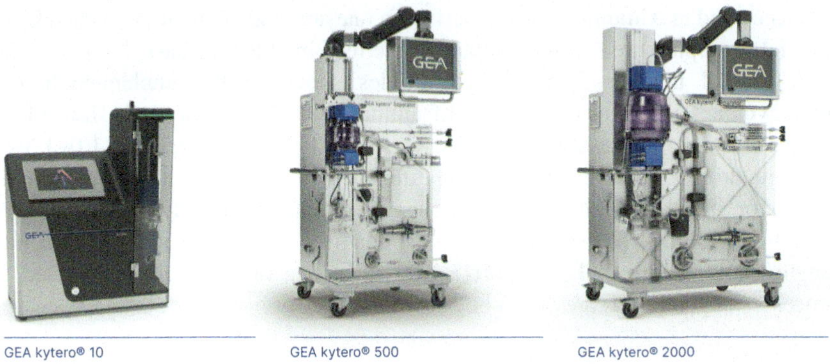

Fig. 3 GEA kytero® series available at lab scale (GEA kytero® 10), pilot scale (GEA kytero® 500), and industrial scale (GEA kytero® 2000). Relative sizes not true to scale

industrial scale (500–2,000 L), providing high separation efficiency while maintaining cell viability for HEK293 and CHO cells. Disc stack separators, such as the GEA kytero® series, operate continuously with single-use bowls at lab scale (5–20 L), pilot scale (50–500 L), and industrial scale (500–2,000 L). The cell suspension enters the separator from the top, the supernatant passes through stacked discs, and the cells are collected and recirculated to the bioreactor. Stainless-steel predecessors achieved perfusion with hybridoma and HeLa cells at densities up to 2×10^7 mL^{-1} and >95% viability in 6–100 L bioreactors [74–76]. Continuous operation reduces downtime compared to intermittent systems and supports higher overall productivity.

Key factors influencing separation efficiency and cell viability include feed flow rate, rotation speed, *g*-force, and clarification area. External recirculation loops, often driven by pumps, may induce shear stress; therefore, pump type and operating conditions must be adapted for sensitive cell lines. While peer-reviewed data on current single-use systems for industrial-scale perfusion are limited, centrifugal separators are a promising technology due to their scalability and established role in downstream processing (Fig. 3).

Hydrocyclones

Hydrocyclones are based on similar separation principles as centrifuges, but do not contain any rotating parts. Hydrocyclones consist of a chamber with a cylindrical section and a conical bottom section. Cell suspension is introduced through a tangential inlet into the cylinder, creating a double vortex that separates heavier/denser particles (e.g., cells) from the lighter/less dense liquid phase (i.e., supernatant). As a result of centrifugal force, drag force, and pressure gradient force, cells move towards the wall as they exit the hydrocyclone through a bottom pipe, known as the underflow. At the same time, the supernatant is pushed in an upward stream to the top of the hydrocyclone, creating the so-called overflow. The separation efficiency and maximum attainable perfusion rate are fundamentally impacted by the flow rate and pressure gradient applied at the feed inlet (0.5–4 bar).

A limited number of published articles showcase the implementation of hydrocyclones in perfusion cultures of animal cells. Bettinardi et al. showed that CHO cells were successfully cultured to 1.5×10^7 mL^{-1} applying 1.0–1.3 bar pressure drops, yielding a separation efficiency of >90% in a 40 L single-use bioreactor bag [77]. Although it is suggested that hydrocyclones sized 3 mm underflow, 2 mm overflow, 10 mm swirl chamber diameter, and 71 mm conical length could theoretically support a 500 L bioreactor [78], empirical data for processes beyond 40 L are yet to be shown.

Inclined Settlers

Inclined settlers are packed with numerous closely stacked inclined plates or lamellae. The settler is externally connected with the bioreactor via a pump-driven recirculation loop. Cell suspension enters the packed plates from the lower part of the settler. Cells sediment in the laminar flow regime on the lower plate, forming a countercurrent flow that enhances the settling efficiency (Boycott effect) [79, 80]. Sedimented cells are then returned to the bioreactor, while supernatant is removed through the top part of the settler. Dead cells and cell debris are preferentially removed from the retained cell concentrate, resulting in a predominantly viable culture in the bioreactor. The sedimentation area is increased to raise perfusion rates. Long residence times and a tendency for cell adhesion to the plates are the major drawbacks of this technology. Different groups have applied coating, vibration, and low temperatures to the lamellae to mitigate the cell adhesion effect and improve the sedimentation efficiency [81, 82]. The simplicity and robustness of this settler facilitate extended perfusion times lasting up to 5 months [83] and support industrial-scale bioreactors of 3,000 L at 1 VVD [84]. Coronel et al. achieved successful growth of avian suspension cells up to 5×10^7 mL^{-1} in a lab-scale settler system connected to a 1 L bioreactor [85].

Acoustic Settlers

Acoustic settlers operate via a recirculation loop connecting the bioreactor to a chamber exposed to an acoustic resonance field. Cells reversibly aggregate in the pressure nodes of the ultrasonic wave, sediment at the conical chamber bottom upon wave deactivation, and are returned as a cell concentrate, while supernatant is removed from the top. Efficiency depends on input power, cell density, and perfusion rate, and high perfusion can cause a sudden efficiency drop [86]. Backflushing with supernatant [87, 88] or increased power improves separation, but higher amplitudes generate heat (up to 1.3°C min^{-1}) [89], reducing viability. Cooling is applied by air (lab scale) or water (pilot scale).

Densities up to 5.5×10^7 mL^{-1} at >95% viability and ~95% efficiency have been achieved in CHO [87], Sf9, and hybridoma cultures [90, 91]. Continuous operation over 110 days at 1×10^7 mL^{-1} and 95% viability has been demonstrated [92]. Acoustic settlers have supported 100 L bioreactors at perfusion rates of 2 VVD, with units available for 200 L [93]. However, further scale-up is limited by the input power needed for retention, which may compromise cell viability.

2.7 *Process Analytical Technologies*

The implementation of process analytical technologies (PAT) was facilitated as part of the initiative "Pharmaceutical Current Good Manufacturing Practices (cGMPs) for the 21st Century – A Risk Based Approach" (August 2002). The goal was to improve continuous, real-time quality assurance and higher productivity through robust process analytics and control schemes for pharmaceutical products. The expectations regarding the use of PAT also include user-friendliness, robustness, real-time monitoring of the bioreactor system, and low cost. The challenges lie in communication between multiple, diverse process systems, a certain degree of flexibility in the event of changes in process sequences and/or software/hardware updates, as well as the required skills of the operators. The PAT scheme comprises various elements, including sensors, analytical tools, process models, data analysis techniques, and control strategies.

2.7.1 Sensors

Sensors are essential for characterizing, monitoring, and controlling bioreactors, providing insights into physical, chemical, and biological behavior. They can be classified as:

1. **In-line** (e.g., temperature, pH, DO): Integrated in bioreactor vessel, direct contact with the process liquid/gas, real-time data for process control.
2. **Online** (e.g., off-gas analysis, mass/volume flow controller): Integrated in piping; direct contact with the process liquid/gas, enables indirect, time-delayed measurements which provide information about the state of the reaction volume and/or can be used to indirectly adjust desired process parameters.
3. **At-line**: Bypass connection, sample analyzed outside the bioreactor, delayed results, often via analytical chemistry methods.
4. **Offline**: Manual sampling and analysis, highly delayed, not linked to the process control system, and only manual adjustments.
5. **Soft sensors**: Also known as software sensors, based on mathematical models implemented as computer programs (software) and provide information similar to that of a hardware sensor (sensor), classified as either model-driven or data-driven.

Collecting and integrating such data within IoT (Internet of Things), AI, and digital twins frameworks enables automated, digitalized bioreactor platforms [94, 98].

For animal cell cultures in stainless-steel STRs, the following PAT hardware sensors are already commercially available for in-line or at-line use, while others remain under development. These sensors are tailored for the food industry, unlike the highly complex and costly systems common in biopharma.

Spectroscopy

Spectroscopy generally refers to the interaction between light and matter. Spectroscopic sensors are used to identify specific substances (such as glucose and lactate) or cells and, in the case of bioreactor processes, to subsequently make quantitative statements. They differ in the various wavelength ranges of the infrared region (near-infrared (NIR), mid-infrared (MIR), and far-infrared (FIR)) as well as the visible and ultraviolet (UV) region.

Transmitted Light Measurement (Optical Density, Turbidity Sensors)

In this method, a light source emits light, and the transmitted light is detected by a detector. Light scattering arises from gas bubbles, solid particles, or cells. The use of turbidity sensors requires calibration for each bioprocess run, as the composition of the culture medium can vary between batches and influence turbidity. This is particularly relevant for microbial applications, whereas animal cell media usually show fewer raw material fluctuations and comparable concentrations of absorbing substances. The dimensionless parameter optical density (OD) describes the extent of light absorption or scattering of a sample at a defined wavelength and path length. OD values can be correlated with total cell count via offline measurements, enabling quantitative assessment of the culture. For animal cell cultures, however, the number of viable cells is more important than the total count. Furthermore, while the size of animal cells, filamentous fungi, and bacteria can vary, it remains constant for microorganisms [37]. Therefore, this method is generally more suitable for microbial processes.

Near-Infrared Absorption

Depending on the application, NIR absorption sensors operate either with a single specific wavelength (monochromatic) for targeted analysis (identification and quantification) of one individual substance, or with a wavelength spectrum (polychromatic) for simultaneous detection of several different substances based on their characteristic absorption patterns. In NIR absorbance sensors, a light source emits light of wavelengths (specific or spectrum) in the NIR range, which, on its way to the detector, encounters particles and molecules located between the light source and the detector. NIR wavelengths induce vibration of C–H, N–H, C–O, and O–H bonds, leading to the absorption of a concentration-dependent portion of the light, so that not all the emitted light reaches the detector.

In both cases, the detected signal depends on substance concentration and path length. The dimensionless parameter AU (absorbance unit) indicates how much light is absorbed by a sample and can be correlated with substrate or metabolite concentrations (e.g., glucose, lactate) or biomass, thereby providing quantitative information on the culture state. The main advantage of these methods is their ability to deliver specific concentration data on individual medium components or cells [99].

Near-Infrared Backscatter

A light source emits light of a specific wavelength in the NIR range, with the detector positioned in close proximity to the source. The emitted light is scattered at the surface of molecules and particles, so that a portion of the scattered light reaches the detector. If the energy is released at the same frequency as the original incident radiation, this is referred to as Rayleigh scattering (elastic scattering). The intensity of the received signal enables the calculation of particle and molecule density, as it is directly related to the amount of detected light [100].

The dimensionless parameter AU (arbitrary unit) indicates how much light is scattered by a sample at a specific wavelength. This value can be correlated with substrate or metabolite concentrations (e.g., glucose or lactate) or biomass, allowing for a quantitative statement to be made about their status within the bioreactor system. The advantage of these methods is that specific information about the concentrations of individual medium components or cells can be obtained. This method is particularly suitable for very high cell densities, which cannot be properly detected by transmission anymore.

Raman Spectroscopy

A light source emits light of a specific wavelength, which, on its way to the detector, encounters molecules and particles located between the light source and the detector. The emitted light is scattered at the surface of the particles, so that a portion of the scattered light reaches the detector. If the energy is released at a different frequency than that of the incident radiation, this is referred to as Raman scattering (inelastic scattering), which occurs much less frequently than elastic or Rayleigh scattering. The frequency can be higher (Stokes) or lower (Anti-Stokes) than the excitation light. To generate the Raman spectrum, these Raman shifts are measured, revealing the so-called chemical fingerprint of each substance, which allows them to be identified and quantified [101].

Raman spectroscopy can be used to identify and quantify individual substances in a cell culture medium. However, the signal from Raman scattering is very weak and can easily be superimposed by background fluorescence signals. In addition, the calibration and analysis of the Raman spectrum is very complex and time-consuming. In particular, a stable process with constant medium properties is necessary for calibration.

Capacitive Measurement

Capacitive measurement systems are a subgroup of impedance measurement systems, where the focus is on the capacitive component of the measurement signal. The method is suitable for both conductive and non-conductive materials. In this process, the cell culture medium containing the cells is located between the plates of a plate capacitor. When an alternating voltage is applied, current flows between the plates. The magnitude of the resulting current depends on the individual dielectric constant of the medium (including the cells), the distance between the capacitor plates, and their size.

The higher the value of the dielectric, the greater the capacitance of the capacitor and thus the current flow through the medium. With this method, only the viable cells should be detected (the plasma membrane is non-conductive); however, the signal is influenced by gas bubbles and potentially high conductivities of the cell culture medium itself. The values obtained by the sensor can be correlated with offline data in order to make a statement about the number of viable cells [37, 102, 103].

Biosensors

Biosensors are mostly used as at-line sensors (flow injection analysis systems) and are categorized into electrochemical, optical, or whole-cell sensors. They comprise a biological component, such as enzymes, antibodies, or oligonucleotides, which convert the substance to be detected with high specificity, and a physical sensor (transducer), which transforms the corresponding biological-chemical signal (e.g., pH value, oxygen consumption) into an electrical or optical signal. The biological component and the transducer are in direct contact with each other. The so-called multi-use sensors, which allow automated quasi-continuous measurements, are preferred for bioreactor processes. These can be used multiple times and allow (intermediate) calibrations if necessary. A disadvantage is that biosensors cannot be sterilized and, therefore, cannot be used in situ in the bioreactor systems. This creates an additional potential source of contamination [37, 98, 104].

2.7.2 Process Models, Data Analysis Techniques, and Control Strategies

Bioreactor Process Models

A distinction is made between mechanistic and data-driven models. Mechanistic models rely on established scientific principles, such as mathematical, physical, chemical, or biological relationships, while data-driven models are derived from experimental or observational data, often using statistical methods or AI. The combination of these two approaches gives rise to hybrid models, which integrate mechanistic and data-driven elements to provide more accurate and reliable process prediction models, ultimately supporting advanced process control optimization (hybrid controllers) and hybrid control strategies. Mechanistic models can further be classified into steady-state models, which use empirical models with constant parameters, and dynamic models, which capture time-dependent variations in the system [95, 96, 105].

Data Analysis Techniques

The main challenge in data analysis lies in selecting the "right" data from large, complex, and variable datasets, commonly referred to as "big data," and in interpreting it correctly across different domains. Big data is characterized by six V's: volume, variety, veracity, velocity, variability, and value.

Multivariate data analysis offers a solution to this challenge. It encompasses statistical methods that consider multiple variables simultaneously, such as sensor data from diverse sources, with the aim of extracting relevant information, revealing hidden relationships, and uncovering interactions within the dataset, thereby deepening the understanding of the process [96, 98, 106, 107].

In addition to, or alongside, multivariate data analysis, machine learning algorithms are increasingly employed. Unlike traditional process models or multivariate analyses, machine learning enables computers to identify patterns and relationships directly from collected big data, without explicit programming. This is particularly useful for detecting complex or hidden patterns that cannot be captured by process models or multivariate analyses. Successful machine learning relies on extensive training and the availability of large datasets. The ultimate goal is to generate reliable process predictions, which can then inform or optimize process control strategies [98, 108].

Process Control Strategies

The combination of sensor data, process models, and data analysis enables the creation of digital twins, which facilitate the rapid testing of process scenarios without using materials. This supports adjustments to process control strategies and provides insights from process runs, which can be used to predict and optimize system behavior. Real-time control and decision-making can be performed by the operator or autonomously by the system.

Two main strategies are commonly used: model predictive control (MPC) and hybrid control. MPC uses a process model to predict future system behavior with several advanced control methods and regulate it according to defined objectives, typically for process optimization [109]. Hybrid control combines mechanistic and data-driven approaches, leveraging scientific relationships and large-scale data analysis for process optimization [96].

Integration of prediction modeling of the bioprocess allows adjustment of the production schedule according to the predicted behavior of the bioreactor as the key process step in cultivated meat production. Although predictive modeling is more challenging in biological systems than in other industries, its investigation is evolving, e.g., for cell culture processes in biotherapeutics manufacturing [110]. For example, a prediction model for commercial-scale production of biotherapeutics in a cell culture perfusion bioreactor has been developed that predicts the bioreactor potency over a period of 5 days, bridging the lag of offline analytics to assess cell culture productivity [111].

3 Prototyping of Bioreactors at Lab Scale

In the context of bioreactors, prototyping refers to the iterative process of designing, building, and testing laboratory-scale systems that meet the biological and technical requirements for cell cultivation. The primary goal is to develop a functional prototype that realistically simulates critical process conditions, such as mixing, oxygen transfer, and temperature control. This allows different design options to be evaluated, sensors and control systems to be integrated, and the influence of process parameters such as agitation speed and aeration on cell growth and product formation to be studied [112]. Thus, lab-scale bioreactor prototyping serves multiple purposes: it enables early-stage functional testing, informs biological feasibility, supports iterative design improvements, and generates essential process data for later scale-up. In contrast to industrial-scale systems, prototypes offer higher flexibility, faster turnaround, and the evaluation of novel configurations without the high investment and operational costs of full-scale production [113, 114]. This substantially reduces financial risk, as capital-intensive full-scale infrastructure is not required at a stage where the feasibility of the design is still unknown. Prototyping is an essential step between theoretical design and large-scale production, enabling early identification of critical design parameters, reducing technical risks, and optimizing process control before substantial resource investments are made. The objectives of prototyping include validating design concepts (e.g., stirring mechanisms and reactor geometries) and ensuring that biological conditions are favorable for cell cultivation. In addition, it is used to integrate and test sensor and automation technologies, perform parameter studies (e.g., pH, oxygen supply), and prepare for scale-up by developing transferable concepts and models. Through prototyping, rapid feedback on reactor design and operation can be obtained, allowing adjustments without the costs and complexity of full-scale systems. For example, prototyping can reveal whether the chosen mixing strategies induce harmful shear stress or whether oxygen delivery is sufficient, enabling early troubleshooting and improving the reliability of subsequent scale-up [115, 116].

However, translating lab-scale findings into industrial processes is not straightforward. Critical bioprocess parameters such as oxygen transfer, nutrient gradients, shear stress, and cell distribution often behave nonlinearly during scale-up. Therefore, careful design of lab-scale prototypes that reflect key characteristics of the target system is essential to reduce the risk of scale-up failure [117].

This section provides an overview of the current strategies and technologies used in the prototyping of bioreactors for cultivated meat. Subsections cover the trade-offs between downscaling and upscaling, advanced manufacturing techniques and material selection, as well as the integration of process monitoring and automation tools. Together, these aspects highlight how lab-scale prototyping serves as a critical foundation for engineering robust and scalable bioreactor systems in cellular agriculture.

3.1 *Trade-Offs Between Downscaling and Upscaling*

Downscaling involves the simulation of large-scale process conditions in smaller lab-scale reactors to perform parameter studies and evaluate culture performance. Its advantages lie in speed, cost efficiency, and flexibility, allowing for fast iterations with minimal material use. However, downscaling faces limitations such as altered mass transfer dynamics, different shear force profiles, and geometric changes that affect mixing and flow patterns. These scale-dependent effects mean that not all parameters measured at a small scale can be directly transferred to larger reactors without careful consideration [118]. Conversely, upscaling entails transferring results from lab-scale prototypes to pilot- or production-scale bioreactors. The benefits of upscaling include achieving suitable production conditions, enabling direct application of findings to industrial settings, and revealing scale-dependent phenomena such as gradients in oxygen or nutrients that only arise at larger volumes. Although resource-intensive and technically demanding, upscaling represents an essential de-risking step, enabling identification of scalability limitations before committing to capital-intensive industrial deployment. A central challenge in bioreactor development is the transferability of process parameters across scales. Parameters such as the k_La, mixing time, cell distribution, and oxygen or nutrient limitation are critical. For instance, k_La reflects how efficiently oxygen is delivered to the culture, directly influencing cell growth, but it depends strongly on reactor volume and agitation type, among other factors. Similarly, mixing time and uniform cell distribution affect culture homogeneity and product quality. Due to the fact that these parameters often behave differently in down-scaled systems, scaling laws and measurement techniques must be carefully applied to ensure relevance [119].

3.2 *Prototyping Technologies for Bioreactor Development: Comparison of Additive and Conventional Manufacturing Processes*

Key aspects of bioreactor prototyping include designing the reactor, selecting materials, and using appropriate manufacturing techniques. These factors

Table 3 Comparison of additive and conventional manufacturing processes

Method	Type	Main advantage	Main limitation
3D printing (FDM, SLA, SLS)	Additive	Rapid iteration, complex geometry	Limited chemical/biological resistance
CNC machining laser cutting	Subtractive	High precision, material flexibility	Material waste, limited for complex internals
Injection molding, vacuum forming	Form-based	High reproducibility, scalable	High upfront cost (e.g., tooling)
Modular assembly systems	Modular	Flexible, low-cost, quick setup	Limited process realism
Digital simulation (CFD, CAD)	Virtual	Fast, no material cost	Dependent on model accuracy
Functional lab prototypes	Physical/ wet lab	Realistic testing of biology and control	Time- and cost-intensive

significantly influence the system's mechanical and thermal stability as well as chemical and biological compatibility. The properties and surface characteristics of materials determine how cells react to the culture environment, while the geometry of the reactor and the mixing mechanisms influence the flow profile, the supply of oxygen, and the formation of shear stress. The chosen manufacturing method also plays a role, as it determines tolerances, surface finish, and the possibility of making rapid adjustments during the prototyping cycle. Optimizing these aspects specifically can result in the development of a functional prototype that simulates realistic process conditions and provides reliable data for later scaling [120, 121].

In recent years, additive manufacturing (3D printing) has become a key tool for bioreactor prototyping due to its ability to produce complex geometries without the need for expensive tooling [120, 122]. At the same time, subtractive methods like the computer numerical control (CNC) machining remain important where high precision and robust materials are needed [123]. Form-based methods, such as injection molding, are less commonly used during early-stage development due to their higher upfront costs; however, they become valuable for later stages, particularly those involving reproducibility and scalability [124]. Besides these physical manufacturing techniques, modular systems and digital prototyping approaches, including computational simulations, support the rapid iteration and virtual testing of critical parameters such as fluid dynamics and mass transfer [125]. An overview and comparison of these prototyping approaches, including their main advantages and limitations, is provided in Table 3.

3.2.1 Additive Manufacturing (3D Printing)

Additive manufacturing (AM), commonly known as 3D printing, has become an increasingly relevant tool for the rapid prototyping of bioreactor components. This technology enables the layer-by-layer fabrication of complex three-dimensional geometries directly from digital models, offering high design flexibility, minimal material waste, and fast iteration cycles. Since its introduction through stereolithography (SLA) in the 1980s, a variety of 3D printing technologies have emerged, including fused deposition modeling (FDM), selective laser sintering (SLS), inkjet-based printing, and contour crafting. These techniques differ in printing resolution, speed, material compatibility, and post-processing requirements, making them suitable for different prototyping needs [122, 126].

The first step in creating a 3D object is to design a computer-aided design (CAD) model, which is then converted into a *.stl* (standard tessellation language or stereolithography) file. This format remains one of the most important interfaces between CAD software and AM devices to this day. The .stl file is then processed using a slicer software, which converts the three-dimensional model into two-dimensional layer data [127]. These layers are then used by the AM device to build up the 3D object. Several factors must be taken into account when selecting the appropriate AM technology, including required accuracy, object size, material requirements, and costs. Post-processing is usually necessary after manufacturing, such as removing excess material or support structures, polishing, painting, dyeing, or infiltrating [120].

A diverse range of materials and AM technologies is currently available, each exhibiting distinct advantages depending on the intended application. Commonly used materials include polymers, metals, ceramics, photopolymers, and polymer plasters. For bioreactor prototyping, two parameters are particularly critical: thermal stability, which is relevant for potential heat sterilization, and achievable dimensional accuracy. Rapid prototyping techniques, such as binder jetting (3DP) [128] and FDM [129], allow for fast model fabrication but are generally limited in mechanical stability and precision. In contrast, laser-based technologies, including SLS [130] and selective laser or electron beam melting (SLM/EBM) [131], enable the production of functional components with enhanced mechanical strength, biocompatibility, and sterilizability. SLA and PolyJet™ printing provide high resolution and mechanically robust parts [132], whereas two-photon polymerization (2PP) offers the highest achievable resolution for microscale structures. Additional methods, such as robocasting, selective heat sintering, and direct ink writing, further expand the range of possibilities for producing complex, application-specific bioreactor prototypes [120].

Overall, AM offers a versatile and powerful approach for the rapid prototyping of bioreactor components. It enables the fabrication of geometrically complex and application-specific structures, including channels for nutrient flow or integrated sensors, which are critical for optimizing cell growth and scaffold design [133]. Despite remaining challenges related to material selection, achievable resolution, and post-processing, ongoing advances in AM technologies and materials continue to expand their potential [134]. Through interdisciplinary collaboration between engineers and biotechnologists, AM can accelerate the development of tailored bioreactor systems, supporting more efficient, scalable, and innovative approaches in cultivated meat bioprocess engineering.

3.2.2 Subtractive Manufacturing: CNC Machining

CNC machining is a subtractive manufacturing process in which material is removed from a solid workpiece using computer-controlled cutting tools. By automating the movement of the worktable and the spindle, CNC systems convert digital CAD models directly into physical parts with high-dimensional accuracy and excellent reproducibility. This automation reduces operator variability, improves repeatability, and enables the rapid fabrication of complex geometries within the limits of the cutting tool's reach. Several machining operations can be performed on CNC systems. In milling, a rotating cutting tool attached to a spindle removes material from a securely positioned workpiece, following toolpaths defined by the CAD model until the final part geometry is achieved. Turning is employed for the production of cylindrical components, in which the workpiece is rotated in a chuck while a single-point cutting tool shapes it to the desired dimensions. This approach is particularly suited for shafts, threaded connectors, or round fittings. Grinding uses a rotating abrasive wheel to refine the surface finish of a part, which is especially valuable for bioreactor components requiring exceptionally smooth, cleanable surfaces. Drilling, in contrast, involves using a rotating drill bit to produce round holes in a stationary workpiece, e.g., to accommodate screws, ports, or fluidic connections [135].

In bioreactor prototyping, CNC machining is widely applied for producing functional components from metals such as stainless steel, aluminum, and titanium, as well as engineering plastics like polyetheretherketone (PEEK), polycarbonate, or acrylics. These materials are chosen for their mechanical stability, chemical resistance, and compatibility with sterilization methods such as autoclaving or chemical cleaning, properties essential for components in direct contact with process media. Typical examples include housings, flanges, impellers, baffles, and customized sensor ports [136].

While traditional CNC machining is optimized for macro-scale features, advances in stage precision and spindle speed have enabled "micromilling," a high-resolution variant capable of producing features down to the micron scale. In bioprocess applications, micromilling is used for creating fine channels, scaffold interfaces, or precision molds for replication processes. CNC machining offers superior surface finish, tight tolerances, and the ability to process durable, sterilization-resistant materials. However, its subtractive nature imposes limitations on certain internal geometries that are more easily achievable with AM, and it generally produces more material waste. For this reason, CNC machining is often combined with rapid prototyping techniques, using additive methods for early-stage conceptual testing and CNC for final, functional parts that require long-term mechanical integrity and process compatibility [123].

3.2.3 Form-Based Methods

Form-based prototyping methods, such as injection molding and vacuum forming, rely on shaping materials using predefined molds. These techniques are well-established in industrial manufacturing and can also play a role in bioreactor development, particularly when functional components or casings are required in larger quantities or in specific geometries [137].

Injection molding involves injecting molten thermoplastics into high-precision molds, enabling the production of complex and repeatable parts with excellent surface finish and tight tolerances. This method is particularly useful for fabricating standardized components such as vessel housings, sensor mounts, or modular inserts. However, due to the high cost and long lead time for mold fabrication, it is better suited for later development stages or scaled production. For prototyping, rapid tooling options such as 3D-printed or soft molds can reduce this barrier [138].

Vacuum forming, on the other hand, offers a low-cost and fast alternative for creating simpler parts by shaping heated plastic sheets over a mold using vacuum pressure. It is ideal for producing lightweight covers, trays, or outer shells in early prototyping phases. While the geometric complexity is limited, vacuum forming enables quick iteration and is compatible with many biocompatible thermoplastics [139].

Together, these form-based methods provide a practical solution for producing components with consistent quality, especially when combined with rapid mold fabrication techniques. Their main limitations lie in the reduced flexibility for design changes and higher initial effort compared to fully digital or modular prototyping approaches.

3.2.4 Modular Assembly Systems

Modular assembly systems utilize standardized, reusable components that can be flexibly combined to construct lab-scale bioreactors. Typical modules include tubing, fittings, pumps, connectors, and vessels in standardized formats (e.g., from laboratory hardware suppliers). This approach facilitates rapid prototyping without the need for custom fabrication, reducing both time and cost during early development phases [140].

The key advantage of modular systems is their adaptability: Design iterations can be quickly implemented by replacing or rearranging individual modules. This makes them ideal for parameter screening, sensor integration trials, or early-stage bioprocess testing. However, modular systems often lack the precision or long-term robustness of custom-fabricated reactors, and their scalability can be limited. Despite this, they are valuable tools in iterative development and early validation [141].

3.2.5 Digital Prototyping and Simulation

Digital simulations are increasingly used in bioreactor prototyping to model physical, chemical, and biological processes before physical construction. Techniques such as computational fluid dynamics (CFD) or finite element analysis (FEA) allow researchers to evaluate fluid flow, mixing behavior, shear stress, oxygen distribution, and heat transfer under different design scenarios [142, 143].

This in silico approach helps identify potential problems, compare design variants, and optimize critical parameters, such as impeller geometry or baffle placement. Digital prototyping can reduce the number of physical prototypes required and significantly shorten development timelines. However, the accuracy of simulations depends on the quality of input data and underlying assumptions, and real-world validation remains essential [143].

3.3 Integration of Sensor Systems for Process Monitoring and Control

The integration of sensor technologies is a crucial component in the development and operation of bioreactor systems, especially when aiming for controlled and reproducible cell culture environments. Although process monitoring and automation are well established in industrial-scale bioprocessing, their implementation in lab-scale setups remains limited. This is largely due to technical complexity, higher relative costs, and spatial restrictions in small-scale systems [144].

To ensure optimal culture conditions and timely intervention, several critical process parameters must be continuously monitored and adjusted. Among the most important are pH and DO, which are essential for maintaining cell viability and function. Additionally, monitoring metabolite concentrations such as glucose and lactate provides insights into the cells' metabolic state, while measurements of cell density and viability help determine process performance and timing for harvesting or media exchange [95].

Various sensor technologies are available to capture these parameters. A general overview of in-line and offline monitoring approaches, including electrochemical and optical sensor systems, is provided in Sect. 2.7.

To make use of sensor data for automated control, integration with hardware and software systems is necessary. This includes control units such as programmable logic controllers (PLCs), feedback loops, pumps for media addition or removal, and interfaces for process visualization and control. In perfusion systems or continuous culture, these systems can maintain homeostasis by automatically adjusting media flow, gas exchange, or temperature based on real-time measurements. However, while such control systems are standard in laboratory-scale bioreactors, their translation to industrial-scale systems with the same level of flexibility is generally not feasible. Industrial bioreactors feature control systems, but these are typically custom-designed and integrated specifically for the process, rather than being a direct scale-up of laboratory controls [145].

Overall, the integration of sensor systems, automation technologies, and intelligent data analysis holds great promise for enhancing reproducibility, efficiency, and scalability in lab-scale bioreactor development. Continued innovation in this area is expected to significantly narrow the gap between research- and production-scale bioprocessing.

4 Bioreactor and Process Design at Industrial Scale

The following sections will briefly introduce bioreactors and process design at industrial manufacturing scale, including scale-up challenges and general design aspects of large bioreactors, as well as key considerations for designing full production lines. Since bioreactor systems for adherent cells have not yet been established at an industrial manufacturing scale, this section will focus on bioreactors for suspension cells, particularly on the most common STR.

4.1 Vessel and Skid Design

The cost of production will be a primary driver in determining the economic viability of cultivated meat. While a significant portion of the production costs is associated with the culture medium, the capital investment will greatly rely on the quantity, sizing, and functionality of production bioreactors [46–48]: the larger the reactors, the lower the costs per kg of product for both CAPEX and OPEX [47].

To date, monoclonal antibody production using animal cells has been established in bioreactors up to 25 m^3 [146]. From an economic perspective, however, there is interest in using significantly larger bioreactors for cultivated meat production. For example, Humbird identified in an extensive analysis of scale-up economics of cultivated meat an optimal bioreactor volume of 50 m^3 for a fed-batch cultivation in an STR [46, 147]. Negulescu et al., however, expect significantly decreasing production costs on an even larger scale, from 33 m^3 STRs to 169 m^3 STRs and further to 209 m^3 ALBs. In 2022, a US-based cultivated meat company announced its plan to build a large-scale cultivated meat facility equipped with ten 250 m^3 bioreactors [148]. A conceptual study of Li et al. revealed that a 300 m^3 ALB could meet adequate mass transfer and mixing demands [28]. Although such bioreactor sizes are common for microbial applications [35], their applicability for animal cell culture has not yet been proven. Today's largest bioreactor for mammalian cell culture, with 30,000 L working volume (40,000 L total volume), was recently established by the US CDMO company Boston Institute of Biotechnology in their GMP facility in Shanghai [149].

From a technical perspective, scaling up bioreactors presents several challenges. Steam sterilization above 2 bar requires compliance with the respective pressure codes, while large agitators introduce higher forces and torque. These factors impact vessel wall thickness, reinforcements, lid and bottom type, double-jacket design (e.g., half-pipe coils, pillow plates), agitator flange and seals, nozzle configuration, and overall aseptic design. Larger liquid volumes demand longer shafts, which must be thicker, aseptically dividable, and sometimes supported to prevent bending and vibration. Impellers are typically assembled inside the vessel due to manhole size and weight limits. As cleaning spray balls may not cover all impeller areas, additional lateral nozzles or jet cleaners are often required to ensure effective cleaning.

Finally, cooling becomes more challenging at larger vessel volumes. While heat induced by metabolic activity and power input by the agitator is linear to the volume, doubling in volume only results in a roughly 60% increase in available heat exchange area at typical H/D ratio of 2:1. Therefore, additional measures such as increasing the cooling area by internal or external heat exchangers, or changing the heat flux by decreasing the cooling medium temperature and thereby enlarging the temperature difference may become necessary. For better homogeneity and higher efficiency of the cooling, separation of the temperature control jacket into multiple departments with separate inlets and outlets is beneficial on a large scale. This also allows for better energy efficiency, e.g., in fed-batch processes, as only the compartments covering the current filling level need to be active.

Next to the vessel, the units with the process equipment have to undergo changes in large-scale equipment. First of all, large-scale vessels are installed on multiple floors/levels, resulting in a split of the process units. This results in additional engineering and on-site work to build interconnecting piping and testing the finalized installation. Also, equipment used in aseptic systems might not be available in the necessary dimensions. One example is aseptic valve that is limited by most suppliers to sizes of 6″ or DN150. Another example is filter housing, for which, depending on aeration rate and maximum pressure drop/built up in the vessel, larger cartridges or multi-cartridge housings are needed. Other challenges may arise with equipment like mass flow meters, control valves, pumps, and aseptic heat exchangers. In such cases, changes to the process, technology, or functionality may be necessary to build a suitable unit.

Furthermore, the size of bioreactors may depend on the scalability of related process equipment, as seen with perfusion technologies [48]. Established systems in the biopharmaceutical industry, such as ATF and TFF systems, are limited in capacity. They scale almost linearly in terms of cost and effort as volumes increase. Although it is possible to connect multiple of these systems to the same bioreactor, the accumulating overall equipment costs are likely to be higher than installing a second bioreactor instead.

When looking at large-scale bioreactors and units, transport and logistics also become a challenge. For most transport ways, there are standard sizes and weights, e.g., 40-ft containers for ships or a standard inner width of 2.4 m for road transport in Europe. Any unit exceeding these standards will require special permits and/or special packaging, which will significantly increase costs. Due to its size, this type of equipment cannot be transported through standard doors and openings, so it must be incorporated into the building itself. This requires coordination with the overall construction progress to allow installation while the building is open or specially planned installation openings that can be closed after equipment installation. In rare cases, equipment might exceed even the maximum possible size limits for transport, e.g., due to bridges or tunnels. Therefore, on-site manufacturing can be considered as

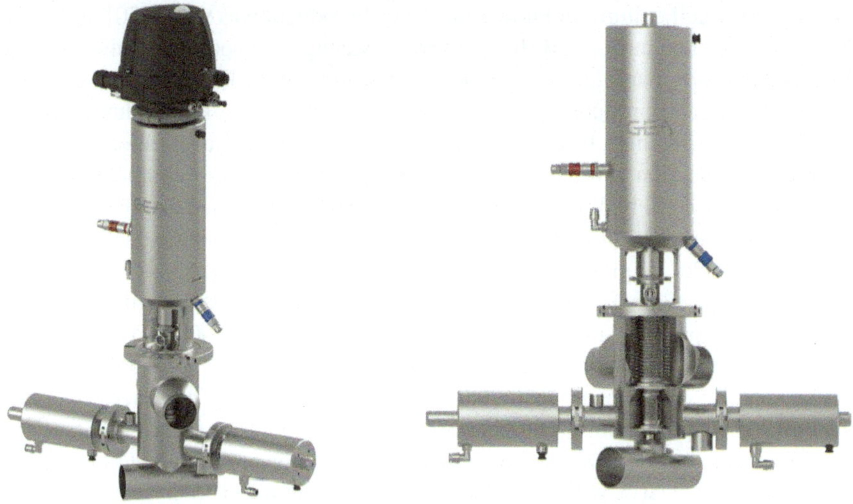

Fig. 4 GEA Aseptomag® double-chamber valve type DK designed for demanding aseptic processes (left) and cross-sectional view visualizing the two valve chambers and functionality of the valve (right). The steam-sterilizable sterile chamber in the middle acts as an integrated steam barrier enhancing the aseptic containment

an option, but it brings further challenges, especially for the vessel regarding welding, cleanliness, or pressure testing.

Industrial bioreactor systems for cultivated meat production may be associated with prolonged cell cultivation cycles, which lead to two additional requirements for bioreactor design. One is to maintain axenic conditions during this period, which requires more sophisticated solutions in terms of sterile barriers, e.g., block and bleed valves with integrated steam barriers (Fig. 4). The other one is the need for redundancy in case one piece of equipment fails during the process, such as broken sensors or blocked filter cartridges. In the case of cultivated meat, this also means that these kinds of equipment need an option for separate cleaning and sterilization, so that the redundant unit can be in operation while the original unit is exchanged or maintained. Both requirements are not only valid for longer production cycles but also for large volumes in general, even during short processes, as the cost of failure rises exponentially with the bioreactor's total volume. This is not only related to direct costs such as the costs of raw materials, energy, labor, as well as additional cleaning and sterilization, but also includes consequential damages, e.g., loss of earnings if the product cannot be sold, delivery commitments are not fulfilled, and damages to the company's image.

While almost exclusively stainless-steel bioreactors[3] are used for large-scale production in microbial applications, single-use bioreactors (SUBs) are frequently

[3] Special applications such as photobioreactors may require completely different bioreactor designs and materials.

used in animal cell culture for biopharmaceutical production as well [150]. Not only limited to cultivated meat, SUBs have several advantages compared to stainless-steel systems and are particularly useful for research and development, contract manufacturing, and production of small product quantities of high-value products [151]. They provide high flexibility in terms of the production of different products and enable production without cross-contamination in the "same" equipment. SUBs are also characterized by lower cleaning and sterilization costs and efforts, as well as lower investment costs [49]. To date, there are many different single-use bioreactor systems available, ranging from a few milliliters to 2,000 L [152, 153], covering research and development as well as small-scale biopharma manufacturing. A few companies also provide larger single-use bioreactors, e.g., the 5,000 L DynaDrive by Thermo Fisher Scientific or ABEC's Custom Single Run (CSR®) product line [154] with a working volume of up to 6,000 L. In an evaluation by Thermo Fisher Scientific of the production of biopharmaceuticals, 15 m^3 stainless-steel bioreactors were determined to be the more economical solution for product quantities above 75–150 kg $year^{-1}$ at 5 g L^{-1} product titer, when compared to 2 or 5 m^3 SUBs [151]. Cultivated meat, however, will require drastically larger quantities in order to become profitable. From this perspective, current SUBs do not seem to be economically applicable. Additionally, the advantages of higher flexibility and lower CAPEX are not only related to higher OPEX, but also create dependencies on the single-use equipment provider in terms of delivery time and whether the specific model is manufactured and provided in the future. Warehousing of single-use equipment can mitigate delivery risks, but storage capacities and related costs must be considered, as well as the limited shelf life of single-use equipment.

The combination of different plastic materials in SUBs makes them difficult to recycle, posing a risk that they will end up in landfills or pollute the oceans [7]. On the other hand, properly designed stainless-steel bioreactors can operate for decades and can be more easily recycled at the end of their lifetime. With single-use systems, flexibility is clearly a benefit during process development or for the production of various products in the same facility, as is common in biopharmaceutical and contract manufacturing. However, flexibility and manual handling for changing and reconnecting the single-use parts may increase the risk of failure through human error. Nevertheless, SUBs have relevance in process development and may also be used in the seed train in cultivated meat production to generate the inoculum for larger stainless-steel production bioreactors [155].

4.2 *Scale-up Challenges and Considerations*

Due to its great importance, the following section provides a brief introduction to bioreactor scale-up, even though a detailed examination of this topic would go beyond the scope of this chapter. For a more detailed introduction, the reader is referred to the pertinent literature. For example, reviews from Lara et al., Alavijeh et al., and Guan et al. discuss general scale-up aspects (inhomogeneities) of bioreactors for different applications [156], digitally enabled approaches for scale-up of mammalian cell culture bioreactors [157], and general scale-up challenges for cultivated meat [158], respectively.

As the volume-specific costs of bioreactors decrease with size [46], bioreactors for large-scale production must be as large as possible to make them economically feasible. However, they must provide comparable process conditions to the lab, pilot, and demo scales to ensure that the bioprocess can be successfully transferred to the larger scale. As the size of the bioreactor increases, a variety of challenges emerge, which are collectively referred to as scale-up challenges. These are not only limited to bioreactors for animal cell cultivation, but differ from the challenges in microbial processes with fermentation volumes that can be 10× larger and beyond. The scale-up of the cultivation is an inevitable but sometimes exhausting step when scaling a bioprocess from lab-scale to commercial production. In general, the scale-up of a bioprocess is a multidisciplinary challenge that typically combines the proper design and characterization of the large-scale bioreactor and experimental validations at both laboratory and pilot scales using the so-called scale-down models [159].

A typical scale-up challenge is the increasing formation of spatial gradients within the bioreactor [156]. Most of them can be attributed to "insufficient" mixing, leading to axial (e.g., nutrients, oxygen, and pH) and radial (temperature) gradients. However, some gradients are an inherent property of the bioreactor, such as spatial differences in shear rates or hydrostatic pressure. Looking at STRs, shear rates are highest at the impeller tip and strongly depend on the impeller design [160]. A common approach to assess, among others, hydrodynamic stress distribution caused by the agitator is the use of CFD, which allows, e.g., comparisons of different impellers and across scales [161]. The hydrostatic pressure can hardly be influenced, as it depends on the liquid height, which is determined by the aspect ratio of the vessel that is usually kept constant or at least similar across scales. However, the hydrostatic pressure directly affects the gas solubility and hence the OTR in a positive way and the carbon dioxide removal in a negative way. The accumulation of CO_2 negatively affects pH and osmolality in cell cultures [161], and, in contrast to the oxygen supply, significantly depends on the volumetric aeration rate rather than the specific power input [146]. Sufficient CO_2 depletion will become particularly challenging in the context of high cell densities enabled by perfusion, as further oxygenation is typically achieved by increasing the oxygen concentration in the sparging gas [161].

Keeping the same aspect ratio of the vessel, the volume-to-surface ratio decreases, making temperature control via the double jacket more demanding. However, this is for animal cell cultivation, usually less challenging than for microbial applications due to the much slower growth and heat formation. Hence, bioreactors for animal cells are typically flatter and wider than their microbial pendants, ranging from about 1.3 to 1.8 (liquid height to diameter) [146]. Although not yet tested at these scales, perfusion-driven high cell density cultures and further enlargement of the bioreactors may require adjustments to accommodate the increasing cooling requirements.

Constant mixing times, tip speed, specific power input, or k_La values are frequently applied as scale-up criteria for bioreactors. However, not all of them can be kept constant simultaneously, and keeping one or the other characteristic identical can significantly impact the others [156]. For example, maintaining the same Reynolds number would reduce the specific power input tremendously, whereas maintaining the same impeller speed would result in much higher specific energy input. Yang et al. investigated different criteria scaling from 3 to 2,500 L, showing that a constant tip speed would lead to a very low impeller speed of only 9 rpm, whereas a constant mixing time would result in a very high agitation speed (190 rpm). Hence, compromises must be made during scale-up. A common scale-up criterion in animal cell culture is a constant-specific power input, which is a good indicator of culture homogenization, mixing, and shear stress [157]. However, it leads to decreased mixing with increasing size, favoring the formation of spatial gradients as discussed above.

In general, the spatial gradients cause constantly changing environmental conditions for the cells moving across the bioreactor, e.g., in terms of shear, pressure, oxygen availability, and nutrient concentration, that may affect the metabolism, yield, and product quality [156]. Since it is technically not feasible to avoid them completely during the scale-up, scale-down models provide the possibility to evaluate the impact of these gradients on a small scale [162]. To mention a few, scale-down models have been developed to simulate changing oxygen concentrations (e.g., through changing oxygen concentrations in the sparging gas [163], cyclic pressure changes [164], or internal [165] or external [166] compartmentalization), for pH values [167, 168] and hydrodynamic stress [169].

4.3 Design Aspects of Industrial Production Lines

Cell cultivation is typically the time- and throughput-limiting step in all bioprocesses, but especially relevant for cultivated meat due to the high doubling time of animal cells compared to microbial growth. This becomes even more pronounced for processes that include a further differentiation step. Therefore, a full production line is usually designed around the bioreactors that are needed to fulfill the annual production target. Due to the limiting nature of an individual bioreactor, one production line typically consists of several bioreactors that share the same upstream and downstream equipment. This not only ensures a higher degree of utilization, but also reduces throughput-specific costs as equipment costs typically do not scale linearly in price with higher capacity [170]. In general, high yields and productivity are crucial for large-scale production. For bioreactors, overall process productivity includes productivity during the actual cultivation, but also non-productive times such as preparation, cleaning, sterilization, and maintenance. Fewer cleaning and sterilization cycles also reduce the costs for water, energy, and chemicals and can reduce the number of CIP distribution lines in a production facility. More advanced cultivation types than simple batch approaches clearly offer higher productivity in both aspects, but they involve more complex and flexible systems, which are consequently more costly.

Each of the various cultivation modes has different advantages and disadvantages, e.g., in terms of process times, productivity, and complexity of the bioreactor (Table 2). Additionally, the choice of the operation mode has implications for the entire process line, as it affects both the upstream and downstream processes. Some effects on the upstream and downstream equipment and equipment utilization have already been discussed in Sect. 2.5.6 with regard to a single bioreactor and are now examined in the context of a large-scale production line. To design an economic production line, thorough production scheduling, including parallelization and staggering (time-shifted operation of parallel equipment) of equipment, as well as precise matching of equipment sizes, is needed to ensure the highest equipment utilization, minimal idle times, and an overall continuous production of the line. A well-understood, robust, and reproducible process at the pilot scale is crucial for doing this work properly.

While batch processes only need preparation and sterilization of one medium and only once before starting the cultivation, other operation modes require different media and repeated or even continuous supply of fresh media throughout the cultivation. Repeated or continuous addition of fresh and sterile medium requires either repeated in-time or even continuous preparation and sterilization of fresh medium or aseptic medium storage to accommodate varying medium demands and production capabilities, e.g., during CIP of the non-aseptic part of media preparation. Although the latter requires additional aseptic equipment, which is therefore more expensive, it is usually the preferred solution for large-scale production, as it provides more flexibility and process safety. Several bioreactors can then be fed from the same aseptic media storage via an aseptic media distribution line, making this approach feasible for all cultivation strategies. Additionally, the buffer capacity of the storage vessels allows the preparation of different media, such as basal medium, fed-batch medium, and differentiation medium, with the same medium preparation system. Again, this needs thorough production scheduling and sizing of the storage vessels according to the specific demands during production.

Besides the actual production bioreactors, a cultivated meat production line requires several smaller seed bioreactors to generate enough biomass to seed the final production bioreactor. The volume typically increases by a factor of 5–10 from one bioreactor to the next. As for the medium preparation, the different cultivation modes can have an impact on the seed train design. Batch and fed-batch processes often require the same number of seed trains as production bioreactors because the cultivation times are quite similar across different bioreactor sizes. Considering staggered bioreactors with continuous or partial harvests and hence extended periods of cultivation time, one seed train can supply several production bioreactors, thereby reducing the number of comparably expensive seed reactors. Independent of the operation mode of the production bioreactor, particularly the first stages of the seed train often rely on (fed-) batch processes due to their simplicity and reliability [49].

The cultivation types also differ with respect to the harvesting process. The entire bioreactor may be harvested all at once at the end of the cultivation, but may also be harvested partially and repeatedly or fully continuously. Looking into a single bioreactor, it means that the downstream equipment would have long idle times for batch, fed-batch, and partial harvest processes, with a negative impact on the total economics. In contrast, fully continuous harvesting has the highest equipment utilization, but typically requires an additional harvest vessel to buffer for cleaning times, especially for non-sterile downstream equipment. On an industrial scale, however, the downstream equipment is typically shared by several bioreactors, leading to a similar design of the downstream line for both continuous and discontinuous harvests. Regardless of the cultivation type, the downstream process of a production line is usually designed for continuous or quasi-continuous operation to maximize equipment usage. For batch, fed-batch, and partial harvest processes, this means parallel bioreactors with time-shifted operation. One bioreactor or part of one bioreactor, in case of partial harvest operation, is harvested into an intermediate storage vessel feeding the further downstream process. An (aseptic) intermediate vessel helps to increase the operational time and therefore the productivity of the bioreactor for batch and fed-batch processes and to maintain axenic conditions in case of partial harvests. Once the batch is processed, the next bioreactor is harvested to optimize equipment utilization, highlighting the need for robust and reproducible processes and thorough production scheduling. In fully continuous production processes, the continuous outflow can be combined from several parallel running bioreactors to the same downstream line. This enables a more economic downstream process, as the equipment typically does not increase linearly in price with higher capacity [170], resulting in a lower throughput-specific investment. As for the discontinuous harvests, an intermediate storage vessel is typically used to compensate for harvest flow fluctuations and to buffer for cleaning times of the downstream equipment. Hence, the downstream processes of production lines are often similar, independent of the actual cultivation type in one individual bioreactor.

When deciding on a cultivation strategy for an industrial production process, several aspects must be considered. As described above, this includes economic aspects such as CAPEX considerations, as well as productivity and medium conversion that need to be compared in a techno-economic analysis. Investigations into the robustness and reliability of the process, as well as risk analysis, are also needed to assess the impact of potential failures (e.g., stopped or reduced growth, equipment failures such as sensor malfunctions, contamination) on the entire production process. Another aspect to consider is batch tracing. This means that a specific final product can be assigned to the raw materials included in its production process. This is simple in batch and fed-batch processes, but hardly possible for continuous operations, whether fully continuous or based on partial harvests. For repeated (fed-) batch and continuous cultivations, changing product properties and quality over time are further aspects that must be considered.

Regardless of the cultivation type, the number and size of the bioreactors and the capacity of the upstream and downstream processes must be precisely matched to ensure a robust production process and an economic production line design with a high degree of equipment utilization. This requires well-known, reliable, and reproducible bioprocesses and thorough mass balancing, production scheduling, and equipment sizing. Additionally, industrial-scale production lines are highly automated, e.g., to minimize human errors.

4.4 Software and Automation

Compared to benchtop or pilot bioreactors, production bioreactors are typically less flexible, as the process has been developed and optimized at lab, pilot, and demo scale. Instead, industrial-scale production lines are highly automated to guarantee process security. In the context of automation, the term process reliability refers to both the process itself and the aspect of safety when handling the equipment. An automated system supports the operator in monitoring the process parameters and, depending on the degree of automation, regulates itself. The less manual intervention is required, the safer the system operation is for the operator and for the process itself. Every manual intervention carries the risk of human error and incurs additional costs for the plant operator in terms of personnel expenses.

The overall automation concept for industrial plants must allow for automated processes with manual interventions, including manual control, and take into account a uniform concept for emergency shutdowns and alarm systems. This type of process integration involves networking sub-processes and individual plant components, usually from different manufacturers. At the start of the project, interface communication or data communication between the various supervisory control and data acquisition (SCADA) systems should therefore be defined. A so-called module type package (MTP) interface should be used in this case. MTP provides all the necessary information, from module properties and status descriptions to alarm handling for a subsystem or system component, to the higher level automation system (process orchestration layer, POL). This means that the various plant components use the same interface language, which reduces the effort involved in complex system integration and thus saves time and money, and leads to maximum flexibility in the event of future plant modifications (modular automation concept). MTP also enables a standardized platform for digital data management [171].

A so-called manufacturing execution system (MES) ensures that, e.g., media are always available in sufficient amounts or that equipment, vessels, and lines are cleaned and sterilized according to need. It optimizes production flows, collects real-time data, and checks the status of performance at the production management level. The overarching, unified automation system POL works as an integration and coordination layer above different SCADA and MES systems.

Process display and control are implemented via a human-machine interface (HMI). In recent years, there has been a positive shift from "colorful, overloaded images" to information focused on the essentials to ensure efficient plant operation. This can be accomplished by restricting visual effects to abnormal situations, employing contrast and color schemes that minimize eye strain, and utilizing a hierarchical screen layout along with the visualization of automated sequences to support normal operation. To optimize the overall process, predictive process control strategies can be established not only at the level of individual units but also at the plant-wide level. As previously described, digital twins can be used to simulate scenarios in advance and adapt the process control strategy accordingly. A digital twin can encompass the entire process plant or at least the critical process steps. The more process steps are included in the digital twin, the more complex and demanding it becomes to create a reliable model. Here, the cost-benefit aspect should be prioritized. Nevertheless, this approach opens entirely new possibilities in terms of energy and resource efficiency, minimizing downtime, and maximizing overall plant productivity.

5 Scaffold Integration in Bioreactor Systems

Scaffolds play a crucial role in structured cultivated meat production as well as in various tissue engineering applications by providing three-dimensional frameworks that support cell attachment, growth, and tissue formation [172]. However, the successful integration of these scaffolds into bioreactor systems goes beyond their standalone properties and requires a harmonized approach that connects scaffold design with bioreactor operation, process control, and scalability. Integrating scaffolds into bioreactor systems presents a complex challenge that requires a close alignment of material properties, cellular biology, and process parameters [173]. This section provides an overview of the key requirements for scaffolds used in bioreactors, introduces different types of scaffolds and manufacturing methods, and discusses cell seeding techniques alongside the adaptation of bioreactor configurations to scaffold designs. Furthermore, strategies for process monitoring during cultivation and challenges related to scaffold harvesting and analysis are examined. The goal is to highlight the interfaces between scaffold design and bioreactor technology, offering practical approaches for successful integration into cultivated meat production processes.

5.1 Requirements for Scaffolds in Bioreactor Applications

The successful integration of scaffolds into bioreactor systems requires balancing biological, structural, and process engineering considerations. In cultivated meat production, scaffolds are not merely passive supports for anchorage-dependent cells but active components that influence cell attachment, proliferation, differentiation, and tissue maturation under dynamic bioprocess conditions. Their design must therefore simultaneously address the biological needs of the cultivated cells and the operational demands of the bioreactor environment [174]. A key technical requirement is sufficient mechanical stability to withstand shear forces, pressure gradients, and repeated handling, as encountered in STR or perfusion systems, without compromising structural integrity over the cultivation period [175, 176]. Inadequate stability can restrict medium flow, reduce oxygen and nutrient delivery, and hinder tissue formation, particularly in perfusion systems where deformation can block channels and impair mass transfer [177]. At the same time, a highly porous and interconnected architecture is essential to enable efficient nutrient and gas exchange, particularly oxygen supply to dense tissue constructs, while also facilitating homogeneous cell seeding and the removal of metabolic waste [178, 179]. The porosity and pore geometry must be tailored to the specific cell type to optimize both diffusion and mechanical performance [180]. Equally important is ensuring biocompatibility, which requires scaffold materials to provide a non-toxic, stable environment that preserves cell viability and function under the variable physical and chemical conditions within a bioreactor. To promote appropriate cellular behavior and tissue development, their structure and surface chemistry should ideally recapitulate key features of the extracellular matrix (ECM), providing both the physical cues and biochemical signals necessary for adhesion, proliferation, and differentiation [181]. Functionalization with bioactive motifs, such as RGD (Arg-Gly-Asp) peptides or ECM-derived proteins, can enhance cell adhesion and signaling, but in cultivated meat applications, these modifications should meet food-grade, animal-free, and regulatory standards [182, 183]. As a result, plant-based or recombinant bioactive components are increasingly used to achieve both functional and ethical compatibility [184], and when scaffolds are part of the final product, all materials must be edible, safe, and preferably plant-derived [183, 185]. In contrast, temporary scaffolds may be designed to degrade during cultivation, allowing cells to replace the structure with their own ECM, provided that degradation occurs in a controlled manner without structural collapse or flow obstruction [186, 187]. The physical design of scaffolds must also be tailored to the specific bioreactor configuration. In packed-bed or fixed-bed systems, scaffolds need to be form-stable to ensure uniform perfusion of the culture medium. In contrast, STRs require scaffolds that are either immobilized or sufficiently robust to remain suspended without fragmentation or causing clogging [188]. Finally, scalability is a decisive factor, as scaffolds optimized for small-scale static cultures may not directly translate to automated large-scale production. Mass manufacturability through reproducible methods such as 3D printing, extrusion, or freeze-drying, combined with compatibility for aseptic processing, sterilization, and automation, is essential to bridge the gap between biological performance and engineering feasibility [189, 190].

5.2 Overview of Different Types of Scaffolds and Materials

The diversity of scaffold types reflects the wide range of structural and functional demands in cultivated meat production. Depending on the target tissue architecture, mechanical properties, and nutrient transport requirements, scaffolds can be broadly

Fig. 5 Comparison of different scaffold types (Created in BioRender. Treinen, C. (2025) https://BioRender.com/3ux9o64)

categorized into porous matrices, hydrogels, fiber-based constructs, and microcarriers (Fig. 5) [10]. Each type offers unique advantages and challenges that influence cell behavior and the overall quality of the final product.

Porous 3D scaffolds, composed of highly porous and interconnected polymer networks, are vital for cultivated meat production as they provide an environment conducive to cell growth, differentiation, and tissue formation within bioreactors [191]. Their sponge-like structure enables the efficient transportation of nutrients and oxygen, which is crucial in dynamic cultivation systems for maintaining cell viability and function. These scaffolds mimic natural ECMs, promoting proper tissue integration and mechanical stability under bioreactor conditions [174]. Techniques such as freeze-drying and water annealing enable precise control of pore size and mechanical properties to suit specific cell types and bioreactor setups [192]. Porous scaffolds are typically fabricated via lyophilization (freeze-drying), where the frozen solvent sublimates to create interconnected pores, or through spray drying and water annealing methods [193]. Sterilization methods such as UV treatment (for suitable structures and conformations), ethylene oxide (EtO), or supercritical CO_2 (sCO_2) are applied to ensure biocompatibility without compromising scaffold structure [194]. While porous scaffolds are promising for scalable bioreactor cultivation, challenges remain in balancing costs, scalability, and optimal pore architecture for efficient mass transfer and tissue development [195].

Hydrogels represent another major class of scaffolds for cultivated meat, providing a hydrated and biocompatible environment that supports cell attachment, proliferation, and differentiation [196]. They are commonly fabricated through molding combined with chemical or physical crosslinking techniques, such as ionic or covalent crosslinking, which stabilize the polymer network while preserving the hydrogel's high water content and softness [197]. Their soft, ECM-like network can be engineered to match the mechanical and structural requirements of different tissue types and to perform reliably under specific bioreactor operating conditions. In dynamic cultures, their viscoelastic properties help reduce mechanical stress on cells, while tailored microstructures can encourage organized tissue formation [198, 199]. However, ensuring long-term stability under shear forces, maintaining nutrient and oxygen transport in thicker constructs, and achieving cost-efficient large-scale production remain key challenges for their integration into bioreactor-based systems.

Nanofibrous scaffolds hold significant promise for integration into bioreactor systems used in cultivated meat production due to their ability to closely replicate the ECM, thereby promoting muscle and fat cell growth and maturation [179]. Their high surface-to-volume ratio promotes enhanced cell attachment and differentiation, which is essential for developing meat-like textures [174]. Fabrication methods like electrospinning allow for precise control over scaffold properties, but scaling these processes to meet industrial bioreactor demands remains a challenge [200, 201].

Microcarriers are small spherical beads, typically 100–500 μm in diameter, widely used in bioreactors for adherent cell expansion due to their high surface-area-to-volume ratio and the ability to enhance scalability in stirred systems [174]. They are commonly fabricated through techniques such as emulsion polymerization, spray drying, or solvent evaporation, allowing control over bead size, surface texture, and porosity to optimize cell attachment and growth. In bioreactor systems such as STRs or wave bioreactors, microcarriers facilitate the expansion of high-density cell cultures by enabling cells to grow on their surfaces while being suspended in the culture medium. This suspension culture approach improves nutrient and oxygen distribution compared to static cultures and supports large-scale production. The cellularized microtissues formed on microcarriers can either be harvested by separating cells from the beads or used directly as components in the final cultivated meat product, potentially reducing the number of downstream processing steps. However, challenges remain in ensuring uniform suspension without damaging cells, optimizing harvesting methods to maximize yield, and controlling costs related to microcarrier production and recovery. Additionally, the choice of edible and food-safe microcarrier materials is essential to meet regulatory and consumer requirements [202]. Careful optimization of bioreactor parameters such as agitation speed, microcarrier concentration, and feeding strategies is necessary to balance cell growth, viability, and product quality for scalable cultivated meat manufacturing [203, 204]. Together, these scaffold types and fabrication approaches form a versatile toolbox that can be tailored to meet the complex structural, mechanical, and biological requirements of cultivated meat production within bioreactor systems.

The selection of materials for scaffold fabrication in cultivated meat production is crucial for meeting biological functionality and food safety requirements. These biomaterials can be broadly classified into natural and synthetic categories, each offering distinct advantages and challenges [205]. Natural biomaterials are typically derived from protein or polysaccharide sources. Protein-based materials such as collagen, gelatin, silk, and fibrin closely mimic the native ECM, promoting cell adhesion and growth. Polysaccharide-based materials include substances like alginate, agarose, chitosan, and hyaluronic acid, which provide biocompatibility and favorable physical properties such as hydration and porosity [206]. Their natural origin generally supports biodegradability and cell-friendly environments, making them attractive as scaffolds for cultivated meat [206]. Synthetic biomaterials primarily consist of polymer and ceramic biomaterials. Polymers such as polyethylene glycol, polyglycolide, and poly-lactic-co-glycolic acid enable precise tuning of mechanical strength, degradation rates, and scaffold architecture [207]. Ceramic biomaterials like hydroxyapatite and bioactive glass are less commonly used in cultivated meat but are prominent in other tissue engineering fields due to their mechanical robustness [208]. The choice of biomaterial impacts scaffold fabrication methods, mechanical properties, nutrient diffusion, and cellular responses in bioreactor systems. Achieving an optimal balance between biological performance, structural integrity, costs, and regulatory compliance remains a key focus for advancing scaffold technologies in large-scale cultivated meat production.

5.3 Methods for Cell Seeding of Scaffolds

Cell seeding onto scaffolds is a critical step in tissue engineering that directly impacts the quality and functionality of the engineered tissue. Methods for cell seeding vary depending on timing, technique, and cell concentration, with the goal of achieving uniform cell distribution and strong cell adhesion on the scaffold.

Various seeding techniques have been developed to optimize cell infiltration and viability, and these can be broadly categorized into static seeding, dynamic seeding (flow/perfusion-based and agitation-based), and assisted seeding methods involving physical forces [209, 210].

Static seeding is the simplest and most widely used technique, where a concentrated cell suspension is directly pipetted onto the scaffold or the scaffold is immersed in the cell suspension, followed by incubation under static conditions to allow cell attachment. While this method is straightforward and requires minimal equipment, it often results in uneven cell distribution, with cells predominantly adhering to the scaffold surface. The efficiency of static seeding strongly depends on scaffold properties such as porosity, wettability, and pore interconnectivity, as poor liquid infiltration can trap air pockets and limit cell penetration. Typical cell densities for static seeding range from 10^5 to 10^6 cells cm^{-3} scaffold, with incubation times commonly between 1 and 4 h [211]. The actual number of cells required depends on the total scaffold volume and porosity, as these factors influence cell attachment and distribution. While applicable for initial screening and small-scale setups, static seeding is often insufficient for large-scale bioreactor cultivation due to limited cell infiltration and nonuniform colonization.

To overcome the limitations of static seeding, dynamic seeding methods have been developed. These approaches apply fluid flow or mechanical agitation to improve cell distribution throughout the scaffold volume. In flow- or perfusion-based dynamic seeding, a cell suspension is actively pumped through the scaffold pores, facilitating deep cell infiltration and simultaneous nutrient supply. Perfusion bioreactors exemplify this technique and have been shown to enhance uniform cell colonization and viability. Cell densities that are applied typically range from approximately 10^6 to 10^7 cells cm^{-3} scaffold, with perfusion superficial velocities adjusted between 0.1 and 10 mm s^{-1}, depending on scaffold type and pore size. However, optimizing flow rates is crucial to minimize shear-induced cell damage [212, 213]. Conversely, agitation-based dynamic seeding involves mechanical movement such as rotation, shaking, or rolling of the scaffold immersed in cell suspension, promoting cell dispersion through constant mixing. Within agitation-based approaches, rotational seeding systems encompass a diverse range of setups where scaffolds are rotated or spun in a cell suspension. Rotational speeds typically range from low speeds around 0.2–500 rpm, with seeding durations of 12–72 h, resulting in seeding efficiencies between 38% and 90% [214]. Generally, lower rotation speeds tend to preserve cell morphology better but require longer incubation times and may lead to lower seeding efficiency, especially at lower cell concentrations. Higher speeds improve cell distribution and attachment by increasing circulation, but risk detaching loosely adhered cells. Finding the right balance between speed and incubation time is therefore crucial [215]. Compared to perfusion-based seeding, agitation methods are technically simpler but can impose mechanical stress on cells and often produce less uniform and controlled cell attachment.

Lastly, assisted seeding methods such as vacuum-assisted and magnetic-assisted seeding employ external physical forces to drive cells more rapidly and deeply into scaffolds, especially those with complex or tortuous pore architectures. These methods improve both the speed and uniformity of cell penetration although they may introduce increased mechanical stress on cells or require additional cell modifications. The choice of seeding strategy depends on the properties of the scaffolds, the cell types, and the tissue engineering objectives.

In vacuum-assisted seeding, the pressure difference created by a vacuum is used to efficiently draw cell suspensions into the pores of scaffolds. This method enables rapid cell colonization, typically achieving efficiencies of between 60% and 90%. The vacuum devices used are simple, inexpensive, and often disposable, which minimizes the risk of contamination [214]. In practice, the scaffold is placed in the syringe, and a defined volume of cell suspension is drawn into it by briefly applying a vacuum. The distribution of cells depends on the porosity of the scaffold, as air in the pores can affect how far the cells penetrate. In scaffolds with low porosity, the vacuum helps cells to penetrate the interior, whereas in scaffolds with high porosity, the suspension is absorbed more easily through air expulsion. Cell adhesion predominantly occurs on the exterior of the scaffolds, and cell viability is not significantly impaired by the vacuum process [216].

In magnetic-assisted seeding, cells are labeled with magnetic particles, either bound to the cell surface or internalized, and guided into scaffolds using an external magnetic field. This method enables rapid and uniform cell distribution, with reported seeding efficiencies often exceeding 90% [217]. Magnetic labeling can be achieved through polymer-based particles or liposome-encapsulated iron oxide nanoparticles. While short-term cell viability is generally maintained at low particle-to-cell ratios, higher concentrations may impair proliferation. The technique supports reproducible scaffold colonization and cell organization resembling native tissue [218]. However, potential impacts of residual magnetic particles on cell behavior, scaffold biocompatibility, and the safety and quality of the cultivated meat product require further investigation before broad application.

While advanced seeding methods offer significant improvements in cell distribution and speed, their implementation in cultivated meat bioreactors requires careful consideration of scalability, regulatory compliance, and product safety. Physical forces used in assisted seeding may stress cells or introduce foreign materials, which could affect cell function or consumer acceptance. Therefore, integrating these techniques into bioreactor workflows demands further optimization and comprehensive evaluation of their impact on tissue maturation and product quality.

5.4 Process Monitoring and Harvesting of Scaffolds

Effective process monitoring is critical to ensure optimal cell growth and tissue development within scaffolds during bioreactor cultivation. Traditional monitoring approaches in bioreactors often focus on bulk parameters such as DO, pH, temperature, and CO_2 [219]. However, the spatial heterogeneity introduced by scaffolds necessitates advanced monitoring strategies capable of providing localized information within the scaffold matrix. This is particularly important for assessing nutrient diffusion, waste accumulation, and cell viability in different scaffold regions.

Cell growth is associated with the consumption of carbon sources, amino acids, vitamins, and other essential nutrients, alongside the production of metabolic by-products such as lactate and ammonia. While the intrinsic effect of lactate on cell growth and productivity remains a topic of ongoing research, elevated ammonium (NH_4^+) levels have been extensively documented as toxic and inhibitory to mammalian cell cultures [220]. Furthermore, monitoring nutrient levels, like glucose and amino acids, is vital not only for maintaining optimal bioprocess conditions but also for enabling efficient medium recycling strategies [221]. Therefore, the availability of sensitive sensors capable of detecting both nutrient and metabolite concentrations, combined with appropriate filtration techniques, could facilitate medium reuse and substantially reduce the overall costs of cultivated meat production. In addition to nutrient and metabolite sensors, biomass monitoring plays a crucial role in assessing cell proliferation and viability within scaffold-based cultures. Conventional biomass estimation methods in bioreactors include direct techniques such as manual cell counting, NIR spectroscopy, and dielectric spectrometry, as well as indirect methods based on gas exchange, nutrient uptake, or redox potential [222]. Noninvasive spectroscopic methods, such as optical and NIR sensors, are widely applied for this purpose due to their ability to provide continuous, real-time measurements without disturbing the culture. These approaches can, in some cases, also give insights into cell viability. However, their accuracy can be affected by non-cellular particles, bubbles, or morphological changes in cells, which may alter the signal and lead to misinterpretation [219].

Electrochemical biosensors (ECBs) have emerged as promising tools for real-time, in situ monitoring of key metabolites and nutrients directly within the bioreactor-scaffold system. Their small size and high sensitivity enable integration into scaffold structures or reactor compartments, allowing continuous measurement of critical parameters, such as glucose and lactate. This facilitates early detection of metabolic shifts, localized nutrient depletion, or toxic metabolite accumulation, which are critical for process optimization [223].

When coupled with optical imaging techniques (e.g., fluorescence-based viability assays) and automated control systems, these integrated sensors support dynamic adjustment of feeding strategies, oxygenation, and perfusion rates to maintain an optimal microenvironment for cell cultivation. Implementing feedback loops between sensor data and bioreactor control systems can further enhance culture stability and product consistency. For example, Raman spectroscopy has been used for real-time glucose monitoring, enabling automated medium feeding via feedback control loops without manual intervention [224]. Integrated optical sensors for pH and DO in SUBs facilitate real-time adjustments and support automated control strategies [222]. Smart sensor systems incorporating adaptive algorithms have demonstrated early event detection and process stability through automated responses in cell cultivation systems [225]. To complement biochemical and physical sensing, live-cell imaging techniques offer an additional layer of information, enabling direct visualization of cell distribution, morphology, and viability within scaffolds under near-physiological conditions. For example, a multi-channel bioreactor has been shown to support online confocal imaging of fluorescently labeled cells embedded in 3D constructs under flow conditions [226]. Similarly, a low-cost 3D-printed microfluidic bioreactor doubles as an imaging chamber for continuous monitoring of organoid growth [227], while simple microfluidic devices have enabled precise, real-time live-cell imaging with multiparametric environmental control [228]. Coupling these imaging platforms with electrochemical and spectroscopic sensors provides a multidimensional understanding of scaffold-cell interactions, thereby supporting more precise process control.

6 Summary and Conclusion

To enable efficient cultivated meat production, high volumes and low production costs with little environmental impact are anticipated. As part of the biotechnological paradigm shift, this requires suitable bioreactor concepts and thus closes interdisciplinary collaboration. Although STRs are currently the most common bioreactors for cultivating animal cells, with volumes of up to 25 m^3, alternative bioreactor designs could provide benefits when increasing size while simultaneously reducing complexity and costs. However, it has to be considered that the current experience of large bioreactors for animal cell cultivation is mainly based on biopharmaceutical production, as large-scale cultivated meat production plants are scarce to date. Nevertheless, the requirements to ensure efficient and robust production are very similar although the revenue of cultivated meat is much lower than that from biopharmaceutical products.

A relevant step towards new bioreactor concepts is to refrain from rapid scale-up and instead focus on downscaling to investigate potential designs within a laboratory setup. The prototyping of bioreactors is therefore vital for effective planning in bioreactor development. It allows early detection and resolution of biological and engineering challenges, supports realistic simulation of process parameters, and improves the transferability of findings to larger scales. Without thorough prototyping, scale-up efforts risk failure due to unforeseen issues. Thereby, prototyping can bridge the gap between design concepts and industrial applications.

Besides the bioreactor itself, there is a need for smart processing with more advanced and reliable process equipment, enabling long-duration continuous production with high productivity. The latter is getting more attractive with the development of scalable perfusion technologies and a drastic reduction in the cost of growth media through media and cell development. Another factor to consider is integrating and harvesting a suitable scaffold in combination with the bioreactor. Advanced process monitoring, incorporating electrochemical biosensors, spectroscopic methods, and live-cell imaging, represents a key enabling technology for successful scaffold incorporation in bioreactor systems, enhancing the understanding of scaffold-cell interactions and supporting process control to improve product quality, reproducibility, and scalability in cultivated meat production. These aspects mark the highly interdisciplinary nature of cultivated meat research, which is essential for developing efficient bioreactor concepts that foster sustainable cultivated meat production towards global food security.

Acknowledgments The work at TUM/Cellular Agriculture was supported by the TUM WACKER Institute for Industrial Biotechnology. The authors would like to thank Dr. Laura Lindenthal and Dr. Carsten Bornhövd (Wacker Chemie AG) for their support and constructive discussions as part of the "ArtMeat" project of the TUM WACKER Institute for Industrial Biotechnology.

Author Contribution Statement

Marie-Luise Schlieker: Conceptualization, Writing – Original Draft, Writing – Review and Editing, Visualization.
David Vorländer: Conceptualization, Writing – Original Draft, Writing – Review and Editing, Visualization.
Laura Pasitka: Writing – Original Draft, Writing – Review and Editing, Visualization.
Sebastian Bode: Writing – Original Draft, Writing – Review and Editing.
Tatjana Krampitz: Writing – Original Draft, Writing – Review and Editing.
Chantal Treinen: Conceptualization, Writing – Review and Editing, Visualization.
Marius Henkel: Conceptualization, Writing – Original Draft, Writing – Review and Editing, Visualization.

Declaration on the Use of Generative AI and AI-Assisted Tools During the writing and editing process, Grammarly AI writing assistant (Grammarly, Inc., San Francisco, USA), ChatGPT (OpenAI, San Francisco, USA) and DeepL Translator (DeepL SE, Köln, Germany) were used to improve language and correct grammar, stylistic, and spelling errors.

Declaration of Interest David Vorländer, Laura Pasitka, Sebastian Bode, and Tatjana Krampitz are employed by GEA Liquid Technologies Germany GmbH, which provides process know-how and equipment, including bioreactors, to various industries. The information provided in this chapter represents their own personal opinions and work experience but is not prescribed by GEA.

References

1. Kantarci N, Borak F, Ulgen KO (2005) Bubble column reactors. Process Biochem 40:2263–2283. https://doi.org/10.1016/j.procbio.2004.10.004
2. Nienow AW (2014) Stirring and stirred-tank reactors. Chem Ing Tech 86:2063–2074. https://doi.org/10.1002/cite.201400087
3. Nienow AW (2006) Reactor engineering in large scale animal cell culture. Cytotechnology 50: 9–33. https://doi.org/10.1007/s10616-006-9005-8
4. Guralp SA, Gubbuk IH, Kucukkolbasi S, Gulari E (2015) Universal cell capture by immobilized antimicrobial peptide plantaricin. Biochem Eng J 101:18–22. https://doi.org/10.1016/j.bej.2015.04.018
5. Martin I, Wendt D, Heberer M (2004) The role of bioreactors in tissue engineering. Trends Biotechnol 22:80–86. https://doi.org/10.1016/j.tibtech.2003.12.001
6. Kawecki NS, Chen KK, Smith CS, Xie Q, Cohen JM, Rowat AC (2024) Scalable processes for culturing meat using edible scaffolds. Annu Rev Food Sci Technol 15:241–264. https://doi.org/10.1146/annurev-food-072023-034451
7. Allan SJ, De Bank PA, Ellis MJ (2019) Bioprocess design considerations for cultured meat production with a focus on the expansion bioreactor. Front Sustain Food Syst 3:44. https://doi.org/10.3389/fsufs.2019.00044
8. Stefaniak JJ, Gailey FB, Brown CS, Johnson MJ (1946) Pilot plant equipment for submerged production of penicillin. Ind Eng Chem 38:666–671. https://doi.org/10.1021/ie50439a010
9. Noll P, Henkel M (2020) History and evolution of modeling in biotechnology: modeling & simulation, application and hardware performance. Comput Struct Biotechnol J 18:3309–3323. https://doi.org/10.1016/j.csbj.2020.10.018
10. Schlicker M-L, Köhne L, Brenner KJ, Linnemann CT, Henkel M (2025) Technology and processes for cultivated meat. Ernahrungs Umschau 72:114–121. https://doi.org/10.4455/eu.2025.025
11. Eagle H (1955) Nutrition needs of mammalian cells in tissue culture. Science 122:501–504. https://doi.org/10.1126/science.122.3168.501
12. Harrison RG, Greenman MJ, Mall FP, Jackson CM (1907) Observations of the living developing nerve fiber. Anat Rec 1:116–128. https://doi.org/10.1002/ar.1090010503
13. Enders JF, Weller TH, Robbins FC (1949) Cultivation of the Lansing strain of poliomyelitis virus in cultures of various human embryonic tissues. Science 109:85–87. https://doi.org/10.1126/science.109.2822.85
14. Van Wezel AL (1967) Growth of cell-strains and primary cells on micro-carriers in homogeneous culture. Nature 216:64–65. https://doi.org/10.1038/216064a0
15. Knazek RA, Gullino PM, Kohler PO, Dedrick RL (1972) Cell culture on artificial capillaries: an approach to tissue growth in vitro. Science 178:65–67. https://doi.org/10.1126/science.178.4056.65
16. Bielser J-M, Wolf M, Souquet J, Broly H, Morbidelli M (2018) Perfusion mammalian cell culture for recombinant protein manufacturing – a critical review. Biotechnol Adv 36:1328–1340. https://doi.org/10.1016/j.biotechadv.2018.04.011
17. Eibl R, Kaiser S, Lombriser R, Eibl D (2010) Disposable bioreactors: the current state-of-the-art and recommended applications in biotechnology. Appl Microbiol Biotechnol 86:41–49. https://doi.org/10.1007/s00253-009-2422-9
18. Post MJ, Levenberg S, Kaplan DL, Genovese N, Fu J, Bryant CJ, Negowetti N, Verzijden K, Moutsatsou P (2020) Scientific, sustainability and regulatory challenges of cultured meat. Nat Food 1:403–415. https://doi.org/10.1038/s43016-020-0112-z
19. Huh D, Matthews BD, Mammoto A, Montoya-Zavala M, Hsin HY, Ingber DE (2010) Reconstituting organ-level lung functions on a chip. Science 328:1662–1668. https://doi.org/10.1126/science.1188302

20. Todhunter ME, Jubair S, Verma R, Saqe R, Shen K, Duffy B (2024) Artificial intelligence and machine learning applications for cultured meat. Front Artif Intell 7:1424012. https://doi.org/10.3389/frai.2024.1424012
21. Nie M, Shima A, Yamamoto M, Takeuchi S (2025) Scalable tissue biofabrication via perfusable hollow fiber arrays for cultured meat applications. Trends Biotechnol 43:1938–1960. https://doi.org/10.1016/j.tibtech.2025.02.022
22. Weissenbach J-L, Sylvia R, von der Brelie A, Brandl M, Fesenfeld M (2022) Consumable tissue-like structure generated with muscle cells grown on edible hollow fibers. Merck Patent GmbH, WO 2022/038240 A2
23. Tzimorotas D, Solberg NT, Andreassen RC, Moutsatsou P, Bodiou V, Pedersen ME, Rønning SB (2023) Expansion of bovine skeletal muscle stem cells from spinner flasks to benchtop stirred-tank bioreactors for up to 38 days. Front Nutr 10. https://doi.org/10.3389/fnut.2023.1192365
24. Hanga MP, De La Raga FA, Moutsatsou P, Hewitt CJ, Nienow AW, Wall I (2021) Scale-up of an intensified bioprocess for the expansion of bovine adipose-derived stem cells (bASCs) in stirred tank bioreactors. Biotechnol Bioeng 118:3175–3186. https://doi.org/10.1002/bit.27842
25. Norris SCP, Kawecki NS, Davis AR, Chen KK, Rowat AC (2022) Emulsion-templated microparticles with tunable stiffness and topology: applications as edible microcarriers for cultured meat. Biomaterials 287:121669. https://doi.org/10.1016/j.biomaterials.2022.121669
26. Sardroud HA, Shokhrollahi Barough M, Askari E, Akbari M (2025) Microcarrier-seeded muscle cells exhibit delayed differentiation in simulated microgravity compared to a terrestrial bioreactor. npj Sci Food 9:153. https://doi.org/10.1038/s41538-025-00498-5
27. Kim M, Harris DM, Cimpeanu R (2025) A simulation modeling framework for fluid motion and transport in a rocking bioreactor with application to cultivated meat production. https://doi.org/10.48550/arXiv.2504.05421
28. Li X, Zhang G, Zhao X, Zhou J, Du G, Chen J (2020) A conceptual air-lift reactor design for large scale animal cell cultivation in the context of in vitro meat production. Chem Eng Sci 211:115269. https://doi.org/10.1016/j.ces.2019.115269
29. Zhou X, Zheng H, Wu Y, Yin H, Mao X, Li N, Guo H, Chang Y, Jiang X, Ai Q, Xue C (2025) Scalable production of muscle and adipose cell-laden microtissues using edible macroporous microcarriers for 3D printing of cultured fish fillets. Nat Commun 16:1740. https://doi.org/10.1038/s41467-025-57015-1
30. Morel A, Wirth SC, Razza N, Hübscher J, Eberle S (2024) Method for operating a bioreactor for cultivated meat and corresponding bioreactor. Mirai Foods AG, EP4328296A1
31. Vila M, Lull T (2021) Process and device for production of cultured meat. BioTech Foods S. L., WO/2021/250292
32. Stanbury PF, Whitaker A, Hall SJ (2017) Microbial growth kinetics. In: Principles of fermentation technology. Butterworth-Heinemann. https://doi.org/10.1016/B978-0-08-099953-1.00002-8
33. Zeta GmbH. Zeta magnetic agitators. https://zetabiosystem.com/wp-content/uploads/2025/11/ZETA_Folder_Magnetic-Agitators_A4_052024_EN_Print.pdf
34. PRG GmbH. PRG Magnetrührwerk MRLL. https://www.prg-gmbh.de/fileadmin/user_upload/Downloads_Gesamt/Flyer/PRG_Produktflyer_Magnetruehrwerk-MRLL-von-unten_DIN-A4_DE_V05_Web_200dpi.pdf
35. Wittmann C, Liao JC (2017) Industrial biotechnology: products and processes. 1st edn. Wiley. https://doi.org/10.1002/9783527807796.fmatter1
36. Al-Rubeai M (2015) Animal cell culture. Cell engineering, vol 9. Springer, Cham. https://doi.org/10.1007/978-3-319-10320-4
37. Chmiel H (2011) Bioprozesstechnik. Spektrum Akademischer Verlag, Heidelberg. https://doi.org/10.1007/978-3-8274-2477-8
38. Sharma R, Harrison STL, Tai SL (2021) Advances in bioreactor systems for the production of biologicals in mammalian cells. ChemBioEng Rev 9:42–62. https://doi.org/10.1002/cben.202100022

39. Grote K-H, Feldhusen J (2014) Dubbel: Taschenbuch für den Maschienenbau. Springer Vieweg, Berlin, Heidelberg. https://doi.org/10.1007/978-3-642-38891-0
40. Yonsel S, Deckwer WD (1991) Design and operation of a highly instrumented airlift/bubble column reactor in pilot scale. Construction and start-up (Hochinstrumentierter Airlift-/Blasensaeulen-Reaktor im Pilotmassstab. Konstruktion und Inbetriebnahme). Bioeng Forschung Praxis 7:193
41. Varley J, Birch J (1998) Reactor design for large scale suspension animal cell culture. Cytotechnology 3:177–205. https://doi.org/10.1023/A:1008008021481
42. Pörtner R, Platas OB, Fassnacht D, Nehring D, Czermak P, Märkl H (2007) Fixed bed reactors for the cultivation of mammalian cells: design, performance and scale-up. Open Biotechnol J 1:41–46. https://doi.org/10.2174/1874070700701010041
43. Catapano G, Czermak P, Eibl R, Eibl D, Pörtner R (2009) Bioreactor design and scale-up. In: Cell and tissue reaction engineering. Springer, Berlin, Heidelberg, pp 173–259. https://doi.org/10.1007/978-3-540-68182-3_5
44. Davis JM, Hanak JAJ (2007) Hollow-fiber cell culture. In: Springer protocols, vol 75.
45. Da Gama Ferreira R, Petrides D (2023) Cultured (cultivated) meat production – process modeling and techno-economic assessment (TEA) using SuperPro designer. https://doi.org/10.13140/RG.2.2.33578.90560
46. Humbird D (2021) Scale-up economics for cultured meat. Biotechnol Bioeng 118:3239–3250. https://doi.org/10.1002/bit.27848
47. Negulescu PG, Risner D, Spang ES, Sumner D, Block D, Nandi S, McDonald KA (2023) Techno-economic modeling and assessment of cultivated meat: impact of production bioreactor scale. Biotechnol Bioeng 120:1055–1067. https://doi.org/10.1002/bit.28324
48. Pasitka L, Wissotsky G, Ayyash M, Yarza N, Rosoff G, Kaminker R, Nahmias Y (2024) Empirical economic analysis shows cost-effective continuous manufacturing of cultivated chicken using animal-free medium. Nat Food 5:693–702. https://doi.org/10.1038/s43016-024-01022-w
49. Pörtner R (2015) Bioreactors for mammalian cells. In: Al-Rubeai M (ed) Animal cell culture. Springer, Cham. https://doi.org/10.1007/978-3-319-10320-4_4
50. Sahm H, Antranikian G, Stahmann K-P, Takors R (2013) Bioverfahrenstechnik. In: Industrielle Mikrobiologie. Springer, Berlin, Heidelberg. https://doi.org/10.1007/978-3-8274-3040-3_2
51. Sonnleitner B, Chmiel H (2011) Wachstum: Kinetik und Prozessführung. In: Chmiel H (ed) Bioprozesstechnik. Spektrum Akademischer Verlag, Heidelberg. https://doi.org/10.1007/978-3-8274-2477-8_4
52. Krull R, Hempel DC, Wucherpfennig T (2014) Bioverfahrenstechnik. In: Grote KH, Bender B, Göhlich D (eds) Dubbel. Springer Vieweg. https://doi.org/10.1007/978-3-662-54805-9_92
53. Chotteau V (2015) Perfusion processes. In: Al-Rubeai M (ed) Animal cell culture. Cell engineering, vol 9. Springer, Cham. https://doi.org/10.1007/978-3-319-10320-4_13
54. Martens DE, van den End EJ, Streefland M (2014) Configuration of bioreactors. In: Pörtner R (ed) Animal cell biotechnology. Methods in molecular biology, vol 1104. Humana Press, Totowa. https://doi.org/10.1007/978-1-62703-733-4_19
55. O'Neill EN, Ansel JC, Kwong GA, Plastino ME, Nelson J, Baar K, Block DE (2022) Spent media analysis suggests cultivated meat media will require species and cell type optimization. npj Sci Food 6:46. https://doi.org/10.1038/s41538-022-00157-z
56. Zhang S, Handa-Corrigan A, Spier RE (1993) A comparison of oxygenation methods fro high-density perfusion culture of animal cells. Biotechnol Bioeng 41:685–692. https://doi.org/10.1002/bit.260410702
57. Maiorella B, Dorin G, Carion A, Harano D (1991) Crossflow microfiltration of animal cells. Biotechnol Bioeng 37:121–126. https://doi.org/10.1002/bit.260370205
58. Clincke M, Mölleryd C, Zhang Y, Lindskog E, Walsh K, Chotteau V (2013) Very high density of CHO cells in perfusion by ATF or TFF in WAVE bioreactor™. Part I. Effect of the cell density on the process. Biotechnol Prog 29:754–767. https://doi.org/10.1002/btpr.1704

59. Clincke M, Mölleryd C, Samani PK, Lindskog E, Fäldt E, Walsh K, Chotteau V (2013) Very high density of Chinese hamster ovary cells in perfusion by alternating tangential flow or tangential flow filtration in WAVE bioreactor™ – part II: applications for antibody production and cryopreservation. Biotechnol Prog 29:768–777. https://doi.org/10.1002/btpr.1703
60. Pasitka L, Cohen M, Ehrlich A, Gildor B, Reuveni E, Ayyash M, Wissotsky G, Herscovici A, Kaminker R, Niv A, Bitcover R, Dadia O, Rudik A, Voloschin A, Shimoni M, Cinnamon Y, Nahmias Y (2022) Spontaneous immortalization of chicken fibroblasts generates stable, high-yield cell lines for serum-free production of cultured meat. Nat Food 4:35–50. https://doi.org/10.1038/s43016-022-00658-w
61. Pollock J, Ho SV, Farid SS (2013) Fed-batch and perfusion culture processes: economic, environmental, and operational feasibility under uncertainty. Biotechnol Bioeng 110:206–219. https://doi.org/10.1002/bit.24608
62. Lang Z, Yan S, Xiong Q, Chen G (2023) WAVE-based intensified perfusion cell culture for fast process development. Biotechnol Lett 45:1117–1131. https://doi.org/10.1007/s10529-023-03405-8
63. Frick T, Noack U, Greller G, Adams T, Fenge C (2011) Increasing efficiency in protein and cell production by combining single-use bioreactor technology and perfusion. BioPharm Int 2011(Suppl 4) https://www.biopharminternational.com/view/increasing-efficiency-protein-and-cellproduction- combining-single-use-bioreactor-technology-and-per
64. Eibl R, Löffelholz C, Eibl D (2013) Disposable bioreactors for inoculum production and protein expression. Methods Mol Biol 1104:265–284. https://doi.org/10.1007/978-1-62703-733-4_18
65. Deo YM, Mahadevan MD, Fuchs R (1996) Practical considerations in operation and scale-up of spin-filter based bioreactors for monoclonal antibody production. Biotechnol Prog 12:57–64. https://doi.org/10.1021/bp950079p
66. Iding K, Lütkemeyer D, Fraune E, Gerlach K, Lehmann J (2000) Influence of alterations in culture condition and changes in perfusion parameters on the retention performance of a 20 μm spinfilter during a perfusion cultivation of a recombinant CHO cell line in pilot scale. Cytotechnology:141–150. https://doi.org/10.1023/A:1008183508439
67. Avgerinos GC, Drapeau D, Socolow JS, Mao J, Hsiao K, Broeze RJ (1990) Spin filter perfusion system for high density cell culture: production of recombinant urinary type plasminogen activator in CHO cells. Nat Biotechnol 8:54–58. https://doi.org/10.1038/nbt0190-54
68. Bierau H, Perani A, Al-Rubeai M, Emery AN (1998) A comparison of intensive cell culture bioreactors operating with Hybridomas modified for inhibited apoptotic response. J Biotechnol 62:195–207. https://doi.org/10.1016/S0168-1656(98)00064-9
69. Vallez-Chetreanu F, Fraisse Ferreira LG, Rabe R, Von Stockar U, Marison IW (2007) An on-line method for the reduction of fouling of spin-filters for animal cell perfusion cultures. J Biotechnol 130:265–273. https://doi.org/10.1016/j.jbiotec.2007.04.007
70. Castilho LR, Medronho R (2002) Cell retention devices for suspended-cell perfusion cultures. Adv Biochem Eng Biotechnol 74:129–169. https://doi.org/10.1007/3-540-45736-4_7
71. Clapp KP, Casten A, Lindskog EK (2018) Chapter 24: Upstream processing equipment. In: Biopharmaceutical processing. https://doi.org/10.1016/B978-0-08-100623-8.00024-4
72. Chen X-Y, Chen J-Y, Tong X-M, Mei J-G, Chen Y-F, Mou X-Z (2020) Recent advances in the use of microcarriers for cell cultures and their ex vivo and in vivo applications. Biotechnol Lett 42:1–10. https://doi.org/10.1007/s10529-019-02738-7
73. Apelman S (1991) Separation of animal cells in continuous cell culture systems. In: Murakami H, Shirahata S, Tachibana H (eds) Animal cell technology: basic & applied aspects, vol 4. Springer, Dordrecht. https://doi.org/10.1007/978-94-011-2844-5_21
74. Björling T, Dudel U, Fenge C (1995) Evaluation of a cell separator in large scale perfusion culture. In: Animal cell technology: developments towards the 21st century. Springer, Dordrecht, pp 671–675. https://doi.org/10.1007/978-94-011-0437-1_105

75. Jäger V (1992) A novel perfusion system for the large-scale cultivation of animal cells based on a continuous flow centrifuge. In: Animal cell technology, pp 397–402. https://doi.org/10.1016/B978-0-7506-0421-5.50093-3
76. Jäger V (1992) High density perfusion culture of animal cells using a novel continuous flow centrifuge. In: Murakami H, Shirahata S, Tachibana H (eds) Animal cell technology: basic & applied aspects. Springer, Dordrecht, pp 209–216
77. Bettinardi IW, Castan A, Medronho RA, Castilho LR (2020) Hydrocyclones as cell retention device for CHO perfusion processes in single-use bioreactors. Biotechnol Bioeng 117:1915–1928. https://doi.org/10.1002/bit.27335
78. Kundu AM, Hiller GW (2021) Hydrocyclones as cell retention devices for an N-1 perfusion bioreactor linked to a continuous-flow stirred tank production bioreactor. Biotechnol Bioeng 118:1973–1986. https://doi.org/10.1002/bit.27711
79. Boycott AE (1920) Sedimentation of blood corpuscles. Nature 104:532. https://doi.org/10.1038/104532b0
80. Kinosita K (1949) Sedimentation in tilted vessels (1). J Colloid Sci 4:525–536. https://doi.org/10.1016/0095-8522(49)90049-5
81. Hecht V, Duvar S, Ziehr H, Burg J, Jockwer A (2014) Efficiency improvement of an antibody production process by increasing the inoculum density. Biotechnol Prog 30:607–615. https://doi.org/10.1002/btpr.1887
82. Searles JA, Todd P, Kompala DS (1994) Viable cell recycle with an inclined settler in the perfusion culture of suspended recombinant Chinese hamster ovary cells. Biotechnol Prog 10:198–206. https://doi.org/10.1021/bp00026a600
83. Henzler H (2012) Continuous fermentation with animal cells. Part 2. Techniques and methods of cell retention. Chem Ing Tech 84:1482–1496. https://doi.org/10.1002/cite.201200003
84. Pohlscheidt M, Jacobs M, Wolf S, Thiele J, Jockwer A, Gabelsberger J, Jenzsch M, Tebbe H, Burg J (2013) Optimizing capacity utilization by large scale 3000 L perfusion in seed train bioreactors. Biotechnol Prog 29:222–229. https://doi.org/10.1002/btpr.1672
85. Coronel J, Gränicher G, Sandig V, Noll T, Genzel Y, Reichl U (2020) Application of an inclined settler for cell culture-based influenza A virus production in perfusion mode. Front Bioeng Biotechnol 8:672. https://doi.org/10.3389/fbioe.2020.00672
86. Ryll T, Dutina G, Reyes A, Gunson J, Krummen L, Etcheverry T (2000) Performance of small scale CHO perfusion cultures using an acoustic cell filtration device for cell retention: characterization of separation efficiency and impact of perfusion on product quality. Biotechnol Bioeng 69:440–449. https://doi.org/10.1002/1097-0290(20000820)69:4<440::AID-BIT10>3.0.CO;2-0
87. Gränicher G, Coronel J, Trampler F, Jordan I, Genzel Y, Reichl U (2020) Performance of an acoustic settler versus a hollow fiber–based ATF technology for influenza virus production in perfusion. Appl Microbiol Biotechnol 104:4877–4888. https://doi.org/10.1007/s00253-020-10596-x
88. Merten O-W (2015) Advances in cell culture: anchorage dependence. Philos Trans R Soc B 370:20140040. https://doi.org/10.1098/rstb.2014.0040
89. Pui PWS, Trampler F, Sonderhoff SA, Groeschl M, Kilburn DG, Piret JM (1995) Batch and Semicontinuous aggregation and sedimentation of hybridoma cells by acoustic resonance fields. Biotechnol Prog 11:146–152. https://doi.org/10.1021/bp00032a005
90. Dalm MCF, Cuijten SMR, Van Grunsven WMJ, Tramper J, Martens DE (2004) Effect of feed and bleed rate on hybridoma cells in an acoustic perfusion bioreactor: part I. Cell density, viability, and cell-cycle distribution. Biotechnol Bioeng 88:547–557. https://doi.org/10.1002/bit.20287
91. Gorenflo VM, Pfeifer TA, Lesnicki G, Kwan EM, Grigliatti TA, Kilburn DG, Piret JM (2004) Production of a self-activating CBM-factor X fusion protein in a stable transformed Sf9 insect cell line using high cell density perfusion culture. Cytotechnology 44:93–102. https://doi.org/10.1007/s10616-005-0703-4

92. Gorenflo VM, Angepat S, Bowen BD, Piret JM (2003) Optimization of an acoustic cell filter with a novel air-backflush system. Biotechnol Prog 19:30–36. https://doi.org/10.1021/bp025625a
93. Gorenflo VM, Smith L, Dedinsky B, Persson B, Piret JM (2002) Scale-up and optimization of an acoustic filter for 200 L/day perfusion of a CHO cell culture. Biotechnol Bioeng 80:438–444. https://doi.org/10.1002/bit.10386
94. Kadlec P, Gabrys B, Strandt S (2009) Data-driven soft sensors in the process industry. Comput Chem Eng 33:795–814. https://doi.org/10.1016/j.compchemeng.2008.12.012
95. Luttmann R, Bracewell DG, Cornelissen G, Gernaey KV, Glassey J, Hass VC, Kaiser C, Preusse C, Striedner G, Mandenius C (2012) Soft sensors in bioprocessing: a status report and recommendations. Biotechnol J 7:1040–1048. https://doi.org/10.1002/biot.201100506
96. Nikita S, Mishra S, Gupta K, Runkana V, Gomes J, Rathore AS (2023) Advances in bioreactor control for production of biotherapeutic products. Biotechnol Bioeng 120:1189–1214. https://doi.org/10.1002/bit.28346
97. Read EK, Park JT, Shah RB, Riley BS, Brorson KA, Rathore AS (2010) Process analytical technology (PAT) for biopharmaceutical products: part I. Concepts and applications. Biotechnol Bioeng 105:276–284. https://doi.org/10.1002/bit.22528
98. Stanbury PF, Whitaker A, Hall SJ (2017) Instrumentation and control. In: Principles of fermentation technology. Butterworth-Heinemann. https://doi.org/10.1016/B978-0-08-099953-1.00002-9
99. Biechele P, Busse C, Solle D, Scheper T, Reardon K (2015) Sensor systems for bioprocess monitoring. Eng Life Sci 15:469–488. https://doi.org/10.1002/elsc.201500014
100. Biskupek-Korell B (2013) Nahinfrarotspektroskopie als schnelle, zuverlässige und kostengünstige Methode. Wiley Analytical Science
101. Bruker (2025) Anwendungen der Raman-Spektroskopie. Accessed 17 Sept 2025
102. I&L Biosystems (2025) FUTURA Inline Biomasse Sensoren. https://il-biosystems.com/de/aber-instruments/device/futura-inline-biomass-sensors/. Accessed 17 Sept 2025
103. Rechner Sensors (2025) Kapazitiver Sensor: Aufbau und Funktionsweise. Accessed 17 Sept 2025
104. Fraunhofer IGB (2025) Biosensorik. https://www.igb.fraunhofer.de/de/forschung/zell-und-gewebetechnologien/biosensorik.html. Accessed 17 Sept 2025
105. Herwig C, Pörtner R, Möller J (2025) Digitale Zwillinge: Werkzeuge und Konzepte für intelligente Bioproduktion. Springer, Cham. https://doi.org/10.1007/978-3-031-75698-6
106. Goldrick S, Sandner V, Cheeks M, Turner R, Farid SS, McCreath G, Glassey J (2020) Multivariate data analysis methodology to solve data challenges related to scale-up model validation and missing data on a micro-bioreactor system. Biotechnol J 15:1800684. https://doi.org/10.1002/biot.201800684
107. Hair JF, Black WC, Babin BJ, Anderson RE (2019) Multivariate data analysis. Cengage
108. SAP (2025) Machine learning: Die Zukunft der Intelligenz | Definition, Arten und Beispiele. https://www.sap.com/germany/products/artificial-intelligence/what-is-machine-learning.html. Accessed 17 Sept 2025
109. Schwenzer M, Ay M, Bergs T, Abel D (2021) Review on model predictive control: an engineering perspective. Int J Adv Manuf Technol 117:1327–1349. https://doi.org/10.1007/s00170-021-07682-3
110. Eslami T, Jungbauer A (2024) Control strategy for biopharmaceutical production by model predictive control. Biotechnol Prog 40:e3426. https://doi.org/10.1002/btpr.3426
111. Panjwani S, Almazan A, Hille R, Spetsieris K (2024) Predictive modeling for cell culture in commercial manufacturing of biotherapeutics. Biotechnol Bioeng 121:3440–3453. https://doi.org/10.1002/bit.28813
112. Schneider SL, Seidel S, Seefeldt A, Romang M, Kastl S, Gensel J, Neumeyer T, John GT, Eibl D (2023) 3D printed, single-use bioreactor with integrated inline sensors for microbial and mammalian cell cultivation – a case study. Processes 11:3231. https://doi.org/10.3390/pr11113231

113. Amer M, Ramsey JD (2018) Multi-chamber single-use bioreactor – a proof of concept prototype. Biochem Eng J 130:113–120. https://doi.org/10.1016/j.bej.2017.12.002
114. Bareither R, Pollard D (2011) A review of advanced small-scale parallel bioreactor technology for accelerated process development: current state and future need. Biotechnol Prog 27:2–14. https://doi.org/10.1002/btpr.522
115. Camburn B, Viswanathan V, Linsey J, Anderson D, Jensen D, Crawford R, Otto K, Wood K (2017) Design prototyping methods: state of the art in strategies, techniques, and guidelines. Des Sci 3:e13. https://doi.org/10.1017/dsj.2017.10
116. Lim D, Renteria ES, Sime DS, Ju YM, Kim JH, Criswell T, Shupe TD, Atala A, Marini FC, Gurcan MN, Soker S, Hunsberger J, Yoo JJ (2022) Bioreactor design and validation for manufacturing strategies in tissue engineering. Bio Des Manuf 5:43–63. https://doi.org/10.1007/s42242-021-00154-3
117. Delvigne F, Takors R, Mudde R, Van Gulik W, Noorman H (2017) Bioprocess scale-up/down as integrative enabling technology: from fluid mechanics to systems biology and beyond. Microb Biotechnol 10:1267–1274. https://doi.org/10.1111/1751-7915.12803
118. Betts JI, Baganz F (2006) Miniature bioreactors: current practices and future opportunities. Microb Cell Factories 5:21. https://doi.org/10.1186/1475-2859-5-21
119. Tajsoleiman T, Mears L, Krühne U, Gernaey KV, Cornelissen S (2019) An industrial perspective on scale-down challenges using miniaturized bioreactors. Trends Biotechnol 37:697–706. https://doi.org/10.1016/j.tibtech.2019.01.002
120. Krujatz F, Lode A, Seidel J, Bley T, Gelinsky M, Steingroewer J (2017) Additive biotech – chances, challenges, and recent applications of additive manufacturing technologies in biotechnology. New Biotechnol 39:222–231. https://doi.org/10.1016/j.nbt.2017.09.001
121. Salehi-Nik N, Amoabediny G, Pouran B, Tabesh H, Shokrgozar MA, Haghighipour N, Khatibi N, Anisi F, Mottaghy K, Zandieh-Doulabi B (2013) Engineering parameters in bioreactor's design: a critical aspect in tissue engineering. Biomed Res Int 2013:1–15. https://doi.org/10.1155/2013/762132
122. Ngo TD, Kashani A, Imbalzano G, Nguyen KTQ, Hui D (2018) Additive manufacturing (3D printing): a review of materials, methods, applications and challenges. Compos Part B Eng 143:172–196. https://doi.org/10.1016/j.compositesb.2018.02.012
123. Böhme A, Schütze F, Sauer S, Foitzik AH (2018) Fabrication and validation by micromilling for bioreactor prototyping. Mater Sci Forum 941:2448–2453. https://doi.org/10.4028/www.scientific.net/MSF.941.2448
124. Pereira T, Kennedy JV, Potgieter J (2019) A comparison of traditional manufacturing vs additive manufacturing, the best method for the job. Procedia Manuf 30:11–18. https://doi.org/10.1016/j.promfg.2019.02.003
125. Sun W, Starly B, Nam J, Darling A (2005) Bio-CAD modeling and its applications in computer-aided tissue engineering. Comput Aided Des 37:1097–1114. https://doi.org/10.1016/j.cad.2005.02.002
126. Wong KV, Hernandez A (2012) A review of additive manufacturing. ISRN Mech Eng 2012:1–10. https://doi.org/10.5402/2012/208760
127. Kulkarni P, Dutta D (1996) An accurate slicing procedure for layered manufacturing. Comput Aided Des 28:683–697. https://doi.org/10.1016/0010-4485(95)00083-6
128. Lu K, Reynolds WT (2008) 3DP process for fine mesh structure printing. Powder Technol 187:11–18. https://doi.org/10.1016/j.powtec.2007.12.017
129. Kristiawan RB, Imaduddin F, Ariawan D, Ubaidillah AZ (2021) A review on the fused deposition modeling (FDM) 3D printing: filament processing, materials, and printing parameters. Open Eng 11:639–649. https://doi.org/10.1515/eng-2021-0063
130. Shirazi SFS, Gharehkhani S, Mehrali M, Yarmand H, Metselaar HSC, Adib Kadri N, Osman NAA (2015) A review on powder-based additive manufacturing for tissue engineering: selective laser sintering and inkjet 3D printing. Sci Technol Adv Mater 16:033502. https://doi.org/10.1088/1468-6996/16/3/033502

131. Murr LE, Gaytan SM, Ramirez DA, Martinez E, Hernandez J, Amato KN, Shindo PW, Medina FR, Wicker RB (2012) Metal fabrication by additive manufacturing using laser and electron beam melting technologies. J Mater Sci Technol 28:1–14. https://doi.org/10.1016/S1005-0302(12)60016-4
132. Melchels FPW, Feijen J, Grijpma DW (2010) A review on stereolithography and its applications in biomedical engineering. Biomaterials 31:6121–6130. https://doi.org/10.1016/j.biomaterials.2010.04.050
133. Auyeung RCY, Kim H, Mathews S, Charipar N, Piqué A (2017) Laser additive manufacturing of embedded electronics. In: Laser additive manufacturing. Woodhead Publishing, pp 319–350. https://doi.org/10.1016/B978-0-08-100433-3.00012-9
134. Wang J, Xie H, Weng Z, Senthil T, Wu L (2016) A novel approach to improve mechanical properties of parts fabricated by fused deposition modeling. Mater Des 105:152–159. https://doi.org/10.1016/j.matdes.2016.05.078
135. Huda Z (2020) Machining processes and machines: fundamentals, analysis, and calculations. CRC Press. https://doi.org/10.1201/9781003081203
136. Chandu HS (2024) A review on CNC machine tool materials and their impact on machining performance. Inpressco. https://doi.org/10.14741/ijcet/v.14.5.4
137. Beelen MJ, Neerincx PE, Van De Molengraft MJG (2011) Control of an air pressure actuated disposable bioreactor for cultivating heart valves. Mechatronics 21:1288–1297. https://doi.org/10.1016/j.mechatronics.2011.09.003
138. Lee UN, Su X, Guckenberger DJ, Dostie AM, Zhang T, Berthier E, Theberge AB (2018) Fundamentals of rapid injection molding for microfluidic cell-based assays. Lab Chip 18:496–504. https://doi.org/10.1039/C7LC01052D
139. Hong F, Tendera L, Myant C, Boyle D (2022) Vacuum-formed 3D printed electronics: fabrication of thin, rigid and free-form interactive surfaces. SN Comput Sci 3:275. https://doi.org/10.1007/s42979-022-01174-1
140. Schuerlein S, Schwarz T, Krziminski S, Gätzner S, Hoppensack A, Schwedhelm I, Schweinlin M, Walles H, Hansmann J (2017) A versatile modular bioreactor platform for tissue engineering. Biotechnol J 12:1600326. https://doi.org/10.1002/biot.201600326
141. Kırdök O, Çetintaş B, Atay A, Kale İ, Akyol Altun TD, Hameş EE (2022) A modular chain bioreactor design for fungal productions. Biomimetics 7:179. https://doi.org/10.3390/biomimetics7040179
142. Sampson K, Koo S, Gadola C, Vasiukhina A, Singh A, Spartano A, Gollapudi R, Duley M, Mueller J, James PF, Yousefi AM (2021) Cultivation of hierarchical 3D scaffolds inside a perfusion bioreactor: scaffold design and finite-element analysis of fluid flow. SN Appl Sci 3:884. https://doi.org/10.1007/s42452-021-04871-3
143. Werner S, Kaiser SC, Kraume M, Eibl D (2014) Computational fluid dynamics as a modern tool for engineering characterization of bioreactors. Pharmaceut Bioprocess 2:85–99. https://doi.org/10.4155/pbp.13.60
144. Buttkewitz MA, Heuer C, Bahnemann J (2023) Sensor integration into microfluidic systems: trends and challenges. Curr Opin Biotechnol 83:102978. https://doi.org/10.1016/j.copbio.2023.102978
145. Chan ZX, Chelvam SP, Sin W-X, Teo DBL, Abdul Rahim AAB, Wu YY, Liu D, Birnbaum ME, Yong D, Ram RJ (2025) Automated, aseptic sampling with small-volume capacity from microbioreactors for cell therapy process analysis. Front Bioeng Biotechnol 13:1612648. https://doi.org/10.3389/fbioe.2025.1612648
146. Sieblist C, Jenzsch M, Pohlscheidt M (2016) Equipment characterization to mitigate risks during transfers of cell culture manufacturing processes. Cytotechnology 68:1381–1401. https://doi.org/10.1007/s10616-015-9899-0
147. Humbird D (2020) Scale-up economics for cultured meat: techno-economic analysis and due diligence. Biotechnol Bioeng 118:3239. https://doi.org/10.17605/OSF.IO/AJSU9
148. GOOD Meat (2025) GOOD Meat Partners with ABEC. https://www.goodmeat.co/all-news/good-meat-partners-with-industry-leader-to-build-first-large-scale-cultivated-meat-facility

149. Boston Institute of Biotechnology (2024) BIB Boasts world's largest bioreactor. https://bostonbib.com/the-boston-institute-of-biotechnology-boasts-worlds-largest-bioreactor/. Accessed 17 Sept 2025
150. Rader RA, Langer E s (2019) Single-use Technologies in biopharmaceutical manufacturing: a 10-year review of trends and the future. In: Eibl R, Eibl D (eds) Single-use technology in biopharmaceutical manufacture. Wiley. https://doi.org/10.1002/9781119477891.ch15
151. Smith MT, Larsen LM, Mullen KM, Johnson J (2020) Operation of 5,000-L single-use bioreactors. BioProcess International
152. Jossen V, Eibl R, Eibl D (2019) Single-use bioreactors – an overview. In: Eibl R, Eibl D (eds) Single-use technology in biopharmaceutical manufacture. 1st edn. Wiley, pp 37–52
153. Junne S, Neubauer P (2018) How scalable and suitable are single-use bioreactors? Curr Opin Biotechnol 53:240–247. https://doi.org/10.1016/j.copbio.2018.04.003
154. ABEC (2025) Custom single-run scalable solutions. https://www.abec.com/single-use/. Accessed 17 Sept 2025
155. de Souza Vandenberghe LP, de Mello AFM, Biagini G, Gruening de Mattos PB, Piazenski IN, Candelario JPM, Soccol CR (2024) Bioreactors for cultivated meat production. Springer, Cham. https://doi.org/10.1007/978-3-031-55968-6_6
156. Lara AR, Galindo E, Ramírez OT, Palomares LA (2006) Living with heterogeneities in bioreactors: understanding the effects of environmental gradients on cells. Mol Biotechnol 34:355–382. https://doi.org/10.1385/MB:34:3:355
157. Alavijeh M, Baker I, Lee YY, Gras SL (2022) Digitally enabled approaches for the scale up of mammalian cell bioreactors. Digit Chem Eng 4:100040. https://doi.org/10.1016/j.dche.2022.100040
158. Guan X, Sun W, Ma Z, Du G, Chen J, Zhou J (2025) From lab to industry: technologies and challenges for scaling up bioprocesses in cell-based food production. Trends Food Sci Technol 161:105040. https://doi.org/10.1016/j.tifs.2025.105040
159. Lara AR, Palomares LA, Ramírez OT (2017) Scale-down: simulating large-scale cultures in the laboratory. In: Ind Biotechnol, pp 55–79. https://doi.org/10.1002/9783527807833.ch2
160. Bustamante MCC, Cerri MO, Badino AC (2013) Comparison between average shear rates in conventional bioreactor with Rushton and Elephant ear impellers. Chem Eng Sci 90:92–100. https://doi.org/10.1016/j.ces.2012.12.028
161. Villiger TK, Neunstoecklin B, Karst DJ, Lucas E, Stettler M, Broly H, Morbidelli M, Soos M (2018) Experimental and CFD physical characterization of animal cell bioreactors: from micro- to production scale. Biochem Eng J 131:84–94. https://doi.org/10.1016/j.bej.2017.12.004
162. Arulrajah P, Lievonen AE, Subaşı D, Pagal S, Weuster-Botz D, Heins A-L (2025) Scale-down bioreactors – comparative analysis of configurations. Bioprocess Biosyst Eng. https://doi.org/10.1007/s00449-025-03182-w
163. Serrato JA, Palomares LA, Meneses-Acosta A, Ramírez OT (2004) Heterogeneous conditions in dissolved oxygen affect N-glycosylation but not productivity of a monoclonal antibody in hybridoma cultures. Biotechnol Bioeng 88:176–188. https://doi.org/10.1002/bit.20232
164. Vardar F, Lilly MD (1982) Effect of cycling dissolved oxygen concentrations on product formation in penicillin fermentations. Eur J Appl Microbiol Biotechnol 14:203–211. https://doi.org/10.1007/BF00498464
165. Gaugler L, Hofmann S, Schlüter M, Takors R (2024) Mimicking CHO large-scale effects in the single multicompartment bioreactor: a new approach to access scale-up behavior. Biotechnol Bioeng 121:1243–1255. https://doi.org/10.1002/bit.28647
166. Anane E, Knudsen IM, Wilson GC (2021) Scale-down cultivation in mammalian cell bioreactors – the effect of bioreactor mixing time on the response of CHO cells to dissolved oxygen gradients. Biochem Eng J 166:107870. https://doi.org/10.1016/j.bej.2020.107870
167. Brunner M, Braun P, Doppler P, Posch C, Behrens D, Herwig C, Fricke J (2017) The impact of pH inhomogeneities on CHO cell physiology and fed-batch process performance – two-

compartment scale-down modelling and intracellular pH excursion. Biotechnol J 12:1600633. https://doi.org/10.1002/biot.201600633
168. Jiang R, Chen H, Xu S (2018) pH excursions impact CHO cell culture performance and antibody N-linked glycosylation. Bioprocess Biosyst Eng 41:1731–1741. https://doi.org/10.1007/s00449-018-1996-y
169. Neunstoecklin B, Stettler M, Solacroup T, Broly H, Morbidelli M, Soos M (2015) Determination of the maximum operating range of hydrodynamic stress in mammalian cell culture. J Biotechnol 194:100–109. https://doi.org/10.1016/j.jbiotec.2014.12.003
170. Harrison RG, Todd PW, Rudge SR, Petrides DP (2015) Bioprocess design and economics. In: Bioseparations science and engineering. Oxford University Press. https://doi.org/10.1093/oso/9780195391817.003.0015
171. Beckhoff New Automation Technology (2025) MTP – Module type package. https://www.beckhoff.com/de-de/produkte/automation/module-type-package-mtp/?pk_campaign=AdWords-AdWordsSearch-mtp_DE&pk_kwd=module%20type%20package%20mtp&gad_source=1. Accessed 17 Sept 2025
172. Chen DXB (2019) Extrusion bioprinting of scaffolds for tissue engineering applications. Springer. https://doi.org/10.1007/978-3-030-03460-3
173. Kulus M, Jankowski M, Kranc W, Golkar Narenji A, Farzaneh M, Dzięgiel P, Zabel M, Antosik P, Bukowska D, Mozdziak P, Kempisty B (2023) Bioreactors, scaffolds and microcarriers and in vitro meat production – current obstacles and potential solutions. Front Nutr 10:1225233. https://doi.org/10.3389/fnut.2023.1225233
174. Fasciano S, Wheba A, Ddamulira C, Wang S (2024) Recent advances in scaffolding biomaterials for cultivated meat. Biomater Adv 162:213897. https://doi.org/10.1016/j.bioadv.2024.213897
175. Chen L, Guttieres D, Koenigsberg A, Barone PW, Sinskey AJ, Springs SL (2022) Large-scale cultured meat production: trends, challenges and promising biomanufacturing technologies. Biomaterials 280:121274. https://doi.org/10.1016/j.biomaterials.2021.121274
176. Ng S, Kurisawa M (2021) Integrating biomaterials and food biopolymers for cultured meat production. Acta Biomater 124:108–129. https://doi.org/10.1016/j.actbio.2021.01.017
177. Yan X, Chen XB, Bergstrom DJ (2012) Modeling of the flow within scaffolds in perfusion bioreactors. Am J Biomed Eng 1:72–77. https://doi.org/10.5923/j.ajbe.20110102.13
178. McClelland RE, Coger RN (2000) Use of micropathways to improve oxygen transport in a hepatic system. J Biomech Eng 122:268–273. https://doi.org/10.1115/1.429657
179. Wang Y, Zou L, Liu W, Chen X (2023) An overview of recent progress in engineering three-dimensional scaffolds for cultured meat production. Foods 12:2614. https://doi.org/10.3390/foods12132614
180. Gleadall A, Visscher D, Yang J, Thomas D, Segal J (2018) Review of additive manufactured tissue engineering scaffolds: relationship between geometry and performance. Burns Trauma 6. https://doi.org/10.1186/s41038-018-0121-4
181. Tibbitt MW, Anseth KS (2009) Hydrogels as extracellular matrix mimics for 3D cell culture. Biotechnol Bioeng 103:655–663. https://doi.org/10.1002/bit.22361
182. Enrione J, Blaker J, Brown D, Weinstein-Oppenheimer C, Pepczynska M, Olguín Y, Sánchez E, Acevedo C (2017) Edible scaffolds based on non-mammalian biopolymers for myoblast growth. Materials 10:1404. https://doi.org/10.3390/ma10121404
183. Singh A, Kumar V, Singh SK, Gupta J, Kumar M, Sarma DK, Verma V (2023) Recent advances in bioengineered scaffold for in vitro meat production. Cell Tissue Res 391:235–247. https://doi.org/10.1007/s00441-022-03718-6
184. Park S-M, Ryoo J-H, Kwon HC, Han SG (2025) Scaffold biomaterials in the development of cultured meat: a review. Food Sci Anim Resour 45:688–710. https://doi.org/10.5851/kosfa.2025.e13
185. Seah JSH, Singh S, Tan LP, Choudhury D (2022) Scaffolds for the manufacture of cultured meat. Crit Rev Biotechnol 42:311–323. https://doi.org/10.1080/07388551.2021.1931803

186. Li Z, Ramay HR, Hauch KD, Xiao D, Zhang M (2005) Chitosan–alginate hybrid scaffolds for bone tissue engineering. Biomaterials 26:3919–3928. https://doi.org/10.1016/j.biomaterials.2004.09.062
187. Modrák M, Trebuňová M, Balogová AF, Hudák R, Živčák J (2023) Biodegradable materials for tissue engineering: development, classification and current applications. J Funct Biomater 14:159. https://doi.org/10.3390/jfb14030159
188. Chen H-C, Hu Y-C (2006) Bioreactors for tissue engineering. Biotechnol Lett 28:1415–1423. https://doi.org/10.1007/s10529-006-9111-x
189. Gu H, Kong Y, Huang D, Wang Y, Raghavan V, Wang J (2025) Scaling cultured meat: challenges and solutions for affordable mass production. Compr Rev Food Sci Food Saf 24: e70221. https://doi.org/10.1111/1541-4337.70221
190. Levi S, Yen F-C, Baruch L, Machluf M (2022) Scaffolding technologies for the engineering of cultured meat: towards a safe, sustainable, and scalable production. Trends Food Sci Technol 126:13–25. https://doi.org/10.1016/j.tifs.2022.05.011
191. Chen Y, Zhang W, Ding X, Ding S, Tang C, Zeng X, Wang J, Zhou G (2024) Programmable scaffolds with aligned porous structures for cell cultured meat. Food Chem 430:137098. https://doi.org/10.1016/j.foodchem.2023.137098
192. Haugh MG, Murphy CM, O'Brien FJ (2010) Novel freeze-drying methods to produce a range of collagen–glycosaminoglycan scaffolds with tailored mean pore sizes. Tissue Eng Part C Methods 16:887–894. https://doi.org/10.1089/ten.tec.2009.0422
193. Kang H-W, Tabata Y, Ikada Y (1999) Fabrication of porous gelatin scaffolds for tissue engineering. Biomaterials:1339–1344. https://doi.org/10.1016/s0142-9612(99)00036-8
194. Dai Z, Ronholm J, Tian Y, Sethi B, Cao X (2016) Sterilization techniques for biodegradable scaffolds in tissue engineering applications. J Tissue Eng 7:2041731416648810. https://doi.org/10.1177/2041731416648810
195. Xiang N, Yuen JSK, Stout AJ, Rubio NR, Chen Y, Kaplan DL (2022) 3D porous scaffolds from wheat glutenin for cultured meat applications. Biomaterials 285:121543. https://doi.org/10.1016/j.biomaterials.2022.121543
196. Chai Q, Jiao Y, Yu X (2017) Hydrogels for biomedical applications: their characteristics and the mechanisms behind them. Gels 3:6. https://doi.org/10.3390/gels3010006
197. El Sayed MM (2023) Production of polymer hydrogel composites and their applications. J Polym Environ 31:2855–2879. https://doi.org/10.1007/s10924-023-02796-z
198. Chaudhuri O (2017) Viscoelastic hydrogels for 3D cell culture. Biomater Sci 5:1480–1490. https://doi.org/10.1039/C7BM00261K
199. Lai W, Geliang H, Bin X, Wang W (2025) Effects of hydrogel stiffness and viscoelasticity on organoid culture: a comprehensive review. Mol Med 31:83. https://doi.org/10.1186/s10020-025 01131-7
200. Kumbar SG, James R, Nukavarapu SP, Laurencin CT (2008) Electrospun nanofiber scaffolds: engineering soft tissues. Biomed Mater 3:034002. https://doi.org/10.1088/1748-6041/3/3/034002
201. Valipouri A (2017) Production scale up of nanofibers: a review. J Text Polym 5:8
202. Bodiou V, Moutsatsou P, Post MJ (2020) Microcarriers for upscaling cultured meat production. Front Nutr 7:10. https://doi.org/10.3389/fnut.2020.00010
203. Berry JD, Liovic P, Šutalo ID, Stewart RL, Glattauer V, Meagher L (2016) Characterisation of stresses on microcarriers in a stirred bioreactor. Appl Math Model 40:6787–6804. https://doi.org/10.1016/j.apm.2016.02.025
204. Lembong J, Kirian R, Takacs JD, Olsen TR, Lock LT, Rowley JA, Ahsan T (2020) Bioreactor parameters for microcarrier-based human MSC expansion under xeno-free conditions in a vertical-wheel system. Bioengineering 7:73. https://doi.org/10.3390/bioengineering7030073
205. Deb P, Deoghare AB, Borah A, Barua E, Das Lala S (2018) Scaffold development using biomaterials: a review. Mater Today Proc 5:12909–12919. https://doi.org/10.1016/j.matpr.2018.02.276

206. Brovold M, Almeida JI, Pla-Palacín I, Sainz-Arnal P, Sánchez-Romero N, Rivas JJ, Almeida H, Dachary PR, Serrano-Aulló T, Soker S, Baptista PM (2018) Naturally-derived biomaterials for tissue engineering applications. In: Novel biomaterials for regenerative medicine, vol 1077. Springer, pp 421–449. https://doi.org/10.1007/978-981-13-0947-2_23
207. Chocholata P, Kulda V, Babuska V (2019) Fabrication of scaffolds for bone-tissue regeneration. Materials 12:568. https://doi.org/10.3390/ma12040568
208. Turnbull G, Clarke J, Picard F, Riches P, Jia L, Han F, Li B, Shu W (2018) 3D bioactive composite scaffolds for bone tissue engineering. Bioact Mater 3:278–314. https://doi.org/10.1016/j.bioactmat.2017.10.001
209. Hollander AP, Hatton PV (2003) Biopolymer methods in tissue engineering, vol 238. Humana Press. https://doi.org/10.1385/159259428X
210. Jouda H, Larrea Murillo L, Wang T (2022) Current progress in vascular engineering and its clinical applications. Cells 11:493. https://doi.org/10.3390/cells11030493
211. Melchels FPW, Barradas AMC, Van Blitterswijk CA, De Boer J, Feijen J, Grijpma DW (2010) Effects of the architecture of tissue engineering scaffolds on cell seeding and culturing. Acta Biomater 6:4208–4217. https://doi.org/10.1016/j.actbio.2010.06.012
212. Melchels FPW, Tonnarelli B, Olivares AL, Martin I, Lacroix D, Feijen J, Wendt DJ, Grijpma DW (2011) The influence of the scaffold design on the distribution of adhering cells after perfusion cell seeding. Biomaterials 32:2878–2884. https://doi.org/10.1016/j.biomaterials.2011.01.023
213. Wendt D, Marsano A, Jakob M, Heberer M, Martin I (2003) Oscillating perfusion of cell suspensions through three-dimensional scaffolds enhances cell seeding efficiency and uniformity. Biotechnol Bioeng 84:205–214. https://doi.org/10.1002/bit.10759
214. Villalona GA, Udelsman B, Duncan DR, McGillicuddy E, Sawh-Martinez RF, Hibino N, Painter C, Mirensky T, Erickson B, Shinoka T, Breuer CK (2010) Cell-seeding techniques in vascular tissue engineering. Tissue Eng Part B Rev 16:341–350. https://doi.org/10.1089/ten.teb.2009.0527
215. Jones G, Cartmell SH (2006) Optimization of cell seeding efficiencies on a three-dimensional gelatin scaffold for bone tissue engineering. J Appl Biomater Biomech 4:172
216. Buizer AT, Veldhuizen AG, Bulstra SK, Kuijer R (2014) Static versus vacuum cell seeding on high and low porosity ceramic scaffolds. J Biomater Appl 29:3–13. https://doi.org/10.1177/0885328213512171
217. Tiwari A, Punshon G, Kidane A, Hamilton G, Seifalian AM (2003) Magnetic beads (Dynabead™) toxicity to endothelial cells at high bead concentration: implication for tissue engineering of vascular prosthesis. Cell Biol Toxicol 19:265–272. https://doi.org/10.1023/B:CBTO.0000004929.37511.ed
218. Shimizu K, Ito A, Arinobe M, Murase Y, Iwata Y, Narita Y, Kagami H, Ueda M, Honda H (2007) Effective cell-seeding technique using magnetite nanoparticles and magnetic force onto decellularized blood vessels for vascular tissue engineering. J Biosci Bioeng 103:472–478. https://doi.org/10.1263/jbb.103.472
219. Djisalov M, Knežić T, Podunavac I, Živojević K, Radonic V, Knežević NŽ, Bobrinetskiy I, Gadjanski I (2021) Cultivating multidisciplinarity: manufacturing and sensing challenges in cultured meat production. Biology 10:204. https://doi.org/10.3390/biology10030204
220. Schneider M (1996) The importance of ammonia in mammalian cell culture. J Biotechnol 46:161–185. https://doi.org/10.1016/0168-1656(95)00196-4
221. Myers GM, Jaros KA, Andersen DS, Raman DR (2023) Nutrient recovery in cultured meat systems: impacts on cost and sustainability metrics. Front Nutr. https://doi.org/10.3389/fnut.2023.1151801
222. Busse C, Biechele P, De Vries I, Reardon KF, Solle D, Scheper T (2017) Sensors for disposable bioreactors. Eng Life Sci 17:940–952. https://doi.org/10.1002/elsc.201700049
223. Ronkainen NJ, Halsall HB, Heineman WR (2010) Electrochemical biosensors. Chem Soc Rev 39:1747. https://doi.org/10.1039/b714449k

224. Berry BN, Dobrowsky TM, Timson RC, Kshirsagar R, Ryll T, Wiltberger K (2016) Quick generation of Raman spectroscopy based in-process glucose control to influence biopharmaceutical protein product quality during mammalian cell culture. Biotechnol Prog 32:224–234. https://doi.org/10.1002/btpr.2205
225. Nettleton DF, Marí-Buyé N, Marti-Soler H, Egan JR, Hort S, Horna D, Costa M, Vallejo Benítez-Cano E, Goldrick S, Rafiq QA, König N, Schmitt RH, R. Reyes A (2023) Smart sensor control and monitoring of an automated cell expansion process. Sensors 23:9676. https://doi.org/10.3390/s23249676
226. Zohar B, Blinder Y, Epshtein M, Szklanny AA, Kaplan B, Korin N, Mooney DJ, Levenberg S (2019) Multi-flow channel bioreactor enables real-time monitoring of cellular dynamics in 3D engineered tissue. Commun Biol 2:158. https://doi.org/10.1038/s42003-019-0400-z
227. Khan I, Prabhakar A, Delepine C, Tsang H, Pham V, Sur M (2021) A low-cost 3D printed microfluidic bioreactor and imaging chamber for live-organoid imaging. Biomicrofluidics 15:024105. https://doi.org/10.1063/5.0041027
228. Babic J, Griscom L, Cramer J, Coudreuse D (2018) An easy-to-build and re-usable microfluidic system for live-cell imaging. BMC Cell Biol 19:8. https://doi.org/10.1186/s12860-018-0158-z

Adv Biochem Eng Biotechnol (2026) 193: 155–200
https://doi.org/10.1007/10_2025_291

Published online: 15 October 2025

Edible Scaffolds for Cultivated Meat Production

Mariana P. Hanga, Simona Fehlmann, Nicole Kleger, Maria Laura Vieri, Travis Callue, Alice Millbank, and Eirini Theodosiou

Contents

M. P. Hanga (✉)
Department of Biochemical Engineering, University College London, London, UK

Quest Meat Ltd., Birmingham, UK
e-mail: m.hanga@ucl.ac.uk

S. Fehlmann and N. Kleger
Sallea AG, Zürich, Switzerland

M. L. Vieri and T. Callue
Department of Biochemical Engineering, University College London, London, UK

A. Millbank and E. Theodosiou
School of Engineering and Innovation, Aston University, Birmingham, UK

Abstract Scaffolds are customisable three-dimensional supports that are compatible with cell culture and thus suitable for a wide range of applications. One emerging application is cultivated meat production. Due to the complexity of cultivated meat products, multiple types of scaffolds would be required for the different manufacturing steps involved. Additionally, as cultivated meat is a food product intended for consumption, there are further requirements in the scaffold's material, properties and method of preparation that are necessary to achieve suitability for use in foods, as well as regulatory requirements for safe use. This chapter focuses on edible scaffolds with applicability in cultivated meat production, exploring established and emerging materials suitable for use in foods, methods for scaffold creation, as well as different types of scaffolds and the diverse roles they play across various stages of the manufacturing process.

Graphical Abstract

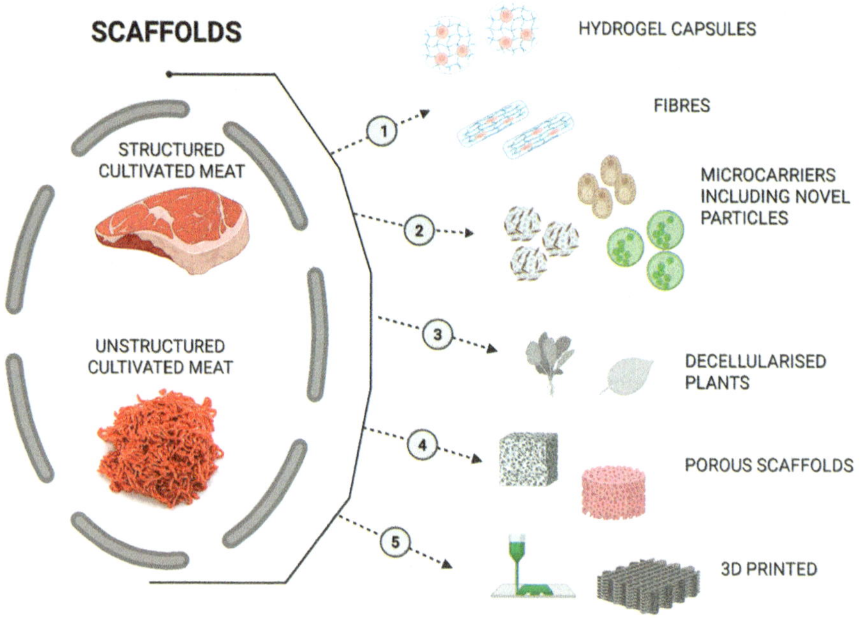

Keywords Biomaterials, Cultivated meat, Food-grade, Scaffolds

Abbreviations

2D	Two-dimensional
3D	Three-dimensional
bAOSMCs	Bovine aortic smooth muscle cells

bASCs	Bovine adipose-derived stem cells
CA	Cellulose acetate
EBP	Extrusion-based printing
ECM	Extracellular matrix
EHDP	Electrohydrodynamic printing
FDA	Food and Drug Administration
FRESH	Freeform reversible embedding of suspended hydrogels
GelMA	Gelatine methacryloyl
GRAS	Generally Recognised as Safe
IBD	Inflammatory bowel disease
IKVAV	Ile-Lys-Val-Ala-Val
iPSCs	Induced pluripotent stem cells
MSCs	Mesenchymal stem cells
PCL	Polycaprolactone
PEO	Polyethylene oxide
pMyoSCs	Porcine myosatellite cells
PolyHIPE	Polymerised high internal phase emulsion
RbSkMCs	Rabbit skeletal muscle myoblasts
RGD	Arginine-glycine-aspartic acid
SBS	Solution blow spinning
SPI	Soybean protein isolate
TVPs	Textured vegetable proteins
WPC	Whey protein concentrate
WPI	Whey protein isolate
ZEM2S	Zebra fish fibroblast stem cells

1 Introduction

Scaffolds are defined as three-dimensional supports with useful characteristics for various applications compatible with cell culture. They are designed to perform one or more of the following functions:

(a) Promote cell-biomaterial interactions, including cell adhesion and extracellular matrix deposition
(b) Facilitate the transport of gases, nutrients, molecular factors, and waste between cells and their environment
(c) Biodegrade over time to enable the formation and maturation of new tissue

Scaffolds have typically been associated with tissue engineering and medical applications [1, 2]. However, more recently, with the development of novel food technologies such as cultivated meat, scaffolds have been found to also play a critical role in the manufacturing of these types of novel products. This book chapter focuses on edible scaffolds with applicability in cultivated meat production, exploring established and novel materials suitable for use in foods, methods for scaffold creation, as well as different types of scaffolds and how they can be utilised throughout the various stages of the manufacturing process.

An important consideration in scaffold design is the nature of the final product that cultivated meat aims to replicate, which can broadly be classified as *unstructured* or *structured*. Unstructured cultivated meat resembles conventional ground meat products (e.g. mince). In this model, tissue architecture is not maintained; instead, a mass of muscle tissue is produced without consideration for factors such as myotube alignment or co-culture with other tissues. The cell mass can then be compressed and formed into burger patties, nuggets, meatballs, or other comparable products. Additional components such as fat (whether plant or animal derived), haem proteins, or related substitutes [3] can also be incorporated into the product after the completion of the cell culture stage. Examples of unstructured cultivated meat products include the cultivated chicken nuggets from JUST and GOOD Meat, which have received regulatory approval in Singapore [4] and the USA, respectively [5]; burgers such as Mosa Meat's proof-of-concept cultivated beef burger [6]; Vow's approved cultivated quail product [7]; and Meatly's UK-approved pet food [8]. Even if the structure and organisation are not necessary for unstructured cultivated meat products, scaffolds still find utility in their production, particularly during the cell proliferation and differentiation stages. They provide surfaces for adherent cells to attach, grow and mature, thereby enhancing cell yields and enabling scalable bioprocessing. Examples of suitable scaffolds include microcarriers, capsules, hydrogels or fibres (Fig. 1). These will be discussed in detail in the later sections.

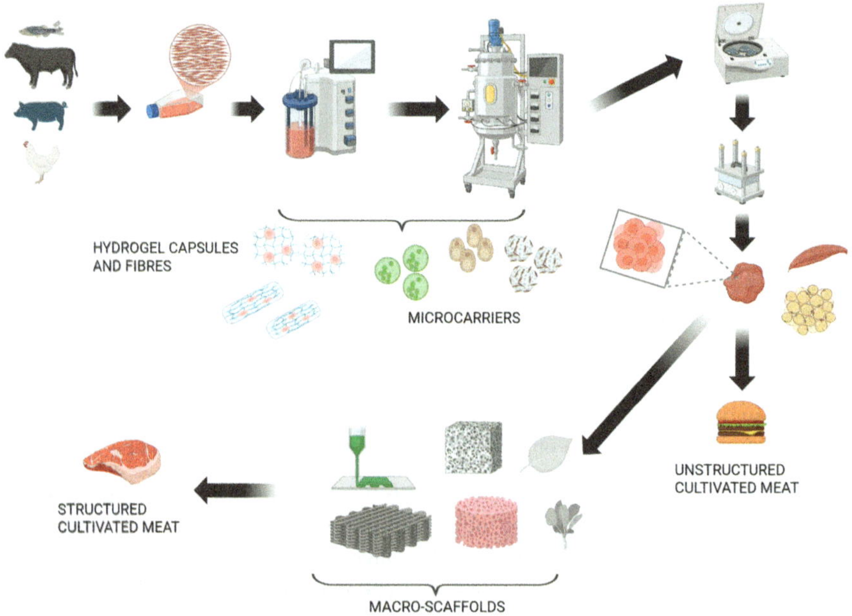

Fig. 1 Scaffold morphologies used throughout the cultivated meat manufacturing process. (Created with BioRender)

On the opposite, a structured product refers to a structurally complete section of muscle tissue, such as a steak, chicken leg or other whole cuts. These products are technically more challenging to manufacture, as their production not only requires the expansion for large numbers of cells but also the reconstruction of complex tissue architecture. This can be carried out using various strategies to create macro-scaffolds (Fig. 1) that support tissue assembly. Challenges include the need for myotube and myofibril alignment and likely the capacity of the scaffold and culture system to allow for contractile and stretch stimulation of the developing tissue construct. Although muscle is the primary tissue in a whole cut, it nevertheless also contains various other cell types, such as adipose (fat), connective and vasculature [9]. This requirement for co-culture of various cell types and/or induced differentiation towards various cell fates simultaneously within the same culture system necessitates broad cytocompatibility of the scaffold. Further, precise and complex media formulations are needed to provide the optimal factors and conditions for all cell types involved. To date, only one structured cultivated meat product has been approved for commercialisation in Israel, and that is the beef steak from Aleph Farms [10]. Despite the increased complexity in manufacturing structured cultivated meat products, there is high demand among consumers for whole cuts.

2 Scaffold Characteristics

Scaffolds for cultivated meat must fulfil multiple roles throughout the bioprocess. Fundamentally, they are required to support cell adhesion, proliferation and, when relevant, differentiation and maturation into specific lineages such as muscle or adipose tissue. Beyond supporting cell growth, scaffold design must also meet a number of application-specific requirements. These include physical and chemical properties compatible with cell culture, as well as characteristics specific to food production such as edibility, safety and contribution to organoleptic properties. Since scaffolds may be incorporated into the final product, their materials and structure must comply with regulatory and consumer standards for food consumption. The following subsections will explore key properties of edible scaffolds in more detail, including their biological function, mechanical performance, material composition and food-grade considerations.

2.1 *Characteristics Specific to Food Production*

When designing scaffolds particularly for cultivated meat, it is crucial that the biomaterial chosen is not only suitable for cell support, but also safe for consumption, sustainable and viable from commercial and technological perspectives. As a novel food technology, cultivated meat and its components must meet food safety regulations, which differ across regions and require formal approval before commercialisation.

A primary regulatory requirement is safety. This includes demonstrating that all materials used and produced, from scaffold materials to the final product and its digestion by-products, are non-toxic and suitable for ingestion. Terms commonly used in the context of food safety include *edible*, *food-grade* and *non-toxic*. In the USA, the FDA also uses the designation *Generally Recognised as Safe* (*GRAS*). In cultivated meat applications, especially when the scaffold is intended to remain in the final product, materials must meet multiple overlapping safety criteria, including biocompatibility during cell culture, food-grade status for regulatory approval and safety of digestion-related breakdown products. These requirements may be evaluated differently depending on the regulatory framework of each jurisdiction [11–13].

For the purposes of this chapter, the term 'edible' is used to encompass any material evidenced as being safely metabolised by the body after oral ingestion, regardless of the current food regulatory status. Materials which are shown to be non-toxic in the body (such as those used for medical implantation) but lacking data pertaining to ingestion are referred to herein as 'potentially edible'.

Opportunities arise when considering that the scaffold can be formulated and designed in ways that add to the final cultivated meat product, such as nutrition or texture. For example, the idea of hybrid products, in which the scaffolding material constitutes an ingredient, has been gaining attention more recently. As such, inspiration has been taken from food science and known food-grade materials, as well as atypical materials with potential including actual foods or plants. The different materials that can be used, common, atypical and emerging, will be covered in Sect. 3.

2.2 Physical Characteristics

The design of a scaffold does not end with the choice of material(s) and morphology. The various physical properties of the scaffold (which in turn are shaped by the specific material, processing and treatment steps, and ultimate morphology of the scaffold) also exert influence over the behaviour of attached cells, and as such must be considered in the context of cell type needs. For example, various specific effects may arise from changes in scaffold stiffness depending on the type of cell being grown on the scaffold. Increased scaffold stiffness has been demonstrated to correlate with an induction of proliferation and reduced capacity for differentiation in various liver-derived stem cells [14]. An increase in collagen scaffold stiffness has also been shown to result in enhanced spreading of myo-progenitor cells, and while myoblast fusion into myotubes was not affected by stiffness, a mid-range value for Young's modulus consistent with that of muscle tissue (approximately 12 kPa) was shown to be optimal for striation in developing myotubes [15]. Myoblastic preference for tissue-like Young's modulus has been supported in the use of other biomaterials such as silk fibroin, where varieties such as silk fibroin derived from

Bombyx mori and *Antheraea mylitta* facilitated enhanced myogenic differentiation, while those varieties with greater elasticity did not [16].

The challenges of fine-tuning the mechanical properties of a scaffold for optimal biocompatibility are evident but deepened when the question of producing a structured cultivated meat product, and thus the need for co-culture of various cell types, is raised. As previously discussed, biocompatibility for skeletal muscle culture is optimal on substrates bearing a Young's modulus similar to that of muscle tissue. However, in an effort to optimise adipogenic differentiation of mesenchymal stem cells (MSCs) [17], noted optimal adipogenic differentiation in softer scaffolds (0.5 vs. 25 kPa). An increase in scaffold stiffness has also been reported to direct MSCs towards an osteogenic differentiation pathway rather than adipogenic, as in the work of Xie et al. [18] though in this instance, the 'stiff' scaffold was substantially stiffer than any discussed thus far. This further highlights the aforementioned need to tailor the mechanical properties of the scaffold to the needs of the specific type of cells used for culture.

Although stiffness and elasticity appear to dominate discussion in the field of mechanical characteristics and biocompatibility, there are other characteristics of relevance. The wettability of a material, denoted by water contact angle, has been noted to exert influence over the behaviour of cultured cells; in the work of Cantini et al. [19], the activity of C2C12 myoblasts was assessed along a wettability gradient ranging from super-hydrophobic to super-hydrophilic. Cells were reported to show increased differentiative potential as hydrophilicity increased, appearing to peak with a water contact angle of approximately 75°, with excessive wettability resulting in a loss of differentiation. These findings are also supported by the work of Feng et al. [20] who noted that an increase in water contact angle (from an original 50–76°) facilitated an increase in fusion index and myotube length, though the latter work combined this regulated hydrophilicity with an increase in conductivity as well as an aligned topography.

Topography is another significant mechanical characteristic with regard to the culture of muscle tissue. For example, the procedure for producing electrospun fibres (see Sect. 4) can be modified to include a rotating collector, resulting in consistent alignment of the nanofibers, which would otherwise be randomly aligned [21, 22]. This change in surface topography served to induce consistent alignment of myotubes cultured on the scaffold surface, better replicating the structure of muscle where fibre alignment enabled longitudinal contractile functions.

Hardness is a property that appears sparsely studied in this space, but existing data has demonstrated the formation of myotubes on PDMS scaffolds with Shore hardness values in the range of 40–70, with qualitative evidence that an intermediate value (Shore hardness 55) best supported myogenic differentiation [23]. It may be argued that the ideal hardness of a scaffold is what mimics the tissue in vivo, as previously demonstrated with stiffness and elasticity. However, that might not be the case. For example, cardiac muscle has been reported as having a Shore hardness value of 20 (±7.5) [24]. It is also important to note that hardness exists as a related, but not interchangeable, characteristic derived from textural profile analysis (TPA), which is relevant to the sensory aspects of consumption rather than biocompatibility.

In this context, the hardness of pork loin has been reported as approximately 70 kPa, a value which increases dramatically to approximately 400 kPa after pan frying [25]. In contrast, cell-laden scaffold constructs as well as blank scaffolds decreased in hardness after cooking, having not been significantly different in hardness from the pork loin samples when raw [25]. This highlights the importance of the mechanical properties of scaffolds not only for biocompatibility, but for consumer acceptance of the final product as well.

2.3 Chemical Characteristics

Most somatic cells are anchorage dependent, meaning that they require a surface to attach to in order to be cultured in vitro. Cell adhesion is a critical and complex part of tissue formation and has many downstream cellular effects. Cells can adhere to each other [26], an essential process in muscle fibre formation [27, 28], or adhere to the extracellular matrix (ECM). ECM-derived proteins are often used in scaffolds; amino acid sequences in protein coatings such as fibronectin and laminin have also been used as cell recognition motifs [29], and the alpha chain of laminin has been demonstrated to play a vital role in myoblastic fusion [30]. By design, scaffolds can mimic the extracellular matrix, thus providing the structural support needed for the cells to attach and stimulate proliferation through specific mechanical and biochemical cues [31]. Examples of commonly used cell recognition motifs include Arg-Gly-Asp (RGD) [32] and Ile-Lys-Val-Ala-Val (IKVAV) [33]. Nevertheless, widely used biomaterials lack these recognition motifs but can still support cell adhesion by other means. For example, silk fibroin bears amino acids with positively charged side chains, which are able to bind to the negatively charged glycocalyx (a surface-expressed layer of glycoproteins) of cells [34]. Furthermore, scaffolds can also support or even direct cell differentiation towards a specific lineage [35].

3 Biomaterials for Scaffold Fabrication

The base materials for the scaffolds must be carefully selected, with foresight given to their current or potential food-grade regulatory status, and consideration given to any potential products arising from the breakdown of the scaffold, ensuring that these too are non-toxic and safe for consumption. Furthermore, careful considerations should also be given to the methods employed for scaffold production with the aim to minimise or completely remove the use of inedible and/or toxic compounds such as solvents or crosslinkers, which are often necessary. Allergens also require consideration as some materials with potential for scaffolding could contain or be listed as major food allergens (e.g. soy, whey, peanut). In such case, the use of those materials is not prohibitive; however, it requires adequate labelling in the end product [35, 36]. Ultimately, the choice of biomaterial for scaffolding needs to not

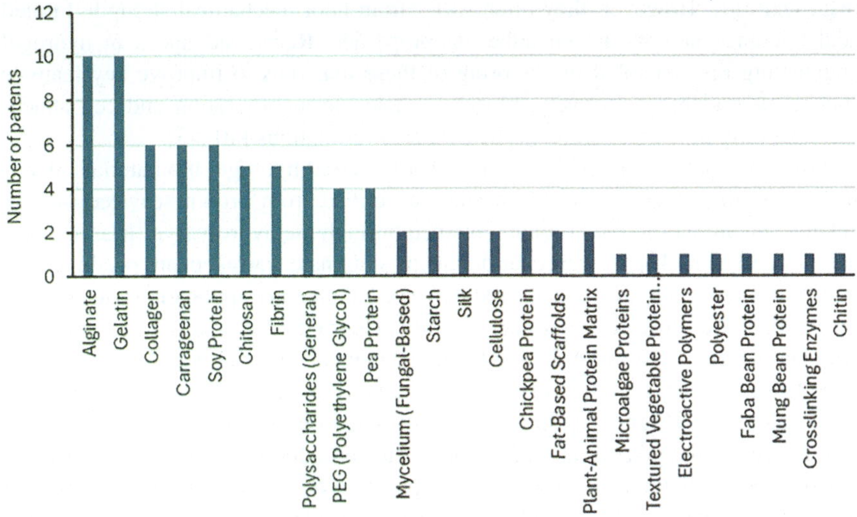

Fig. 2 Scaffolding materials used by the cultivated meat industry. (Data obtained from patent searches)

only fulfil the technical requirements for cultivated meat production, but also be commercially viable, sustainable and safe for consumption, while contributing positively to the appearance, taste, texture and even nutrition of the final cultivated meat products, with the aim to recapitulate the features of traditional meat.

To ascertain which biomaterials are most utilised to date for the production of cultivated meat, a manual review of patents filed with relevance to scaffolds developed for cultivated meat was conducted. A pre-existing compilation of relevant companies was used as a reference [37]. The search was carried out on Google Patents by searching for each company's name individually and using the following inclusion criteria: 'cultivated meat', 'edible', 'scaffold' and 'bio-scaffold', with no exclusion criteria except that only patents written in English were reviewed. From 32 companies, 17 were found to have filed relevant patents. A total of 47 patents were identified and reviewed (Fig. 2).

3.1 Commonly Used Biomaterials

Commonly used biomaterials for scaffold production include hydrogels, which have proven to be promising in tissue engineering due to their biocompatibility, water-rich porous structure and tuneable properties. These characteristics enable them to closely mimic the biomechanical characteristics of native extracellular matrices [38]. Hydrogels are generally made of either natural or synthetic polymers, each possessing distinct advantages and limitations [39]. Natural polymers are generally edible and exhibit low immunogenicity, making them attractive for use in cultivated meat scaffolds. However, they often suffer from poor mechanical strength, limited acid tolerance and weak cell adhesion properties. Recent advances in hydrogel engineering have enabled the tailoring of these materials to improve key features such as degradability, porosity, bioactive molecule incorporation and cell attachment, helping to address some of their inherent limitations [40–43].

From the patent search (Fig. 2), the most preferred hydrogel biomaterials identified were either alginate, a polysaccharide derived from brown seaweed [44] or gelatine, an animal-derived protein obtained through the hydrolysis or heat denaturation of collagen [45]. Polysaccharides such as alginate, carrageenan, pectin, gellan gum and agarose are widely preferred in tissue engineering applications due to their biocompatibility, low cost and degradability. Their edibility and the fact that most are plant derived also make them attractive candidates for cultivated meat production. These materials are non-toxic, generally possess favourable mechanical properties and have a long-standing history of safe use in the food industry [46–50]. However, the general absence of intrinsic cell adhesion motifs on most polysaccharides limits their effectiveness in supporting attachment and proliferation of adherent-dependent cells [51–53]. Despite lacking the cell adhesion motifs, some polysaccharides, such as chitosan, still possess the necessary cytocompatibility and allow cell attachment and proliferation. In the case of chitosan, its biological behaviour is determined by its chemical properties, particularly the degree of deacetylation, as well as the method of processing [54]. Furthermore, as chitosan is typically derived from crustaceans, its consumer acceptance could be limited [50]. Chitosan isolated from fungi [55], on the other hand, could be viewed as a more acceptable material by some consumers. To enhance the adhesive properties of polysaccharides, a typical approach is to chemically modify these materials. Many strategies to achieve this have been successfully implemented, for example by introducing RGD motifs [50], by mixing with other functional biomaterials such as gelatine [56] or by chemical crosslinking with laminin-derived peptides [57, 58]. Unfortunately, most of the modifications introduce additional costs, with the chemical ones requiring the use of toxic chemicals, which for cultivated meat production would introduce safety risks.

Generally, alginate exhibits low immunogenicity; however, this is dependent on factors such as purity, molecular structure and processing methods used [59, 60]. On the other hand, polysaccharides such as carrageenan, despite being widely used as a food additive for its properties as a gelling and thickening agent, emulsifier and stabiliser, were found to activate the innate immune pathways of inflammation while altering the gut microbioma composition and the thickness of the mucus barrier, thus being linked to diseases such as inflammatory bowel disease (IBD) [61]. Despite its wide use in biomedical applications, chitosan has also raised concerns due to its potential immunomodulatory effects [48, 62–64].

Animal-derived proteins, such as collagen and gelatine, were also identified as preferred for scaffold production for cultivated meat (Fig. 2). This was not surprising as it is well known that collagen is an abundant animal-derived protein that plays key roles in moderating cellular processes such as adhesion, migration and differentiation [65]. Similarly, gelatine is derived through the hydrolysis of collagen. Both animal-derived proteins have been widely used in tissue engineering and other biomedical applications due to their versatility in processing into a variety of forms from films to nano- and microparticles, meshes, sponges and even nano- and microfibres [65]. However, their use in cultivated meat production is under scrutiny due to the ethical and sustainability concerns, as well as batch-to-batch variation and risk of zoonotic disease transmission, being derived directly from the animal. Recombinant alternatives have also been developed. However, their production is still relatively expensive due to the low yields achieved and high costs of production [66]. Despite this, both collagen and gelatine have been successfully used for the production of muscle fibres [67–69] with applicability in cultivated meat production. For example, MacQueen et al. [67] used crosslinked gelatine fibres for the culture of bovine aortic smooth muscle cells and rabbit skeletal muscle myoblasts and achieved some of the structural and mechanical features of meat, however, without achieving mature contractile architecture. In another example, type I collagen scaffolds with aligned pores were cultured with C2C12 myoblasts resulting in the formation of myotubes aligned to the pore structures [70].

3.2 *Atypical and Emerging Biomaterials*

Although biomaterials traditionally used in tissue engineering have dominated early scaffold development for cultivated meat, a growing number of atypical and unexpected materials have been trialled in recent years. The search for these materials has been fuelled by the pressing need to achieve cost efficiency in cultivated meat production with the goal to reach cost parity with traditional meat. Additionally, as the idea of the first generation of hybrid cultivated meat products is taking shape, scaffolds are becoming an even more important and integral part of these products, with the intention to not only provide typical scaffold functionality, but also contribute to the organoleptic properties of the final cultivated meat product. A list of edible scaffolding materials chosen by different cultivated meat companies and made public through publications and/or patents is given in Table 1.

Much inspiration in the choice of materials has been drawn from food science and from nature. Among the most promising are plant proteins, which are typically cheap, abundant and sustainable and can be produced in different scaffold morphologies. Proteins such as soy, pea, zein and glutenin have been made into films with appropriate mechanical characteristics for cultivated meat development [71, 72]. Films made from glutenin and zein have been found to effectively promote the growth of aligned cells and to support the development of aligned myotubes [73], highlighting their potential as promising candidates for upcoming research in the production of cultivated meat. Several companies have adopted plant-based proteins as their scaffold material of choice (as identified in Table 1), including DaNAgreen (South Korea) for their porous macro-scaffolds, Gelatex (Estonia) for both microcarriers and macro-scaffolds [74], Sallea (Switzerland) for porous macro-scaffolds [75] and Wild Type (USA) for their fibre-spun macro-scaffold

Table 1 List of edible scaffolds developed by industry for cultivated meat production and made public through publication and/or patents

Company name	Location	Type of scaffold	Technology used (if known)	Material/s	Patents (if available)	Reference
Algimatrix	USA	Macro-scaffold	Lyophilisation	Alginate		[284]
CytoNest	USA	Macro-scaffold	Mechanical fibre drawing/ touch spinning	Zein, polycaprolactone	US2023174928A1	[285]
DaNAgreen	South Korea	Macro-scaffold	Porogen leaching	Vegetable protein	WO2022050733A1	[286]
Gelatex	Estonia	Microcarriers, macro-scaffold	Halospinning	Soy, pea, potato, lupin and hybrids	EP4527920A1 EP3954811A1	[74, 287]
Luyef	Chile	Macro-scaffold	Decellularisation	Algae	WO2023102083A1	[81]
Mirai International	Switzerland	Microcarriers	Not disclosed	Polysaccharides	WO2023104768A1	[288]
MyoWorks	India	Macro-scaffold	Cultivation	Polysaccharide-mycelium hybrid	WO2022208525A1	[289]
Naturbeads	UK	Microcarriers	Membrane emulsification	Cellulose		
Quest Meat	UK	Microcarriers/ macro-scaffold	Not disclosed	Brewing waste; yeast	WO2024100401	[82]
Sallea	Switzerland	Macro-scaffold	Salt leaching	Soy, pea, lentil protein based and hybrids	WO2023001418A1	[75, 290]
Seawith	South Korea	Macro-scaffold	Decellularisation and hydrogel	Algae-derived gel	EP4079838A1	[291]
Smart MCs	Australia	Microcarriers	Not disclosed	Gelatine based		[292]
TruSpin Nanomaterial Innovation	USA	Macro-scaffold	Electrospinning	Hyaluronic acid		[293]
Spiderwort	Canada	Macro-scaffold/ microcarriers	Decellularisation	Fruits and vegetables; fungal	WO2025059779A1	[259]
Wild type[a]	USA	Macro-scaffold	Fibre-spinning technique	Plant based	WO2020123876A1	[76]

[a] These companies have developed scaffolds in-house; however, these are not commercially available for other parties

[76]. Decellularised plants and textured vegetable proteins (TVPs) have also been successfully used for cultivated meat applications. For example, decellularised spinach leaves have been tested with cells and have been found to have comparable cell culture performance to gelatine-coated glass [77]. Similarly, Spiderwort (Canada) uses plant tissues stripped of cellular content, including fruits and vegetables, as scaffolds [78]. In addition, rice grains have been used successfully as scaffolds for cell cultivation and differentiation with potential in alternative hybrid foods [79].

Another example of an atypical scaffold material is the use of different species of algae, which have been identified as promising candidates for cultivated meat production due to their low cost, abundance, sustainability and nutritional properties [80]. Algae have been the material of choice for Luyef (Chile) for the fabrication of their decellularised scaffolds [81].

Yeast and mycelium are also emerging as potential materials to be integrated in different scaffold morphologies in cultivated meat production. For example, Quest Meat Ltd. (UK) are utilising spent brewing yeast to make irregularly shaped microcarriers and macro-scaffolds that can be incorporated in the final cultivated meat product where it adds to texture and nutritional profile of the final product [82]. Mycelium is the vegetative growth form of filamentous fungi; it develops thread-like hyphae that form a dense, interconnected 3D network. The natural fibrous structure combined with good mechanical properties and a range of beneficial bioactive compounds (e.g. polysaccharides, phenols and carotenoids) found in edible fungal species make mycelium an appealing material for creating sustainable scaffolds. These attributes have already attracted interest in tissue engineering fields, such as wound healing, where their ability to support the attachment of mammalian cells has been demonstrated [83, 84]. Given their scalability and inherent edibility, mycelium-based scaffolds may also have significant potential for applications in cultivated meat production. In addition, edible mycelia could be bioengineered to enhance their nutritional profile and sensory characteristics [85] opening new avenues for more versatile and appealing cultivated meat products.

Silk fibroin, a food-grade fibrous protein derived from silkworm cocoons, is also emerging as a material with applicability in cultivated meat. For example, electrospun silk fibroin demonstrated the successful attachment and proliferation of bovine adipose-derived stem cells (bASCs) and porcine myosatellite cells (pMyoSCs) [86, 87].

4 Scaffold Morphologies

As well as the choice of material itself, the scaffold morphology plays a pivotal role in facilitating cell attachment, proliferation and differentiation during cultivated meat production. As native cellular environments are inherently three-dimensional (3D), conventional two-dimensional (2D) culture systems fail to accurately replicate critical aspects such as cell polarity, intercellular interactions and complexity of the extracellular matrix. Due to these limitations, 3D culture systems such as scaffolds have gained increasing attention for their ability to provide a more physiologically relevant microenvironment that mimics more closely the in vivo conditions [88]. Preadipocytes, for example, tend to mature more efficiently in 3D spheroids than in 2D monolayers [89].

As each biomaterial type has unique intrinsic physical and chemical properties, different materials may be particularly suited or unsuited to certain scaffold morphologies. For example, alginate is easily made into hydrogel scaffolds, 3D printed scaffolds, microcarriers or capsules for cell encapsulation by crosslinking with divalent ions and tuning its mechanical properties by adjusting concentrations [90–92]. On the other hand, it is poorly suited to form electrospun fibres due to its low structural rigidity and its polymer chains being insufficiently long and flexible enough to achieve optimal molecular entanglement, thus leading to interruptions in the fibre jet and the formation of beads instead [93]. Zein, in contrast, is well suited to electrospinning due to its molecular entanglement and hydrophobic nature. It also dissolves readily in organic solvents that evaporate well, which is crucial to the fibre formation process [94]. Conversely, zein's hydrophobic nature makes it inapplicable to hydrogel scaffold formation, while alginate is not only water soluble, but also easily cross-linkable in aqueous solutions in the presence of divalent cations, thus giving great ease of sol-gel transformation [93].

Comparing biomaterials' applicability to scaffold morphologies is not always straightforward. For example, wet spinning relies on the phase exchange between a solvent and a non-solvent, for achieving polymer precipitation into a fibre morphology [95]. In this case, the efficiency of the wet-spinning process relies more on the dynamic between these two liquid phases and their compatibility (or lack thereof) with the polymer, rather than the properties of the polymer itself.

4.1 Microcarriers

Microcarriers are micrometre-sized scaffolds typically with diameters of 100–200 μm used in bioreactor cultures as attachment substrates for adherent-dependent cells [27, 96–98]. On microcarriers, cells attach and grow on the outside surface except for macro-porous microcarriers where cell infiltration into the pores is possible [99]. Microcarriers are an established technology used successfully in biotherapeutics manufacturing [100–105] due to the significant advantages that they introduce over 2D culture such as high surface area-to-volume ratio, scalability and versatility [27].

Current commercially available microcarriers are diverse, made from a variety of materials such as polystyrene, dextran, cellulose and gelatine [27, 97]. They are typically spherical or oval shaped and vary in terms of surface coatings and charge properties [99, 106, 107]. This variety allows commercially available microcarriers to cater to a broad range of cell types and applications. However, the materials currently used in commercially available microcarriers are not fit for purpose for cultivated meat production as they raise either safety concerns or in some cases even ethical concerns.

For cultivated meat to become scalable and economically viable, the efficient expansion of mammalian cells must be achieved under suspension culture conditions in bioreactors. Microcarriers are essential to achieving this. To date, there are multiple examples of commercially available microcarriers used for the successful expansion of cells relevant to cultivated meat production in spinner flasks [97, 104, 108–111] also in litre-scale bioreactors [98, 109, 111]. However, commercially available microcarriers such as Plastic and Cytodex1 or 3 are not edible, and as such, they would require removal from the final cell mass which can only be achieved through a combination of harsh enzymatic treatments and intense agitation followed by a filtration step [112].

In recent years, with the emergence of cultivated meat, edible microcarriers have gained a lot of interest with multiple companies generating innovations in the microcarrier space (Table 1). The choice of materials for edible microcarriers will not be discussed in this section. Instead, the innovative approaches in the edible microcarrier space will be covered.

The cellular organisation and tissue architecture formed during expansion directly influence the texture and organoleptic properties of the final cultivated meat product [113, 114]. To this end, edible microcarriers can be engineered with controlled surface topography that can direct cell alignment and fusion, mimicking muscle fibre formation. For example, Norris et al. [115] developed an embossing technique to create grooved edible spherical gelatine microcarriers. These microcarriers showed modest but consistent improvements in the proliferation and alignment of myogenic cells compared to smooth spherical carriers, demonstrating that surface microarchitecture can guide tissue-like organisation even in suspension.

Unlike typical microcarriers, innovative approaches can also offer opportunities for improved nutrient and oxygen delivery, supplemented with efficient waste removal. For instance, microalgae-based carriers have been investigated as live, metabolically active platforms. These microalgae were able to simultaneously absorb toxic metabolic by-products (e.g. ammonia) and supply oxygen via photosynthesis [116]. Moreover, as microalgae are highly nutritious, if included in the final cultivated meat product, they would be contributing valuable proteins, lipids, antioxidants and other nutritionally relevant components [117].

While the majority of commercially available microcarriers are spherical, other shapes such as fibrous or irregularly shaped could have more promise for mimicking the extracellular matrix and achieving better physical properties (e.g. topography) to support cell attachment and even differentiation. For example, electrospun fibrous microcarriers could better support attachment and differentiation of myosatellites. Recent work by Dagès et al. [87] demonstrated the successful use of silk fibroin electrospun mats with bovine adipose-derived stem cells suggesting that the next step in their work to achieve the full potential of these fibrous scaffolds would be to develop them in irregularly shaped, fibrous microcarriers. Another example of irregularly shaped microcarriers with potential is Quest Meat's microcarriers derived from yeast [82]. Recently, out-of-the-box innovative approaches to carriers for cultivated meat inspired from food science have been taken. For example, rice grains have been shown to be a suitable carrier for beef cells with applicability in cultivated meat. The rice grains were transformed into a hybrid product containing beef and were found not only to be able to support cell attachment and differentiation, but also to contribute nutritionally. However, the rice grains required an additional coating of fish gelatine to achieve cell attachment and proliferation [79].

Another emerging innovative microcarrier with potential is based on naturally forming edible fungal particles [118]. Certain food-grade mycelial strains produce spherical pellets during submerged fermentation. These structures can be directly harvested and autoclaved for use as microcarriers. Their fibrous, porous matrix emulates ECM-like features, while their high surface area and hyphal surface layers promote nutrient and oxygen diffusion. Importantly, they are highly customisable in terms of size, porosity, texture and colour. Although autoclaving may reduce volatile aromas, it can also promote the release of umami-enhancing compounds, potentially contributing to the savoury flavour profile of the final product [119]. Additionally, the fungal kingdom is a rich source of bioactive and adhesive compounds. Notably, a recent computational analysis by [120] revealed that over 98% of fungal species in a curated database exhibit a high density of arginine-glycine-aspartic acid (RGD) motifs which are short peptide sequences known to facilitate integrin-mediated cell adhesion. This suggests that fungal microcarriers may offer intrinsic bio-adhesive functionality without the need for chemical surface modifications that otherwise would be required when materials like polysaccharides are used [51–53].

4.2 Casted Hydrogel Scaffolds

Certain edible biomaterials, particularly hydrogels, can be shaped by casting into carefully designed moulds and then solidified through chemical or physical crosslinking. This straightforward yet versatile method allows precise control over the final geometry, size and surface features of the scaffold. By adjusting the composition and the crosslinking conditions, it is possible to tailor the mechanical strength, porosity and nutrient diffusion properties of the resulting construct [121–124]. When combined with complementary techniques such as sacrificial templating, casting can also produce more complex structures, including interconnected pores or embedded microchannels [125–127]. While casting and sacrificial templating remain practical and cost-effective for producing bulk structures with defined external shapes, they offer less flexibility for creating sophisticated internal microarchitectures. For applications that demand finer structural control, internal alignment or precise spatial distribution of bioactive compounds, alternative methods such as 3D printing, salt leaching, freeze-drying, emulsion templating, electrospinning or decellularisation generally provide clearer advantages. Moreover, casting can be labour-intensive and difficult to scale. For example, Furuhashi et al. [128] demonstrated that casting hydrogel modules with internal striped structures and embedded pillars enabled bovine myoblasts to align and contract in response to electrical stimulation. However, each module had to be manually cast and assembled layer by layer, which is impractical for large-scale production. Moreover, the hydrogels used (fibrin and Matrigel) were not edible or suitable for food production. The authors therefore suggested that automated techniques such as 3D bioprinting could offer a more efficient alternative for producing such structures at scale, ideally in combination with edible, food-safe materials.

A potential challenge with larger 3D scaffolds is limited oxygen and nutrient diffusion. One promising alternative is to use thin, film-like scaffolds that maintain better nutrient and gas exchange and can then be stacked or rolled into larger constructs. For example, Korean researchers demonstrated how casting and layer-by-layer assembly can be used to fabricate edible polysaccharide films for controlled nutrient delivery, significantly improving cost efficiency and scalability for C2C12 cell culture and showing clear potential for cultivated meat applications [129]. Due to its simplicity and affordability, casting is also widely used to fabricate micropatterned film-like scaffolds. In the context of cultivated meat, micro-grooved surfaces impart directional cues that promote muscle cell alignment and differentiation. Several studies have shown successful cell proliferation and alignment with this approach, and the resulting aligned muscle layers can later be assembled together with separate smooth layers designed for fat tissue, helping to recreate the complex structure of real meat [72, 130–135]. These recent results demonstrate the potential of combining patterned, film-like scaffolds with smart nutrient delivery to address some of the scalability challenges of cultivated meat production.

Taken together, these examples illustrate that while casting remains a practical and versatile method for shaping edible biomaterials into simple or patterned structures, it may not always provide the fine internal architectures needed to fully replicate the texture, alignment, and complexity of real meat. To address these limitations, the following sections examine a range of complementary strategies for the fabrication of scaffolds which offer greater control over micro- and macro-scale features essential for structured cultivated meat. Techniques such as electrospinning, spinning, freeze-drying, emulsion templating, gas foaming and decellularisation can generate scaffolds with tailored porosity, fibrous alignment and improved nutrient exchange, all of which are critical for supporting cell infiltration, vascularisation and development of thicker, more realistic tissue constructs.

4.3 Fibrous Scaffolds

Fibrous scaffolds are especially suited for the manufacture of structured cultivated meat products. Fibrous scaffolds come in a wide range of sizes, from nanofibre matrices to micro-fibrous structures and to macro-sized hollow fibre constructs. Nanofibres have the advantage of structurally mimicking the extracellular matrix on which cells typically grow in vivo [136, 137]. Micro-sized fibres have surface area advantages for increased cell attachment [138], while hollow fibres mimic the structure of blood vessels and are advantageous for developing vascularised tissue constructs [139].

4.3.1 Fibrous Mats/Membranes

There is a variety of processes that can be used for the creation of edible fibrous scaffolds including electrospinning, solution blow spinning (SBS), centrifugal spinning, wet spinning and dry spinning, as well as various combinations of the above [140–145]. However, each of these processes will produce different types of fibrous matrices. For example, electrospinning, a technique explored the most to date for tissue engineering applications due to its versatility and cost-effectiveness, is able to produce fibrous membranes that closely resemble the morphology of the extracellular matrix [146–148]. However, its high energy demand and apparent low productivity can potentially hinder its application as a manufacturing platform for the cultivated meat industry. To date, electrospinning has also been employed extensively in other industries (e.g. electronics, catalysis, textiles, sensors, filtration), whereas its use in the food sector is extremely limited, mostly to the production of food packaging [149–151].

Non-woven electrospun membranes hold a lot of promise as scaffolds for cultivated meat because of their ECM-like topography, which can promote cell attachment and drive targeted differentiation, in addition to the rapid diffusion of oxygen and nutrients [87, 136]. One limitation of the electrospun constructs is the maximum thickness of the membranes that can be achieved, which in most cases cannot exceed a couple of millimetres [148]. Consequently, electrospun scaffolds cannot alone enable the engineering of structured cultivated meat. This drawback could be addressed through the combination of different types of electrospinning (e.g. random and aligned) [152], or the use of hybrid manufacturing (e.g. electrospinning and 3D printing) [153], that will result in enhanced support functionality.

There is a plethora of natural, edible and abundant materials derived from various sources (i.e. plants, seaweed, bacteria and animals) that can be electrospun either alone or in blends [154]. Of these, electrospun cellulose acetate (CA) alone or blended with annatto extract, a natural plant-derived food dye with anti-inflammatory properties, has been shown to support the proliferation and differentiation of immortalised mouse myoblasts (C2C12) [136]. The same authors also successfully created 3D structures through stacked layers of 'muscle' sheets comprising randomly aligned CA electrospun fibres (four layers with a total thickness of 300–400 μm) laden with different cell types (C2C12, H9c2 – rat cardiomyoblasts or cSCs – primary chicken satellite cells) [155, 156]. Cellulose acetate is a versatile and low-cost material with applicability in many industries. However, in the food industry, it has predominantly been used as packaging [157] and not as a food component.

Edible biomaterials are not always suitable for electrospinning. However, this challenge can be overcome by using a 'carrier' polymer like polyethylene oxide

(PEO). PEO is hydrophilic, non-toxic, biocompatible and biodegradable, with GRAS status, and it can improve the electrospinnability of various otherwise 'difficult-to-spin' biomaterials [158]. For example, blends of zein (a low-molecular-weight globular protein found in maize) with PEO have been used to electrospin fibres with different morphologies depending on the molecular weight of the PEO used (ribbon-like from 1,000,000 Mw; cylindrical from 100,000 Mw). However, in this particular work, no cell studies were carried out to evaluate the suitability of the electrospun fibres [159]. Da Trindade and co-workers [160] also blended PEO with zein and pea protein to reproduce the fibrous structure of traditional meat in vegetable protein-based products. The electrospun fibres had good thermal stability and therefore the potential to maintain physical characteristics during cooking. Finally, various electrospun combinations of PEO with whey protein concentrate (WPC) were reported by [158], and despite the lack of biological testing, once again this formulation could be suitable for cell culture. Zein electrospun fibre matrices have also been shown to support myogenic cell culture to a degree comparable to that of tissue culture plastic, highlighting their potential for cultivated meat applications [161]. However, challenges remain in that such scaffolds must be formed by 'green electrospinning', with non-toxic crosslinkers such as cinnamaldehyde [162]. Other successfully electrospun fibres for cultivated meat applications were created from modified starch blended with either glycomacropeptide or whey protein isolate (WPI) but not yet tested for cell attachment and proliferation [163].

Solution blow spinning (SBS) is another way to produce fibres in nano- and micro-scales, with a relatively simple set-up and the possibility of large-scale manufacturing [140]. The technology has been used mainly with non-edible polymers, except for porcine gelatine. SBS glucose-gelatine fibrous mats with cotton tuft-like appearance and enriched with either Matrigel matrix, laminin-111 or vitronectin supported the growth of primary hepatocytes [164]. A variation of SBS is HaloSpin™, a technology patented by Gelatex Technologies OÜ [165]. This device enables the continuous production of fibres from a variety of synthetic and natural polymers (including edible ones) with inexpensive and simple equipment requirements. HaloSpin™ has been employed for making fibrous scaffolds for cultivated meat starting from plant proteins (e.g. potato, rice, sunflower) to carbohydrates (e.g. dextrose, fructose and maltitol) and polysaccharides (e.g. pullulan, starch and pectin) [74].

4.3.2 Fibres

One of the benefits of creating fibres is the fact that they can be tailored to different lengths while maintaining their high surface area. This offers the versatility of producing scaffolds that can serve different purposes. For example, short-length (~20 μm average length) gelatine fibres were found to promote aggregation of bovine aortic smooth muscle cells (bAOSMCs) and rabbit skeletal muscle myoblasts (RbSkMCs), and as a result of their shorter length, they could serve as microcarriers inside stirred-tank bioreactors, whereas longer gelatine fibres (>1 cm average length) promoted cell alignment and therefore could be considered for the production of whole cuts of meat [67].

Fibres can be produced using various methods including wet spinning, immersion-rotary jet spinning or solution blow spinning. Fibres produced using wet spinning have typical diameters in the micrometre range and therefore are closer mimics to animal muscle fibres, which measure 1–40 mm in length and 20–100 μm in diameter (depending on species and muscle types) [166]. For example, Cui et al. [167] used wet-spun blends of sodium alginate and soybean protein isolate (SPI) to create fibres with final SPI concentrations between 2 and 10 wt% and diameters of 30–100 μm. The increased SPI content further enhanced porosity, surface roughness and water-holding capacity, all of which could be seen as positive contributors to cell culture performance and texture profile of the final product. Unfortunately, however, the study did not include any biological characterisation of the scaffolds.

Another use of fibres could be as aids to improve texture. For example, wet-spun fibres of recombinant fibroin-inspired protein with diameters of up to 30 μm were employed as additives to improve the texture of a plant-based ground pork analogue product [168]. The use of recombinant proteins instead of animal- or insect-derived ones can be very appealing from sustainability and animal welfare perspectives [169]. Although the authors did not include details about how the fibres were created, they stated that they can be produced at kg scales using safe chemicals and are composed of protein with minimal impurities [168], which therefore makes them suitable candidates as scaffolds for culturing animal cells.

Using a variation method of wet spinning called immersion-rotary jet spinning, gelatine fibres with diameters in the range of 1–10 μm were produced. To preserve their structural integrity, the gelatine fibres had to be crosslinked, and this was achieved using a microbial transglutaminase, which is a common food additive. However, this approach failed to maintain the fibrous morphology during the 6-day culture [67]. Although this enzyme has a GRAS status, there are questions about its food safety, with potential side effects that can induce autoimmune/neurodegenerative diseases [170]. Replacing the transglutaminase with chemical crosslinkers such as EDC and NHS that are typically employed for crosslinking [67] would ensure the structural integrity of the fibres over longer periods, but these chemicals are not food safe and therefore are not suitable for cultivated meat applications.

4.3.3 Hollow Fibres

Hollow fibres can also have applicability in cultivated meat production, though they are distinct from other scaffold morphologies discussed thus far in that they are not always used as scaffolds; hollow fibres are often used as a feature of a bioreactor, such as channels for media supply, rather than being cell-laden structures in and of themselves [171]. As scaffolds, hollow fibres are a potential means of addressing mass transfer obstacles in 3D culture systems using other scaffold morphologies [172], with hollow fibres being cited as having the capacity to facilitate cell densities of up to 2×10^8 cells/mL versus a capacity of up to 2×10^6 cells/mL for microcarriers

[173]. Furthermore, the Japan-based cultivated meat company Integriculture recently unveiled a structured cultivated chicken nugget made using a hollow fibre scaffold system [174]. However, a need for further work is evident to achieve the production of edible hollow fibres.

4.4 *Porous Scaffolds*

This section refers specifically to macro-porous scaffolds with applicability in structured cultivated meat products. Porous scaffolds provide a 3D environment that mimics the natural extracellular matrix, supporting the attachment, proliferation and differentiation of various cell types, including stem cells, muscle and fat cells. Interconnected porous architectures are particularly relevant, as they facilitate key cellular processes such as adhesion, migration and maturation, while also enabling efficient nutrient and waste exchange and supporting scaffold remodelling over time. Structures with pores >50 μm enhance cell infiltration, promote uniform tissue development and encourage vascular-like network formation, all of which are essential for producing thicker, structured cultivated meat constructs. In contrast, scaffolds with smaller pore sizes (<50 μm) tend to be more suitable for the adsorption of proteins such as albumin and insulin, which can enhance cell proliferation and differentiation [175–182]. Moreover, different cell types have been shown to favour specific pore size ranges. For example, porous scaffolds with pore sizes ranging from 50 to 200 μm have been shown to support the attachment, growth, and fusion of skeletal muscle cells, with evidence that these functions are better served by pore sizes at the lower end of this range [181]. In contrast, studies have reported that mesenchymal stem cells can differentiate into adipocytes in scaffolds with pore sizes ranging from 200 to 600 μm, with larger pores generally providing more effective support for adipose tissue formation [183]. However, it is important to note that these findings are primarily based on research focused on human skeletal muscle and adipose tissue engineering, rather than directly on cultivated meat systems.

Finally, it is essential to recognise that porosity also influences the mechanical properties of the scaffold, such as stiffness and stress relaxation [183, 184], the importance of which in scaffold design has already been discussed in the previous section. This relationship should be carefully considered when designing porous scaffolds for cultivated meat, as both structural and mechanical requirements must be balanced to support successful tissue development.

4.4.1 Freeze-Dried Porous Scaffolds

Porous scaffolds with tailored pore size and porosity can be produced using a variety of methods, each leading to specific characteristics in the scaffold. One method that has been successfully employed in tissue engineering due to its versatility and usability with a range of biomaterials is freeze-drying, also known as lyophilisation, a dehydration process used to remove water from frozen materials. This method can produce scaffolds with porosity exceeding 90% and a pore diameter range of 20–400 μm [185]. Freeze-drying is advantageous in tissue engineering applications as it uses water as a solvent, thus avoiding any toxic organics and resulting in biocompatible, environmentally friendly scaffolds. It also allows the incorporation of heat-sensitive additives such as proteins, drugs or growth factors, and when using optimal parameters, the pores and internal alignment can be tuned to suit the application. Additionally, freeze-drying can be combined with other methods such as gas foaming or salt leaching to further enhance the scaffold properties. However, freeze-drying is energy-intensive and time-consuming and can lead to structural and textural damage due to ice crystal formation during the drying process [185–188].

Natural biomaterials such as sodium alginate, cellulose, silk fibroin, collagen, gelatine and chitosan, often blended with synthetic polymers or bioactive components, have been commonly used to fabricate freeze-dried porous scaffolds for tissue engineering [69, 185, 188–190]. The versatility of this method supports its application in food-related fields. For example, freeze-dried porous scaffolds made from natural polymers have been shown to support the aggregation and differentiation of human adipocytes and adipose stem cells [191–193]. There are findings with applicability in cultivated meat.

Through controlled freezing, the directional growth of ice crystals can be guided, thus resulting in aligned pore architectures known as anisotropic porous scaffolds. These types of scaffolds could be particularly useful for supporting myoblast orientation and fusion or for reproducing the anisotropic organisation found in native bone and muscle tissue [194–198]. This approach was employed by Xia et al. who used directional freezing of a soy protein/carrageenan/alginate hydrogel to create aligned porous scaffolds which were found to support C2C12 alignment and differentiation [199]. Similarly, Chen et al. developed freeze-dried collagen scaffolds with aligned porosity and showed that these constructs significantly improved the alignment and maturation of pig musculoskeletal stem cells compared to isotropic controls [200].

Freeze-dried edible biomaterials, including non-mammalian gelatine, chitosan, yeast and various plant-based proteins, have recently been investigated for cultivated meat, with promising outcomes [56, 201–210]. For example, Kim et al. (2024) demonstrated the successful fabrication of polysaccharide–pea protein scaffolds containing a high pea protein concentration (13.4% w/v) and achieving approximately 80% porosity. The high pea protein content not only enhanced cell adhesion but also supported cell differentiation into muscle tissue and had the potential to improve the final product's nutritional value. Interestingly, Wang et al. (2024) demonstrated an innovative approach using proanthocyanidins and dialdehyde chitosan as natural crosslinkers to modify a collagen scaffold, which was then freeze-dried to achieve a high porosity of 90%. This scaffold supported the growth and differentiation of bovine myoblasts effectively. Their results showed that these so-called green crosslinkers made the scaffold edible and food-safe while also enhancing its porosity, swelling capacity, degradation behaviour and mechanical strength. Notably, the natural pigments present in proanthocyanidins gave the final cultured meat a desirable colour, reducing the need for artificial colouring agents. In a follow-up study, the same research group also showed that this approach can be extended by incorporating alternative proteins, such as yeast protein, into the scaffold design [205].

In addition to porous 3D scaffolds, freeze-drying can also be used to fabricate film-like structures, which may be relevant for sheet-based or layered cultivated meat formats. For example, Rubio et al. developed freeze-dried mushroom-derived chitosan films that supported the adhesion and proliferation of insect muscle cells [211]. In a subsequent study, the authors demonstrated that C2C12 and bovine muscle cells could grow on a porous wheat glutenin scaffold produced using freeze-drying [73], reinforcing the potential of cost-effective and widely available plant- and fungi-derived proteins as viable scaffolding materials for cultured meat applications. Additionally, freeze-drying can also be employed when fabricating films with specific surface topographies. For example, Chen et al. demonstrated that freezing collagen in a mould before lyophilisation creates microgroove-like aligned ridges on the scaffold surface, guiding the alignment and elongation of rat myoblasts [212].

4.4.2 Polymerised High Internal Phase Emulsion (PolyHIPE)-Based Porous Scaffolds

Using PolyHIPEs to fabricate 3D porous scaffolds for cell culture is a relatively recent development that is rapidly expanding. It offers precise control over size, distribution and interconnectivity of pores in scaffolds, which are critical for cell infiltration and tissue development. This technology also enables customisation of the scaffold's shape and its physical, chemical and mechanical properties, with the resulting structures often exhibiting good stability. Furthermore, it can be scaled up due to its simplicity, reproducibility and compatibility with standard processing methods. For this reason, PolyHIPEs have been used for a variety of applications in tissue engineering, including as scaffolds for both hard and soft tissues [213–215].

Emulsions are widely used in many fields, including cosmetics, pharmaceuticals and the food industry. However, since emulsions are not thermodynamically stable, they are prone to destabilisation. Maintaining emulsion stability is, therefore, essential and can be achieved by using stabilisers, such as emulsifiers, or by optimising fabrication techniques. Incorporating stabilisers is often the most straightforward and effective choice [213, 215–217]. However, many widely used stabilisers are synthetic and not intended for consumption, which means they often need to be removed following fabrication for food applications. This not only increases processing costs but also presents environmental concerns and complicates regulatory approval, potentially limiting the broader use of emulsions and emulsion templating in the cultivated meat industry. However, edible natural biopolymers and proteins such as pectin, starch, cellulose, xanthan gum or lecithin have emerged as promising alternative stabilisers, thus addressing these limitations [213, 214, 218–

221]. Nonetheless, the choice of stabiliser must be carefully considered, as several emulsifiers commonly used in food have been shown to negatively affect the gut microbiota and are associated with metabolic disorders, including inflammation and glucose intolerance [222, 223].

In the context of cellular agriculture, this technique has begun to show promising applications. For instance, Gao et al. developed a porous scaffold made from pea protein using emulsion templating, which supported the adhesion, growth and moderate differentiation of muscle stem cells derived from the large yellow croaker [224]. Norris et al. applied this technique to produce edible gelatine microcarriers with both smooth and grooved surface features, which supported the proliferation and differentiation of C2C12 cells and bovine satellite muscle cells (bMyoSCs) in suspension culture. Furthermore, they formed a patty using the mature construct, which exhibited browning during cooking, a characteristic feature of the Maillard reaction [115]. However, their approach involved the use of Span 80, an emulsifier shown to significantly disrupt the gut microbiota [222, 223].

4.4.3 Porous Scaffolds Produced Using Gas Foaming

Gas foaming is another promising technique for the fabrication of scaffolds for cultivated meat as it can achieve interconnected porous architectures [177]. A significant advantage of gas foaming is that it does not require the use of toxic organic solvents, which are often used in alternative scaffold fabrication techniques, while offering a simple, cost-effective and potentially scalable approach for the production of porous scaffolds [225, 226]. However, gas foaming is constrained by the need for polymers with suitable rheological properties, which limits the range of materials suitable and the mechanical properties achievable in the resulting scaffolds. Moreover, the process is highly sensitive to specific manufacturing parameters, which can pose challenges for reproducibility and for achieving consistent pore uniformity, cellular distribution and interconnectivity [175, 225–227].

Porous scaffolds based on foamed gelatine and alginate have been successfully applied in tissue engineering and regenerative medicine, suggesting their potential use for cellular agriculture [228–233]. However, to date, no published studies have reported the use of gas foaming to fabricate scaffolds specifically for cultivated meat, seafood or leather sectors. This may reflect several factors, including the limited offering of typical biomaterial choices compatible with foaming processes. Nevertheless, with the development of the cultivated meat industry and the identification of emerging biomaterials, there is an opportunity that gas foaming can become more prevalent. Researchers interested in adopting gas foaming for cultivated meat applications could consider using polysaccharides and plant proteins, derived from sources such as legumes, seeds or cereals, potentially in combination with functional additives like polyphenols. These materials are edible, cost-effective and capable of stabilising foams, making them promising candidates for the development of suitable scaffold structures [234, 235].

4.4.4 Porous Scaffolds Produced Using Sacrificial Template Leaching

Sacrificial template leaching is a widely used method for generating porous scaffolds and is increasingly being explored for applications in cultivated meat and fish. This approach involves embedding a removable space-holding material, such as sugar, sodium chloride (NaCl), calcium carbonate ($CaCO_3$) or paraffin, into a scaffold matrix. Once the matrix is formed, the sacrificial material is removed, typically through solvent leaching, to create a porous structure. Overall, sacrificial templating offers several advantages such as the formation of open, interconnected pores and the ability to control pore size and geometry based on the properties of the sacrificial particles. However, the method often results in random porosity and typically requires a high-volume fraction of the sacrificial material to ensure particle contact and complete removal.

Several sacrificial materials have been investigated for the fabrication of scaffolds. Sugar particles ranging from 50 to 400 μm have been compacted and cast with secalin to produce edible scaffolds. With a cross section of 2 mm, these scaffolds successfully supported the cultivation of porcine adipocytes [236]. Apart from sugar, NaCl is a widely used leachable for tissue engineering applications due to its non-toxicity, broad availability and affordability. For example, NaCl has been used in combination with poly(lactic acid) and nanocellulose to fabricate porous scaffolds for tissue engineering [237]. Calcium carbonate is another interesting candidate for sacrificial templating. As a calcium source, it can simultaneously act as a sacrificial particulate and a source of crosslinking ions for alginate-based scaffolds. For example, Wulf et al. (2022) have successfully grown various cell types (HeLa, HEK293 and L929 cells) on $CaCO_3$-crosslinked alginate scaffolds [238]. Paraffin particles have also been employed to introduce structured porosity in bacterial cellulose, enabling the cultivation of human smooth muscle cells [239].

To improve structural control, sacrificial templating has been combined with additive manufacturing. Three-dimensional printing of carbohydrate glass lattices composed of glucose, sucrose or dextran has enabled the creation of vascular-like networks within gelling matrices such as agarose, alginate and fibrin [240]. In another approach, NaCl-loaded alginate hydrogels have been printed to generate hierarchical porosity, although these constructs have not yet been evaluated for cell compatibility [241]. Commercially, the Zurich-based start-up Sallea has developed a templating strategy by using 3D printed NaCl templates to produce edible, protein-based scaffolds for whole-cut cultivated meat products [75], demonstrating the translational potential of sacrificial templating in food-grade scaffold design.

4.4.5 Textured Vegetable Protein-Derived Scaffolds (TVPs)

Textured vegetable protein-derived scaffolds (TVPs) are gaining significant attention as edible, scalable and food-grade materials for cultivated meat and fish production. Originally developed in the 1960s to replicate the texture and nutritional profile of animal meat using plant proteins [36], TVPs are now being explored as scaffolds that can support cell attachment and tissue formation. Wet extrusion of TVPs typically produces fibrous structures suitable for whole-cut analogues, while dry-extruded TVPs result in porous, sponge-like textures that are particularly attractive for minced meat applications due to their high surface area [215]. Their porosity facilitates cell adhesion and nutrient diffusion, making dry-extruded TVPs especially promising for cultivated meat.

Beyond their structural features, TVPs offer several practical advantages. They are already produced at industrial scale, which supports cost-effective and reproducible manufacturing. Their widespread use as food ingredients may also simplify regulatory approval compared to novel biomaterials. In addition, the diversity of plant protein sources, such as soy, wheat, pea and mung bean, allows for tailoring scaffold properties to specific cell types and product formats. However, several challenges remain. These include limited cell migration in dense or fibrous scaffolds, difficulty achieving cell alignment in dry-extruded formats and suboptimal cell adhesion without surface modification.

To address these limitations, some studies have applied animal-derived coatings such as gelatine or fibrinogen to improve cell attachment [36, 242, 243]. Guan et al. (2023), for example, reported a fourfold increase in cell adhesion on TVPs when using a fibrinogen hydrogel [244]. However, reliance on animal-derived components may undermine the ethical and sustainability goals of cultivated meat and could negatively affect consumer acceptance. Encouragingly, other studies have demonstrated successful cell growth on uncoated TVPs using only foetal bovine serum [245]. This suggests that further optimisation of surface properties, such as reducing hydrophobicity or introducing bioactive motifs, could enhance performance while maintaining a fully plant-based scaffold.

4.5 Decellularised Scaffolds

While scaffold design most commonly involves starting with a biomaterial and building a structure therefrom, an alternative top-down approach is to start with a complete biological structure (of animal or plant origin) and remove cells and genetic material (using chemical, enzymatic or mechanical methods), in a process known as decellularisation [246]. Decellularised tissues and organs (such as cartilage, muscle, heart, liver) have already been used successfully as scaffolds in tissue engineering/regenerative medicine [247], but the lack of standardised processing methods, ethical concerns and risks of pathogenicity and immunogenicity hinder their wider application [248]. Plant-based decellularised materials, on the other hand, bypass the above issues. For example, decellularised plant leaves have been used to produce porous and pre-vascularised scaffolds that were able to support the attachment of endothelial cells along the interior of the vascular structure and of mesenchymal stem cells and induced pluripotent stem cell (iPSC)-derived cardiomyocytes to the exterior [249]. Although most of the decellularised plant scaffolds to date have been applied to the wider field of tissue engineering rather than cultivated meat specifically, they nonetheless show potential for application to the latter, as this pre-existing vasculature, alongside the edibility and sustainable sourcing of plant leaves, could potentially address key challenges for cultivated meat scaffolds [250].

The process of plant decellularisation leaves behind the ECM, which is composed mainly of cellulose (the most abundant natural polymer) [190], hemicellulose and either pectin or lignin [251]. One of the major benefits of these scaffolds is the preservation of the microarchitecture of the original plant, which in turn can give topographical cues and contact guidance to cells [252]. Liu et al. [253] explored the anisotropic structure of decellularised Chinese chive leaves and shitake mushroom stems to promote muscle myogenesis of C2C12 cells, whereas the isotropic nature of decellularised loofah slices enhanced adipogenesis of murine 3T3L1 preadipocytes. Other decellularised plants used as cellular scaffolds in the literature include parsley with C2C12 [200]; banana leaves with zebra fish fibroblast stem cells (ZEM2S) [254]; celery with chicken myoblasts [255]; jackfruit core with porcine myoblasts [256]; white asparagus with C2C12 and porcine adipose-derived stem cells [25]; and algae *Durvillaea antarctica* with C2C12 [81]. It is worth noting that although cellulose, the main component of decellularised plants, is edible, biocompatible and inexpensive, it lacks animal cell adhesion sites, and in most studies (such as those mentioned above), the scaffolds required additional coatings (e.g. gelatine, fibronectin, poly-L-lysine, RGD custom peptides) or extra supplements in the growth media (e.g. Matrigel®) to promote cell attachment. Interestingly, however, Allan and co-workers [250] succeeded in the attachment, proliferation, differentiation and uniaxial alignment of C2C12 cells on decellularised garden-variety grass, relying solely on topographical cues and without the need for functionalisation. Similarly, [257] grew bovine satellite cells and quail myoblasts (QM7) on decellularised agricultural harvest wastes, such as corn husk and jackfruit rind cut in squares, circles and triangles, whereas Jones et al. [77] used just decellularised spinach and bovine satellite cells, although there was a lack of spontaneous cell alignment.

Further drawbacks from the use of decellularised plant scaffolds in the cultivated meat industry arise from the decellularisation process itself. Depending on the method used, there is the possibility of leaving behind cellular residues, introducing chemical odours and damaging the ECM [190]. The most popular method of employing ionic and non-ionic detergents, such as sodium dodecyl sulphate and Triton X-100, is lengthy and uses toxic and hazardous substances [77]. Alternative detergents have been proposed such as replacing Triton X-100 with Polysorbate-20 [258], but they need further evaluation, and so does the nutritional profile of the decellularised plants, following the nutritional losses via the removal of beneficial plant proteins, lipids and polysaccharides [253].

Alternative formulations of decellularised plants and fungi can also exist for food or medical applications, with a prime example being the patent authored by Hickey et al. [259]. The authors describe ways to create 3D structures comprising decellularised plants or fungi from a variety of sources (e.g. cellulose-based, chitin-based, chitosan-based, lignin-based, pectin-based) and edible or non-edible polymeric hydrogels, aerogels or foams acting as carriers. Despite all examples in the patent involving cell culture being relevant to medical applications, the authors claim that their invention can be adapted to suit a food-safe manufacturing process that can be done in a kitchen. The latter is extremely appealing for future developments in the cultivated meat industry, with the aim to decentralise meat production and make it more local to farms, restaurants and even households.

4.6 3D Printed Scaffolds

While random porous scaffolds, such as those produced via electrospinning or foaming, are cost-effective and widely accessible, their inherently disordered pore architecture often impedes cell migration and nutrient diffusion [11, 260]. This limitation restricts tissue formation to thin layers, typically to only a few millimetres in thickness, thereby precluding their use in the production of whole-cut meat analogues [242, 245].

To address these constraints, additive manufacturing technologies, commonly referred to as 3D printing, have emerged as promising alternatives, particularly for the production of structured cultivated meat products due to their ability:

1. To achieve precise spatial control over scaffold geometry, pore size and internal architecture, customised for the targeted biological process
2. To provide textural control of the scaffold (e.g. stiffness, fibrous/anisotropic morphology)
3. To achieve the challenge of generating thick constructs [67]

There are several approaches to 3D printing of scaffolds including *scaffold-first approaches* where the scaffold is printed and subsequently inoculated with cells [261–265] and *bioprinting* which involves the simultaneous deposition of cells and biomaterials [266–272].

Several companies are actively developing 3D printed scaffolds for cultivated meat. Aleph Farms technology involves pioneering multi-material printing for structured steaks.

3D printing technologies can be used on their own. However, often they are combined with other pore-generating techniques in hybrid approaches to create hierarchical porosity across multiple length scales. For example, the integration of 3D printing with freeze-drying has been explored as a method to introduce micro-porosity into already printed scaffolds, to facilitate cell attachment and nutrient diffusion [273, 274]. Another example is the integration of salt leaching in 3D printing to generate microporosity [241]. Such an approach has also been taken by the ETH Zurich spin-off Sallea, which has leveraged indirect printing with salt templates to produce edible scaffolds from food-grade materials [75, 275].

4.6.1 Scaffold-First Extrusion-Based Printed Scaffolds

Among the various 3D printing modalities, extrusion-based printing (EBP), which is a direct printing method, is the most extensively applied in the context of edible scaffolds. This technique involves the layer-by-layer deposition of a viscoelastic ink through a nozzle, guided by a computer-controlled path [262, 276]. Alongside a relatively simple process set-up, this method offers material flexibility, as it is largely independent of the ink chemistry unlike other printing methods that often require specific chemical formulations. Instead, it relies heavily on the rheological properties of the material. For cultivated meat production, shear-thinning hydrogels enriched with plant proteins can provide a viable platform for the design of 3D printed scaffolds, for example, successfully grown C2C12 and ZEM2S cells on an alginate-based EBP scaffold containing 15 wt% plant protein powder (pea or soy protein isolate) [210].

Despite its versatility, extrusion-based printing is constrained by resolution limits. To optimise printing resolution, certain materials may be used to tune the rheological properties of protein-loaded inks and thus to optimise resolution [277]. Examples of these materials could include polysaccharides, such as pectin [262], konjac glucomannan and xanthan gum [277]. Alternatively, electrohydrodynamic printing (EHDP) can also be employed. EHDP combines principles of extrusion and near-field electrospinning, using a high-voltage electric field to draw fine fibres from a polymer solution and placing them in a controlled manner on a substrate. This technique enables the fabrication of micro- to nanoscale fibres with morphologies reminiscent of the extracellular matrix (ECM). For example, EHDP was used to produce edible scaffolds from polycaprolactone (PCL) blended with plant-derived proteins such as zein, which were found to support the proliferation of fibroblasts and myoblasts [264, 278]. However, EHDP remains limited by its low throughput, with printing times exceeding 40 min for modestly sized constructs (6 cm × 6 cm × 0.5 mm (24 layers)).

Another challenge that often occurs in EBP is the limited stability of a scaffold at larger printing volumes. To mitigate gravitational collapse during printing, support-bath strategies have been developed, wherein the ink is deposited into a viscous medium, such as agar gel, thus providing a temporary mechanical support [261]. This approach has been adopted commercially, by Aleph Farms, to fabricate multi-material, steak-like scaffolds with spatially defined mechanical gradients [274].

4.6.2 Bioprinted Scaffolds

While scaffold-first approaches offer structural precision and material flexibility, they require a separate cell inoculation step, which can result in a heterogeneous cell distribution and integration within the scaffold. In contrast, bioprinting enables the simultaneous deposition of cells and biomaterials, combining them into a printable bioink that is solidified in layers into defined architectures [279]. This approach introduces additional complexity as the bioink must not only be printable and edible for cultivated meat, but also support cell viability, differentiation and maturation during and after the bioprinting process. Moreover, the printing conditions must be gentle enough to preserve cell integrity, yet robust enough to yield constructs with desirable mechanical and textural properties [260]. Many bioprinting strategies have been adapted from tissue engineering, where the goal is to replicate functional tissues. However, cultivated meat applications differ in that the tissue does not need to be physiologically functional, but all components must be food-safe and palatable.

Multicellular bioprinting strategies which involve the co-printing or post-assembly of multiple cell types are increasingly explored as they can help replicate the complex composition of whole cuts. While promising, such approaches face challenges in co-cultivation, as different cell types often require distinct culture conditions. As an alternative, tissues may be printed and matured separately and fused only post-harvest. For instance, Kang et al. (2021) demonstrated the separate bioprinting and subsequent assembly of muscle and fat tissues using enzymatic crosslinking (e.g. transglutaminase) to form integrated constructs [280].

Despite significant progress, many current bioinks rely on synthetic polymers (e.g. PEGDA) [272] or animal-derived components (e.g. gelatine, fibrinogen) [270, 281], which are unsuitable for food applications. The development of fully edible, plant-based bioinks that meet the mechanical, rheological and biological requirements of bioprinting remains a key research priority. As the field advances, integrating insights from tissue engineering with food-grade material science will be essential to realise the full potential of bioprinting for cultivated meat production.

Similarly to scaffold-first approaches, extrusion bioprinting allows for spatial control over cell types, densities and matrix compositions, which is essential for mimicking the heterogeneous structure of traditional meat and fish. Notably, the freeform reversible embedding of suspended hydrogels (FRESH) method developed by Shiwarski et al. (2021) enables printing within a support bath, which stabilises soft bioinks during deposition [282]. Similarly, Kang et al. (2021) demonstrated the fabrication of multi-tissue constructs by printing distinct fibres in a support medium, followed by post-printing assembly [280]. However, additional challenges are faced particularly as shear stresses during printing can induce cell alignment [283] but may also compromise cell viability if not carefully controlled.

Light-based bioprinting techniques, such as stereolithography and digital light processing, offer even higher resolution and design flexibility. However, their reliance on photopolymerisable monomers, many of which are not food-safe, has so far restricted their use in cultivated meat applications. For example, Jeong et al. (2022) employed a gelatine methacryloyl (GelMA)-based bioink containing bovine embryonic fibroblasts, which were directed towards myogenic and adipogenic lineages [272]. While GelMA is widely used in biomedical applications due to its low toxicity and tuneable properties, it is not considered edible, inhibiting its applicability in food applications [268].

5 Conclusion

Scaffolds play a vital role in scaling up the cultivated meat production by supporting cell adhesion, proliferation and differentiation of adherent cell types. However, since cultivated meat is intended for consumption, the use of edible scaffolds introduces considerations in addition to biological compatibility. These include their potential to contribute to the texture, structure and nutritional profile of the final food product while also ensuring cost-effectiveness and compliance with regulatory and consumer safety standards. The variability in scaffold material properties, morphologies and fabrication methods means that there is no one-size-fits-all solution. Instead, scaffold design must be tailored to the bioprocessing approaches and cell types used, as well as the product format (e.g. structured vs. unstructured meat).

A wide array of promising materials and strategies have emerged in recent years from plant-based proteins and mycelium to recombinant and hybrid approaches. Scaffold morphologies, such as microcarriers, hydrogel constructs, fibrous mats, porous foams and decellularised materials, each offer unique advantages at different stages of the production process and at different scales. Nevertheless, several challenges remain, including scalability, ensuring food safety, achieving suitable mechanical properties and reproducing the complex architecture of whole cuts. As the field of cultivated meat progresses, the standardisation of scaffold testing protocols and comprehensive characterisation of scaffold properties will be essential to accelerate innovation and enable comparison between cultivated and conventional meat products. Ultimately, the optimisation of edible scaffold technologies will be key to advancing from proof of concept to commercial viability and widespread adoption of cultivated meat.

References

1. Langer R, Tirrell DA (2004) Designing materials for biology and medicine. Nature 428(6982): 487–492
2. Dhandayuthapani B, Yoshida Y, Maekawa T, Kumar DS (2011) Polymeric scaffolds in tissue engineering application: a review. Int J Polym Sci 2011:1–19
3. Rubio NR, Xiang N, Kaplan DL (2020) Plant-based and cell-based approaches to meat production. Nat Commun 11:6276
4. BBC (2020) Singapore approves lab-grown ‘chicken’ meat. BBC News [cited 2025 Jul 30]. https://www.bbc.com/news/business-55155741
5. U.S. Food and Drug Administration (2024) FDA completes second pre-market consultation for human food made using animal cell culture technology. FDA [cited 2025 Jul 31]. https://www.fda.gov/food/hfp-constituent-updates/fda-completes-second-pre-market-consultation-human-food-made-using-animal-cell-culture-technology
6. Post MJ (2014) Cultured beef: medical technology to produce food. J Sci Food Agric 94(6): 1039–1041
7. ANZFSA (2025) A1269 – Cultured quail as a novel food. Food Standards Australia New Zealand [cited 2025 Jul 31]. https://www.foodstandards.gov.au/food-standards-code/applications/A1269-Cultured-Quail-as-a-Novel-Food

8. BBC (2024) Lab-grown meat in pet food: UK to allow sales of meatly product. BBC News [cited 2025 Jul 31]. https://www.bbc.com/news/articles/c19k0ky9v4yo
9. Allen J, Ramsden M, Nisar S (2024) Skeletal muscle structure, function and pathology. Orthop Trauma 38(3):137–144
10. Ministry of Health Israel (2024) First in the world: the Ministry of Health has approved cattle-based cultivated meat Ministry of Health [cited 2025 Jul 31]. https://www.gov.il/en/pages/17012024-02
11. Mariano E, Lee DY, Lee J, Choi Y, Park J, Han D et al (2025) A review on the characterization of edible scaffolds for cultured meat: physical, chemical, biocompatibility, and food safety evaluation methods. Food Chem 469:142493
12. Lanzoni D, Rebucci R, Formici G, Cheli F, Ragone G, Baldi A et al (2024) Cultured meat in the European Union: legislative context and food safety issues. Curr Res Food Sci 8:100722
13. Heo S, Lee G, Jeong DW (2025) A study on safety evaluation system of cultured foods among alternative proteins. Food Sci Biotechnol 34(2):365–371
14. Wells R (2008) The role of matrix stiffness in regulating cell behavior. Hepatology 47:1394
15. Engler AJ, Griffin MA, Sen S, Bönnemann CG, Sweeney HL, Discher DE (2004) Myotubes differentiate optimally on substrates with tissue-like stiffness: pathological implications for soft or stiff microenvironments. J Cell Biol 166(6):877–887
16. Chaturvedi V, Naskar D, Kinnear B, Grenik E, Dye D, Grounds M et al (2017) Silk fibroin scaffolds with muscle-like elasticity support in vitro differentiation of human skeletal muscle cells. J Tissue Eng Regen Med 11:3178–3192
17. Xiang S, Zhong L, Fritch M, Li L, Velankar S, Liu Y et al (2021) Caveolin-1 mediates soft scaffold-enhanced adipogenesis of human mesenchymal stem cells. Stem Cell Res Ther 12:347
18. Xie J, Zhang D, Zhou C, Yuan Q, Ling Y, Zhou X (2018) Substrate elasticity regulates adipose-derived stromal cell differentiation towards osteogenesis and adipogenesis through β-catenin transduction. Acta Biomater 79:83–95
19. Cantini M, Sousa M, Moratal D, Manocd J, Salmerón-Sánchez M (2012) Non-monotonic cell differentiation pattern on extreme wettability gradients. Biomater Sci 1(2):202–212
20. Feng T, Ceroni L, Tromp L, Siebenmorgen C, Casalini S, Menna E et al (2025) Functionalized multi-walled carbon nanotube enhanced myogenic differentiation for aligned topography-induced skeletal muscle engineering. Small 21:e2504992
21. Aviss KJ, Gough JE, Downes S (2010) Aligned electrospun polymer fibres for skeletal muscle regeneration. Eur Cells Mater 19:193–204
22. Choi JS, Lee SJ, Christ G, Atala A, Yoo. (2008) The influence of electrospun aligned poly (3-caprolactone)/collagen nanofiber meshes on the formation of self-aligned skeletal muscle myotubes. Biomaterials 29:2899–2906
23. Qi L, Zhang F, Wang K, Chen B, Li X, Xu J et al (2025) Advancements in skeletal muscle tissue engineering: strategies for repair and regeneration of skeletal muscle beyond self-repair. Regene Biomat 12:rbaf050
24. Tejo-Otero A, Fenollosa-Artés F, Achaerandio I, Rey-Vinoles S, Buj-Corral I, Mateos-Timoneda M et al (2022) Soft-tissue-mimicking using hydrogels for the development of phantoms. Gels 8(1):40
25. Murugan P, Yap WS, Ezhilarasu H, Suntornnond R, Le QB, Singh S et al (2024) Decellularised plant scaffolds facilitate porcine skeletal muscle tissue engineering for cultivated meat biomanufacturing. npj Sci Food 8(1):25
26. Gumbiner BM (2005) Regulation of cadherin-mediated adhesion in morphogenesis. Nat Rev Mol Cell Biol 6(8):622–634
27. Bodiou V, Moutsatsou P, Post MJ (2020) Microcarriers for upscaling cultured meat production. Front Nutr 7:10
28. Hung M, Lo HF, Beckmann AG, Demircioglu D, Damle G, Hasson D et al (2024) Cadherin-dependent adhesion is required for muscle stem cell niche anchorage and maintenance. Development 151(7):dev202387

29. Bajaj P, Schweller RM, Khademhosseini A, West JL, Bashir R (2014) 3D biofabrication strategies for tissue engineering and regenerative medicine. Annu Rev Biomed Eng 16(1): 247–276
30. Chaturvedi V, Dye D, Kinnear B, van Kuppevelt T, Grounds M, Coombe D (2015) Interactions between skeletal muscle myoblasts and their extracellular matrix revealed by a serum free culture system. PLoS One 10:e0127675
31. Cai S, Wu C, Yang W, Liang W, Yu H, Liu L (2020) Recent advance in surface modification for regulating cell adhesion and behaviors. Nanotechnol Rev 9(1):971–989
32. D'Souza SE, Ginsberg MH, Plow EF (1991) Arginyl-glycyl-aspartic acid (RGD): a cell adhesion motif. Trends Biochem Sci 16:246–250
33. Tashiro K, Sephel GC, Weeks B, Sasaki M, Martin GR, Kleinman HK et al (1989) A synthetic peptide containing the IKVAV sequence from the A chain of laminin mediates cell attachment, migration, and neurite outgrowth. J Biol Chem 264(27):16174–16182
34. Leal-Egaña A, Scheibel T (2012) Interactions of cells with silk surfaces. J Mater Chem 22(29): 14330–14336
35. Post MJ, Levenberg S, Kaplan DL, Genovese N, Fu J, Bryant CJ et al (2020) Scientific, sustainability and regulatory challenges of cultured meat. Nat Food 1(7):403–415
36. Levi S, Yen FC, Baruch L, Machluf M (2022) Scaffolding technologies for the engineering of cultured meat: towards a safe, sustainable, and scalable production. Trends Food Sci Technol 126:13–25
37. Mellon J (2020) Moo's law an investor's guide to the new agrarian revolution. Fruitful Publications
38. Geckil H, Xu F, Zhang X, Moon S, Demirci U (2010) Engineering hydrogels as extracellular matrix mimics. Nanomedicine 5(3):469–484
39. Gyles DA, Castro LD, Silva JOC, Ribeiro-Costa RM (2017) A review of the designs and prominent biomedical advances of natural and synthetic hydrogel formulations. Eur Polym J 88:373–392
40. Ho TC, Chang CC, Chan HP, Chung TW, Shu CW, Chuang KP et al (2022) Hydrogels: properties and applications in biomedicine. Molecules 27(9):2902
41. Mano J, Silva G, Azevedo H, Malafaya P, Sousa R, Silva S et al (2007) Natural origin biodegradable systems in tissue engineering and regenerative medicine: present status and some moving trends. J R Soc Interface 4:999–1030
42. Huang J, Zhang Z, Jiang H (2023) Edible hydrogels with shrinkage tolerance in acids and stomach-friendly mechanical moduli. Appl Mater Today 32:101786
43. Le Y, Li H, Liao X, Wu Y, Zhang M, Jiang Y et al (2024) Edible hydrogel with dual network structure for weight management. Food Res Int 190:114560
44. Moxon SR, Corbett NJ, Fisher K, Potjewyd G, Domingos M, Hooper NM (2019) Blended alginate/collagen hydrogels promote neurogenesis and neuronal maturation. Mater Sci Eng C 104:109904
45. Rashid AB, Showva NN, Hoque ME (2023) Gelatin-based scaffolds: an intuitive support structure for regenerative therapy. Curr Opin Biomed Eng 26:100452
46. Zarrintaj P, Manouchehri S, Ahmadi Z, Saeb MR, Urbanska AM, Kaplan DL et al (2018) Agarose-based biomaterials for tissue engineering. Carbohydr Polym 187:66–84
47. Yegappan R, Selvaprithiviraj V, Amirthalingam S, Jayakumar R (2018) Carrageenan based hydrogels for drug delivery, tissue engineering and wound healing. Carbohydr Polym 198: 385–400
48. Manigandan V, Karthik R, Ramachandran S, Rajagopal S (2018) Chapter 15 – Chitosan applications in food industry. In: Grumezescu AM, Holban AM (eds) Biopolymers for food design. Academic Press, pp 469–491. https://www.sciencedirect.com/science/article/pii/B9780128114490000153
49. Sahoo DR, Biswal T (2021) Alginate and its application to tissue engineering. SN Appl Sci 3(1):30

50. Ng S, Kurisawa M (2021) Integrating biomaterials and food biopolymers for cultured meat production. Acta Biomater 124:108–129
51. Hu Y, Li Y, Xu FJ (2017) Versatile functionalization of polysaccharides via polymer grafts: from design to biomedical applications. Acc Chem Res 50(2):281–292
52. Lih E, Oh SH, Joung YK, Lee JH, Han DK (2015) Polymers for cell/tissue anti-adhesion. Prog Polym Sci 44:28–61
53. Chen S, Cui S, Zhang H, Pei X, Hu J, Zhou Y et al (2018) Cross-linked pectin nanofibers with enhanced cell adhesion. Biomacromolecules 19(2):490–498
54. Rodríguez-Vázquez M, Vega-Ruiz B, Ramos-Zúñiga R, Saldaña-Koppel DA, Quiñones-Olvera LF (2015) Chitosan and its potential use as a scaffold for tissue engineering in regenerative medicine. Biomed Res Int 2015:1–15
55. Erdoğmuş SF, Altıntaş ÖE, ÇefiK S. (2023) Production of fungal chitosan and fabrication of fungal chitosan/polycaprolactone electrospun nanofibers for tissue engineering. Microsc Res Techn 86(10):1309–1321
56. Enrione J, Blaker J, Brown D, Weinstein-Oppenheimer C, Pepczynska M, Olguín Y et al (2017) Edible scaffolds based on non-mammalian biopolymers for myoblast growth. Materials 10(12):1404
57. Hsueh YS, Chen YS, Tai HC, Mestak O, Chao SC, Chen YY et al (2017) Laminin-alginate beads as preadipocyte carriers to enhance adipogenesis *in vitro* and *in vivo*. Tissue Eng Part A 23(5–6):185–194
58. Alheib O, Da Silva LP, Da Silva Morais A, Mesquita KA, Pirraco RP, Reis RL et al (2022) Injectable laminin-biofunctionalized gellan gum hydrogels loaded with myoblasts for skeletal muscle regeneration. Acta Biomater 143:282–294
59. Kokova V, Lukova P, Baldzhieva A, Katsarov P, Delattre C, Molinié R et al (2023) Extraction, structural characterization, and in vivo anti-inflammatory effect of alginate from Cystoseira crinita (Desf.) Borry harvested in the Bulgarian Black Sea. Mar Drugs 21(4):245
60. Rahman MM, Shahid MA, Hossain MT, Sheikh MS, Rahman MS, Uddin N et al (2024) Sources, extractions, and applications of alginate: a review. Discov Appl Sci 6(8):443
61. Borsani B, De Santis R, Perico V, Penagini F, Pendezza E, Dilillo D et al (2021) The role of carrageenan in inflammatory bowel diseases and allergic reactions: where do we stand? Nutrients 13(10):3402
62. Fong D, Hoemann CD (2018) Chitosan immunomodulatory properties: perspectives on the impact of structural properties and dosage. Future Sci OA 4(1):FSO225
63. Ghattas M, Dwivedi G, Chevrier A, Horn-Bourque D, Alameh MG, Lavertu M (2025) Chitosan immunomodulation: insights into mechanisms of action on immune cells and signaling pathways. RSC Adv 15(2):896–909
64. Wagener J, Wang X, Becker KL, Aimanianda V, Valsecchi I, Gresnigt MS et al (2025) Immunomodulatory function of chitosan is dependent on complement receptor 3. Cell Surf 14: 100146
65. Brovold M, Almeida JI, Pla-Palacín I, Sainz-Arnal P, Sánchez-Romero N, Rivas JJ et al (2018) Naturally-derived biomaterials for tissue engineering applications. In: Chun HJ, Park K, Kim CH, Khang G (eds) Novel biomaterials for regenerative medicine. Springer, Singapore, pp 421–49 [cited 2025 Jul 30]. http://link.springer.com/10.1007/978-981-13-0947-2_23
66. Jie IWK, Lee KWA, Yoon SE, Song JK, Chan LKW, Lee CH et al (2025) Advancements in clinical utilization of recombinant human collagen: an extensive review. Life 15(4):582
67. MacQueen LA, Alver CG, Chantre CO, Ahn S, Cera L, Gonzalez GM et al (2019) Muscle tissue engineering in fibrous gelatin: implications for meat analogs. npj Sci Food 3(1):20
68. Basurto IM, Mora MT, Gardner GM, Christ GJ, Caliari SR (2021) Aligned and electrically conductive 3D collagen scaffolds for skeletal muscle tissue engineering. Biomater Sci 9(11): 4040–4053
69. Basurto IM, Bandara GC, Boudreau RD, Shriver SB, Muhammad SA, Christ GJ et al (2025) Freeze-dried porous collagen scaffolds for the repair of volumetric muscle loss injuries. ACS Biomater Sci Eng 11(3):1598–1611

70. Kroehne V, Heschel I, Schügner F, Lasrich D, Bartsch JW, Jockusch H (2008) Use of a novel collagen matrix with oriented pore structure for muscle cell differentiation in cell culture and in grafts. J Cell Mol Med 12(5a):1640–1648
71. Dong M, Tian L, Li J, Jia J, Dong Y, Tu Y et al (2022) Improving physicochemical properties of edible wheat gluten protein films with proteins, polysaccharides and organic acid. LWT 154:112868
72. Xiang N, Yao Y, Yuen JSK, Stout AJ, Fennelly C, Sylvia R et al (2022) Edible films for cultivated meat production. Biomaterials 287:121659
73. Xiang N, Yuen JSK, Stout AJ, Rubio NR, Chen Y, Kaplan DL (2022) 3D porous scaffolds from wheat glutenin for cultured meat applications. Biomaterials 285:121543
74. Banerjee A, Martens ME, Liira R, Neivelt KI, Moelder L, Howlader MM et al (2025) Plant-based microfibrous scaffolds for culturing meat. US20250059513A1 [cited 2025 May 7]. https://patents.google.com/patent/US20250059513A1/en
75. Kleger N, Fehlmann S, Lee SS, Dénéréaz C, Cihova M, Paunović N et al (2022) Light-based printing of leachable salt molds for facile shaping of complex structures. Adv Mater 34(32): 2203878
76. Elfenbein A, Kolbeck JL, Larson B (2020) Synthetic food compositions. WO2020123876A1
77. Jones JD, Rebello AS, Gaudette GR (2021) Decellularized spinach: an edible scaffold for laboratory-grown meat. Food Biosci 41:100986
78. Andrew P, Anna C, Cesar T, Ryan H, Michelle B-M, Natalie C et al (2025) Plant-based or fungal-based micro-carrier scaffolds optimized for large-scale cell-culture production and other applications. WO; WO 2025/059779 A1. https://lens.org/155-112-707-128-962
79. Park S, Lee M, Jung S, Lee H, Choi B, Choi M et al (2024) Rice grains integrated with animal cells: a shortcut to a sustainable food system. Matter 7(3):1292–1313
80. Espinosa-Ramírez J, Mondragón-Portocarrero AC, Rodríguez JA, Lorenzo JM, Santos EM (2023) Algae as a potential source of protein meat alternatives. Front Nutr:10 [cited 2025 Jul 31]. https://www.frontiersin.org/journals/nutrition/articles/10.3389/fnut.2023.1254300/full
81. Villanueva A, Sanchez FS, Gerdtzen Z, Blanchard K, Barroso P, Brain S (2023) Methods for creating edible scaffolding using Chilean seaweed (Cochayuyo). WO; WO 2023/102083 A1. https://lens.org/077-309-322-936-514
82. Wall I, Sinden J, Petra H, Miguel F, Debora O (2023) Methods and products for meat cell culture. GB, WO; WO 2024/100401 A1. https://lens.org/090-972-751-138-109
83. Ruggeri M, Miele D, Contardi M, Vigani B, Boselli C, Icaro Cornaglia A et al (2023) Mycelium-based biomaterials as smart devices for skin wound healing. Front Bioeng Biotechnol 11. https://www.frontiersin.org/journals/bioengineering-and-biotechnology/articles/10.3389/fbioe.2023.1225722
84. Antinori ME, Contardi M, Suarato G, Armirotti A, Bertorelli R, Mancini G et al (2021) Advanced mycelium materials as potential self-growing biomedical scaffolds. Sci Rep 11(1): 12630
85. Maini Rekdal V, van der Luijt CRB, Chen Y, Kakumanu R, Baidoo EEK, Petzold CJ et al (2024) Edible mycelium bioengineered for enhanced nutritional value and sensory appeal using a modular synthetic biology toolkit. Nat Commun 15(1):2099
86. Li Y, Liu W, Li S, Zhang M, Yang F, Wang S (2021) Porcine skeletal muscle tissue fabrication for cultured meat production using three-dimensional bioprinting technology. J Future Foods 1(1):88–97
87. Dagès BAS, Fabian JA, Polakova D, Rysova M, Topham PD, Souppez JBRG et al (2025) Edible electrospun materials for scalable cultivated beef production. Food Bioprod Process 149:118–129
88. Wolff A, Frank M, Staehlke S, Peters K (2022) A comparative study on the adipogenic differentiation of mesenchymal stem/stromal cells in 2D and 3D culture. Cells 11(8):1313
89. Endo K, Sato T, Umetsu A, Watanabe M, Hikage F, Ida Y et al (2023) 3D culture induction of adipogenic differentiation in 3T3-L1 preadipocytes exhibits adipocyte-specific molecular expression patterns and metabolic functions. Heliyon 9(10):e20713

90. Hanga MP, Holdich RG (2014) Membrane emulsification for the production of uniform poly-N-isopropylacrylamide-coated alginate particles using internal gelation. Chem Eng Res Des 92(9):1664–1673
91. Falcone G, Schrüfer S, Kuth S, Mazzei P, De Pasquale S, Del Gaudio P et al (2024) Ready-to-print alginate inks: the effect of different divalent cations on physico-chemical properties of 3D printable alginate hydrogels. Carbohydr Polym Technol Appl 7:100524
92. Wang L, Li G, Li X, Zhang Y, Liu G, Xie M et al (2025) Emerging materials in cultivated meat: engineering sustainable food solutions – a review. Adv Funct Mater 35(3): 2413316
93. Zdiri K, Cayla A, Elamri A, Erard A, Salaun F (2022) Alginate-based bio-composites and their potential applications. Funct Biomater 13(3):117
94. Azizi H, Arash K, Ghorani B (2023) Structural elucidation of Gluten/Zein nanofibers prepared by electrospinning process: focus on the effect of zein on properties of nanofibers. Polym Test 128:108231
95. Kariduraganavar M, Kittur A, Kamble R (2014) Polymer synthesis and processing. In: Natural and synthetic biomedical polymers. Elsevier, Burlington, MA, San Diego, CA
96. Badenes SM, Fernandes-Platzgummer A, Rodrigues CAV, Diogo MM, da Silva CL, Cabral JMS (2016) Chapter 4 – Microcarrier culture systems for stem cell manufacturing. In: Cabral JMS, Lobato de Silva C, Chase LG, Margarida Diogo M (eds) Stem cell manufacturing. Elsevier, Boston, pp 77–104 [cited 2025 May 21]. https://www.sciencedirect.com/science/article/pii/B9780444632654000042
97. Hanga MP, Ali J, Moutsatsou P, De La Raga FA, Hewitt CJ, Nienow A et al (2020) Bioprocess development for scalable production of cultivated meat. Biotechnol Bioeng 117(10): 3029–3039
98. Hanga MP, De La Raga FA, Moutsatsou P, Hewitt CJ, Nienow AW, Wall I (2021) Scale-up of an intensified bioprocess for the expansion of bovine adipose-derived stem cells (bASCs) in stirred tank bioreactors. Biotechnol Bioeng 118(8):3175–3186
99. Li B, Wang X, Wang Y, Gou W, Yuan X, Peng J et al (2015) Past, present, and future of microcarrier-based tissue engineering. J Orthop Transl 3(2):51–57
100. Badenes SM, Fernandes TG, Rodrigues CAV, Diogo MM, Cabral JMS (2014) Scalable expansion of human-induced pluripotent stem cells in xeno-free microcarriers. In: Turksen K (ed) Stem cells and good manufacturing practices. Springer, New York, pp 23–9 [cited 2025 Jul 30]. https://link.springer.com/10.1007/7651_2014_106
101. Gupta P, Ismadi MZ, Verma PJ, Fouras A, Jadhav S, Bellare J et al (2016) Optimization of agitation speed in spinner flask for microcarrier structural integrity and expansion of induced pluripotent stem cells. Cytotechnology 68(1):45–59
102. Rafiq QA, Hanga MP, Heathman TRJ, Coopman K, Nienow AW, Williams DJ et al (2017) Process development of human multipotent stromal cell microcarrier culture using an automated high-throughput microbioreactor. Biotechnol Bioeng 114(10):2253–2266
103. Lawson T, Kehoe DE, Schnitzler AC, Rapiejko PJ, Der KA, Philbrick K et al (2017) Process development for expansion of human mesenchymal stromal cells in a 50L single-use stirred tank bioreactor. Biochem Eng J 120:49–62
104. Verbruggen S, Luining D, van Essen A, Post MJ (2018) Bovine myoblast cell production in a microcarriers-based system. Cytotechnology 70(2):503–512
105. Rafiq QA, Ruck S, Hanga MP, Heathman TRJ, Coopman K, Nienow AW et al (2018) Qualitative and quantitative demonstration of bead-to-bead transfer with bone marrow-derived human mesenchymal stem cells on microcarriers: utilising the phenomenon to improve culture performance. Biochem Eng J 135:11–21
106. Chen XY, Chen JY, Tong XM, Mei JG, Chen YF, Mou XZ (2020) Recent advances in the use of microcarriers for cell cultures and their ex vivo and in vivo applications. Biotechnol Lett 42(1):1–10

107. Hanga MP, Nienow AW, Murasiewicz H, Pacek AW, Hewitt CJ, Coopman K (2021) Expansion of human mesenchymal stem/stromal cells on temporary liquid microcarriers. J Chem Technol Biotechnol 96(4):930–940
108. Liu Y, Wang R, Ding S, Deng L, Zhang Y, Li J et al (2022) Engineered meatballs via scalable skeletal muscle cell expansion and modular micro-tissue assembly using porous gelatin microcarriers. Biomaterials 287:121615
109. Tzimorotas D, Solberg NT, Andreassen RC, Moutsatsou P, Bodiou V, Pedersen ME et al (2023) Expansion of bovine skeletal muscle stem cells from spinner flasks to benchtop stirred-tank bioreactors for up to 38 days. Front Nutr 10:1192365
110. Yen FC, Glusac J, Levi S, Zernov A, Baruch L, Davidovich-Pinhas M et al (2023) Cultured meat platform developed through the structuring of edible microcarrier-derived microtissues with oleogel-based fat substitute. Nat Commun 14(1):2942
111. Bodiou V, Cristini N, De Cristofaro L, Pareek T, Rajagopal V, Verrougstraete L et al (2025) Process intensification of cultivated meat production through microcarrier addition strategy optimisation. Sci Rep 15(1):14080
112. Nienow AW, Hewitt CJ, Heathman TRJ, Glyn VAM, Fonte GN, Hanga MP et al (2016) Agitation conditions for the culture and detachment of hMSCs from microcarriers in multiple bioreactor platforms. Biochem Eng J 108:24–29
113. Fish KD, Rubio NR, Stout AJ, Yuen JSK, Kaplan DL (2020) Prospects and challenges for cell-cultured fat as a novel food ingredient. Trends Food Sci Technol 98:53–67
114. Zhang P, Zhao X, Zhang S, Li G, Midgley AC, Fang Y et al (2024) The important role of cellular mechanical microenvironment in engineering structured cultivated meat: recent advances. Curr Res Food Sci 9:100865
115. Norris SCP, Kawecki NS, Davis AR, Chen KK, Rowat AC (2022) Emulsion-templated microparticles with tunable stiffness and topology: applications as edible microcarriers for cultured meat. Biomaterials 287:121669
116. Hopfner U, Schenck TL, Chávez MN, Machens HG, Bohne AV, Nickelsen J et al (2014) Development of photosynthetic biomaterials for in vitro tissue engineering. Acta Biomater 10(6):2712–2717
117. Wu JY, Tso R, Teo HS, Haldar S (2023) The utility of algae as sources of high value nutritional ingredients, particularly for alternative/complementary proteins to improve human health. Front Nutr 10:1277343
118. Ogawa M, Kermani AS, Huynh MJ, Baar K, Leach JK, Block DE (2024) Edible mycelium as proliferation and differentiation support for anchorage-dependent animal cells in cultivated meat production. npj Sci Food 8(1):23
119. Alim A, Yang C, Song H, Liu Y, Zou T, Zhang Y et al (2019) The behavior of umami components in thermally treated yeast extract. Food Res Int 120:534–543
120. Teo YX, Lee KY, Goh CJH, Wang LC, Sobota RM, Chiam KH et al (2023) Fungus derived protein particles as cell-adhesive matrices for cell-cultivated food. npj Sci Food 7(1):34
121. Zhu P, Huang W, Guo X, Chen L (2021) Strong and elastic pea protein hydrogels formed through pH-shifting method. Food Hydrocoll 117:106705
122. McClements DJ (2024) Composite hydrogels assembled from food-grade biopolymers: fabrication, properties, and applications. Adv Colloid Interf Sci 332:103278
123. Hennink WE, van Nostrum CF (2012) Novel crosslinking methods to design hydrogels. Adv Drug Deliv Rev 64:223–236
124. Wollschlaeger JO, Maatz R, Albrecht FB, Klatt A, Heine S, Blaeser A et al (2022) Scaffolds for cultured meat on the basis of polysaccharide hydrogels enriched with plant-based proteins. Gels 8(2):94
125. Li S, Li H, Shang X, He J, Hu Y (2023) Recent advances in 3D printing sacrificial templates for fabricating engineered vasculature. MedComm Biomater Appl 2(3):e46
126. Najaf Zadeh H, Huber T, Nock V, Fee C, Clucas D (2020) Complex geometry cellulose hydrogels using a direct casting method. Bioengineering 7(2):58

127. Xie Y, Lee K, Wang X, Yoshitomi T, Kawazoe N, Yang Y et al (2021) Interconnected collagen porous scaffolds prepared with sacrificial PLGA sponge templates for cartilage tissue engineering. J Mater Chem B 9(40):8491–8500
128. Furuhashi M, Morimoto Y, Shima A, Nakamura F, Ishikawa H, Takeuchi S (2021) Formation of contractile 3D bovine muscle tissue for construction of millimetre-thick cultured steak. npj Sci Food 5(1):6
129. Park S, Jung S, Heo J, Koh WG, Lee S, Hong J (2021) Chitosan/cellulose-based porous nanofilm delivering C-phycocyanin: a novel platform for the production of cost-effective cultured meat. ACS Appl Mater Interfaces 13(27):32193–32204
130. Jaques A, Sánchez E, Orellana N, Enrione J, Acevedo CA (2021) Modelling the growth of in-vitro meat on microstructured edible films. J Food Eng 307:110662
131. Chen Y, Bassey AP, Zhu H, Zhou G (2023) Fabrication of cell cultured meat by hydrogel with topographic microstructures. Food Biosci 55:102910
132. Acevedo CA, Orellana N, Avarias K, Ortiz R, Benavente D, Prieto P (2018) Micropatterning technology to design an edible film for in vitro meat production. Food Bioprocess Technol 11(7):1267–1273
133. Orellana N, Sánchez E, Benavente D, Prieto P, Enrione J, Acevedo CA (2020) A new edible film to produce in vitro meat. Foods 9(2):185
134. Yao Y, Yuen Jr JSK, Sylvia R, Fennelly C, Cera L, Zhang KL et al (2024) Cultivated meat from aligned muscle layers and adipose layers formed from glutenin films. ACS Biomater Sci Eng 10(2):814–824
135. Li CH, Yang IH, Ke CJ, Chi CY, Matahum J, Kuan CY et al (2022) The production of fat-containing cultured meat by stacking aligned muscle layers and adipose layers formed from gelatin-soymilk scaffold. Front Bioeng Biotechnol 10. https://www.frontiersin.org/journals/bioengineering-and-biotechnology/articles/10.3389/fbioe.2022.875069
136. Santos AEAD, Cotta T, Santos JPF, Camargos JSF, Carmo ACCD, Alcântara EGA et al (2023) Bioactive cellulose acetate nanofiber loaded with annatto support skeletal muscle cell attachment and proliferation. Front Bioeng Biotechnol 11:1116917
137. Jeong D, Jang G, Jung WK, Park YH, Bae H (2024) Stretchable zein-coated alginate fiber for aligning muscle cells to artificially produce cultivated meat. npj Sci Food 8(1):13
138. Christopherson GT, Song H, Mao HQ (2009) The influence of fiber diameter of electrospun substrates on neural stem cell differentiation and proliferation. Biomaterials 30(4):556–564
139. Bosch-Rué E, Delgado LM, Gil FJ, Perez RA (2021) Direct extrusion of individually encapsulated endothelial and smooth muscle cells mimicking blood vessel structures and vascular native cell alignment. Biofabrication 13(1):015003
140. Daristotle JL, Behrens AM, Sandler AD, Kofinas P (2016) A review of the fundamental principles and applications of solution blow spinning. ACS Appl Mater Interfaces 8(51): 34951–34963
141. Luo CJ, Stoyanov SD, Stride E, Pelan E, Edirisinghe M (2012) Electrospinning versus fibre production methods: from specifics to technological convergence. Chem Soc Rev 41(13):4708
142. Sarkar K, Gomez C, Zambrano S, Ramirez M, De Hoyos E, Vasquez H et al (2010) Electrospinning to Forcespinning™. Mater Today 13(11):12–14
143. Rohani Shirvan A, Nouri A, Sutti A (2022) A perspective on the wet spinning process and its advancements in biomedical sciences. Eur Polym J 181:111681
144. Stoessel PR, Raso RA, Kaufmann T, Grass RN, Stark WJ (2015) Fibers mechanically similar to sheep wool obtained by wet spinning of gelatin and optional plasticizers. Macromol Mater Eng 300(2):234–241
145. Dai Y, Ahmed J, Edirisinghe M (2023) Pressurized gyration: fundamentals, advancements, and future. Macromol Mater Eng 308(7):2300033
146. Flores-Rojas GG, Gómez-Lazaro B, López-Saucedo F, Vera-Graziano R, Bucio E, Mendizábal E (2023) Electrospun scaffolds for tissue engineering: a review. Macromol 3(3): 524–553

147. Ji D, Lin Y, Guo X, Ramasubramanian B, Wang R, Radacsi N et al (2024) Electrospinning of nanofibres. Nat Rev Methods Primers 4(1):1
148. Xue J, Wu T, Dai Y, Xia Y (2019) Electrospinning and electrospun nanofibers: methods, materials, and applications. Chem Rev 119(8):5298–5415
149. Das KP, Satapathy BK (2024) Highly porous agro-augmenting urea-biochar/polylactic acid-based microfibrous electrospun mats as sustainable controlled release fertilizer carriers. ACS Appl Polym Mater 6(11):6262–6277
150. Han WH, Li X, Yu GF, Wang BC, Huang LP, Wang J et al (2022) Recent advances in the food application of electrospun nanofibers. J Ind Eng Chem 110:15–26
151. Mercante LA, Scagion VP, Migliorini FL, Mattoso LHC, Correa DS (2017) Electrospinning-based (bio)sensors for food and agricultural applications: a review. TrAC Trends Anal Chem 91:91–103
152. Park SH, Kim MS, Lee B, Park JH, Lee HJ, Lee NK et al (2016) Creation of a hybrid scaffold with dual configuration of aligned and random electrospun fibers. ACS Appl Mater Interfaces 8(4):2826–2832
153. Xie Z, Gao M, Lobo AO, Webster TJ (2020) 3D bioprinting in tissue engineering for medical applications: the classic and the hybrid. Polymers 12(8):1717
154. Akshay Kumar KP, Zare EN, Torres-Mendieta R, Wacławek S, Makvandi P, Černík M et al (2021) Electrospun fibers based on botanical, seaweed, microbial, and animal sourced biomacromolecules and their multidimensional applications. Int J Biol Macromol 171:130–149
155. Santos AEAD, Guadalupe JL, Albergaria JDS, Almeida IA, Moreira AMS, Copola AGL et al (2024) Random cellulose acetate nanofibers: a breakthrough for cultivated meat production. Front Nutr 10:1297926
156. Moreira AMS, Nogueira JM, Carceroni J, Guadalupe JL, Dos Santos AEA, Fagundes AMA et al (2024) Acetate cellulose fibrous scaffold is suitable for cultivated fat production. Curr Res Food Sci 9:100903
157. Teixeira SC, Gomes NO, Oliveira TVD, Fortes-Da-Silva P, Soares NDFF, Raymundo-Pereira PA (2023) Review and perspectives of sustainable, biodegradable, eco-friendly and flexible electronic devices and (bio)sensors. Biosens Bioelectron X 14:100371
158. Zhong J, Mohan SD, Bell A, Terry A, Mitchell GR, Davis FJ (2018) Electrospinning of food-grade nanofibres from whey protein. Int J Biol Macromol 113:764–773
159. Johnson JK, Ohst D (2020) Electrospun polymer fibers for cultured meat production. US20200245658A1 [cited 2025 May 7]. https://patents.google.com/patent/US20200245658A1/en
160. Da Trindade LG, Zanchet L, Bonsanto FP, Braga ARC (2024) Electrospun fibers of zein and pea protein to create high-quality fibrous structures in meat analogs. Front Bioeng Biotechnol 12:1483966
161. Callue T (2020) Development of biomaterial scaffolds for cultured meat and engineered muscle. University of Bath. https://researchportal.bath.ac.uk/en/studentTheses/development-of-biomaterial-scaffolds-for-cultured-meat-and-engine
162. Nurul Alam AMM, Kim CJ, Kim SH, Kumari S, Lee EY, Hwang YH et al (2024) Scaffolding fundamentals and recent advances in sustainable scaffolding techniques for cultured meat development. Food Res Int 189:114549
163. Mukka S, Ziegler GR (2024) Aligned electrospun starch-pullulan-protein fibers. Food Hydrocoll 157:110412
164. Klaas M, Möll K, Mäemets-Allas K, Loog M, Järvekülg M, Jaks V (2021) Long-term maintenance of functional primary human hepatocytes in 3D gelatin matrices produced by solution blow spinning. Sci Rep 11(1):20165
165. Martens ME, Enn R, Õun S, Veinla M, Toots KM, Järvekülg M (2023) Device and method for producing polymer fibers and its uses thereof. US11697892B2 [cited 2025 May 7]. https://patents.google.com/patent/US11697892B2/en

166. Sha L, Xiong YL (2020) Plant protein-based alternatives of reconstructed meat: science, technology, and challenges. Trends Food Sci Technol 102:51–61
167. Cui B, Liang H, Li J, Zhou B, Chen W, Liu J et al (2022) Development and characterization of edible plant-based fibers using a wet-spinning technique. Food Hydrocoll 133:107965
168. Chuang R, Naidu A, Galipon J (2025) Enhancing meat analog texture using wet-spun fibroin protein fibers: a novel approach to mimic whole-muscle meat. J Texture Stud 56(1):e70001
169. Bomkamp C, Skaalure SC, Fernando GF, Ben-Arye T, Swartz EW, Specht EA (2022) Scaffolding biomaterials for 3D cultivated meat: prospects and challenges. Adv Sci 9(3): 2102908
170. Lerner A, Benzvi C (2021) Microbial transglutaminase is a very frequently used food additive and is a potential inducer of autoimmune/neurodegenerative diseases. Toxics 9(10):233
171. Ellis M, Dunsford I, Allan S (2022) Method of culturing muscle cells with perfusion bioreactor. WO2024038281A1
172. Ellis M, Chaudhuri J (2006) Poly(lactic-co-glycolic acid) hollow fibre membranes for use as a tissue engineering scaffold. Biotechnol Bioeng 96:177
173. Tuomisto HL, Allan SJ, Ellis MJ (2022) Prospective life cycle assessment of a bioprocess design for cultured meat production in hollow fiber bioreactors. Sci Total Environ 851:158051
174. Minghao N, Shima A, Yamamoto M, Takeuchi S (2025) Scalable tissue biofabrication via perfusable hollow fiber arrays for cultured meat applications. Trends Biotechnol 43:1938
175. Dehghani F, Annabi N (2011) Engineering porous scaffolds using gas-based techniques. Curr Opin Biotechnol 22(5):661–666
176. Choi SW, Yeh YC, Zhang Y, Sung HW, Xia Y (2010) Uniform beads with controllable pore sizes for biomedical applications. Small 6(14):1492–1498
177. Flégeau K, Pace R, Gautier H, Rethore G, Guicheux J, Le Visage C et al (2017) Toward the development of biomimetic injectable and macroporous biohydrogels for regenerative medicine. Adv Colloid Interf Sci 247:589–609
178. Ma Y, Wang X, Su T, Lu F, Chang Q, Gao J (2022) Recent advances in macroporous hydrogels for cell behavior and tissue engineering. Gels 8(10):606
179. Caldwell AS, Aguado BA, Anseth KS (2020) Designing microgels for cell culture and controlled assembly of tissue microenvironments. Adv Funct Mater 30(37):1907670
180. Gao D, Shipman WD, Sun Y, Glahn JZ, Beraki L, Hsia HC (2025) Macroporous scaffolds based on biomass polymers and their applications in wound healing. J Bioresour Bioprod 10(1):14–31
181. Kozan NG, Joshi M, Sicherer ST, Grasman JM (2023) Porous biomaterial scaffolds for skeletal muscle tissue engineering. Front Bioeng Biotechnol 11:1245897
182. Perez RA, Mestres G (2016) Role of pore size and morphology in musculo-skeletal tissue regeneration. Mater Sci Eng C 61:922–939
183. Su T, Xu M, Lu F, Chang Q (2022) Adipogenesis or osteogenesis: destiny decision made by mechanical properties of biomaterials. RSC Adv 12(38):24501–24510
184. Nerger BA, Kashyap K, Deveney BT, Lou J, Hanan BF, Liu Q et al (2024) Tuning porosity of macroporous hydrogels enables rapid rates of stress relaxation and promotes cell expansion and migration. Proc Natl Acad Sci 121(45):e2410806121
185. Fereshteh Z (2018) Chapter 7 – Freeze-drying technologies for 3D scaffold engineering. In: Deng Y, Kuiper J (eds) Functional 3D tissue engineering scaffolds. Woodhead Publishing, pp 151–174. https://www.sciencedirect.com/science/article/pii/B9780081009796000070
186. Ghorbandoust M, Fasihi M, Norouzbeigi R (2024) Tuning pore size and density of rigid polylactic acid foams through thermally induced phase separation and optimization using response surface methodology. Sci Rep 14(1):12395
187. Tchessalov S, Maglio V, Kazarin P, Alexeenko A, Bhatnagar B, Sahni E et al (2023) Practical advice on scientific design of freeze-drying process: 2023 update. Pharm Res 40(10): 2433–2455

188. Katrilaka C, Karipidou N, Petrou N, Manglaris C, Katrilakas G, Tzavellas AN et al (2023) Freeze-drying process for the fabrication of collagen-based sponges as medical devices in biomedical engineering. Materials 16(12):4425
189. Angraini N, Syarifuddin S, Rauf N, Tahir D (2025) Advancements in bone tissue engineering: a comprehensive review of biomaterial scaffolds and freeze-drying techniques from perspective global and future research. Artif Organs [cited 2025 June 14]. https://doi.org/10.1111/aor.14976
190. Tang Y, Shi C, Zhu Y, Yang M, Sheng K, Zhang X (2024) Cellulose as a sustainable scaffold material in cultivated meat production. Curr Res Food Sci 9:100846
191. Krontiras P, Gatenholm P, Hägg DA (2015) Adipogenic differentiation of stem cells in three-dimensional porous bacterial nanocellulose scaffolds. J Biomed Mater Res B Appl Biomater 103(1):195–203
192. Dinescu S, Galateanu B, Lungu A, Radu E, Nae S, Iovu H et al (2014) Perilipin expression reveals adipogenic potential of hADSCs inside superporous polymeric cellular delivery systems. Biomed Res Int 2014(1):830791
193. Sawadkar P, Mandakhbayar N, Patel KD, Owji N, Rajasekar P, Sarama R et al (2024) 3D porous binary composites of collagen, elastin, and fibrin proteins orchestrate adipose tissue regeneration. Macromol Biosci 24(8):2400073
194. Basurto IM, Muhammad SA, Gardner GM, Christ GJ, Caliari SR (2022) Controlling scaffold conductivity and pore size to direct myogenic cell alignment and differentiation. J Biomed Mater Res A 110(10):1681–1694
195. Kozan NG, Caswell S, Patel M, Grasman JM (2023) Aligned collagen sponges with tunable pore size for skeletal muscle tissue regeneration. J Funct Biomater 14(11):533
196. Samolyk BL, Pace ZY, Li J, Billiar KL, Coburn JM, Whittington CF et al (2024) Developing porous fibrin scaffolds with tunable anisotropic features to direct myoblast orientation. Tissue Eng Part C Methods 30(5):217–228
197. Dekkers BL, Boom RM, van der Goot AJ (2018) Structuring processes for meat analogues. Trends Food Sci Technol 81:25–36
198. Dewey MJ, Nosatov AV, Subedi K, Harley B (2020) Anisotropic mineralized collagen scaffolds accelerate osteogenic response in a glycosaminoglycan-dependent fashion. RSC Adv 10(26):15629–15641
199. Xia P, Miyajima H, Fujita S (2025) Development of biomimetic edible scaffolds for cultured meat based on the traditional freeze-drying method for Ito-Kanten (Japanese freeze-dried agar). Gels 11(4):299
200. Chen Y, Zhang W, Ding X, Ding S, Tang C, Zeng X et al (2024) Programmable scaffolds with aligned porous structures for cell cultured meat. Food Chem 430:137098
201. Kim DH, Han SG, Lim SJ, Hong SJ, Kwon HC, Jung HS et al (2024) Comparison of soy and pea protein for cultured meat scaffolds: evaluating gelation, physical properties, and cell adhesion. Food Sci Anim Resour 44(5):1108–1125
202. David S, Ianovici I, Guterman Ram G, Shaulov Dvir Y, Lavon N, Levenberg S (2024) Pea protein-rich scaffolds support 3D bovine skeletal muscle formation for cultivated meat application. Adv Sustain Syst 8(6):2300499
203. Xia P, Miyajima H, Fujita S (2025) Porous edible cryogel inspired by the production of shimitofu (freeze-dried tofu) for high-density cell culture in cultured meat. Food Hydrocoll 163:111155
204. Wang Y, Zhong Z, Munawar N, Wang R, Zan L, Zhu J (2024) Production of green-natural and "authentic" cultured meat based on proanthocyanidins-dialdehyde chitosan-collagen ternary hybrid edible scaffolds. Food Res Int 175:113757
205. Wang Y, Zhong Z, Munawar N, Zan L, Zhu J (2024) 3D edible scaffolds with yeast protein: a novel alternative protein scaffold for the production of high-quality cell-cultured meat. Int J Biol Macromol 259:129134

206. Lee H, Kim D, Choi KH, Lee S, Jo M, Chun SY et al (2024) Animal-free scaffold from brown algae provides a three-dimensional cell growth and differentiation environment for steak-like cultivated meat. Food Hydrocoll 152:109944
207. Zagury Y, Ianovici I, Landau S, Lavon N, Levenberg S (2022) Engineered marble-like bovine fat tissue for cultured meat. Commun Biol 5(1):927
208. Li L, Chen L, Chen X, Chen Y, Ding S, Fan X et al (2022) Chitosan-sodium alginate-collagen/gelatin three-dimensional edible scaffolds for building a structured model for cell cultured meat. Int J Biol Macromol 209:668–679
209. Rao KM, Kim HJ, Won S, Choi SM, Han SS (2023) Effect of grape seed extract on gelatin-based edible 3D-hydrogels for cultured meat application. Gels 9(1):65
210. Kim WJ, Kim Y, Ovissipour R, Nitin N (2024) Plant-based biomaterials as scaffolds for cellular agriculture. Future Foods 10:100468
211. Rubio NR, Fish KD, Trimmer BA, Kaplan DL (2019) In vitro insect muscle for tissue engineering applications. ACS Biomater Sci Eng 5(2):1071–1082
212. Chen S, Nakamoto T, Kawazoe N, Chen G (2015) Engineering multi-layered skeletal muscle tissue by using 3D microgrooved collagen scaffolds. Biomaterials 73:23–31
213. Wang Y, Ai C, Wang H, Chen C, Teng H, Xiao J et al (2023) Emulsion and its application in the food field: an update review. eFood 4(4):e102
214. Aldemir Dikici B, Claeyssens F (2020) Basic principles of emulsion templating and its use as an emerging manufacturing method of tissue engineering scaffolds. Front Bioeng Biotechnol 8:875. https://www.frontiersin.org/journals/bioengineering-and-biotechnology/articles/10.3389/fbioe.2020.00875
215. Zhang T, Sanguramath RA, Israel S, Silverstein MS (2019) Emulsion templating: porous polymers and beyond. Macromolecules 52(15):5445–5479
216. McClements DJ, Jafari SM (2018) Improving emulsion formation, stability and performance using mixed emulsifiers: a review. Adv Colloid Interf Sci 251:55–79
217. Yiu CC, Liang SW, Mukhtar K, Kim W, Wang Y, Selomulya C (2023) Food emulsion gels from plant-based ingredients: formulation, processing, and potential applications. Gels 9(5):366
218. Cui L, Guo J, Meng Z (2023) A review on food-grade-polymer-based O/W emulsion gels: stabilization mechanism and 3D printing application. Food Hydrocoll 139:108588
219. Xia B, Shen Y, Zhao R, Deng J, Wang C (2024) Interactions with soy lecithin regulate the emulsification capacity of pea protein: effects of soy lecithin concentration. Food Hydrocoll 155:110168
220. Shao P, Feng J, Sun P, Xiang N, Lu B, Qiu D (2020) Recent advances in improving stability of food emulsion by plant polysaccharides. Food Res Int 137:109376
221. Robin AL, Sankhla D (2013) European Legislative Framework controlling the use of food additives. In: Saltmarsh M (ed) Essential guide to food additives. The Royal Society of Chemistry [cited 2025 June 17]. https://doi.org/10.1039/9781849734981-00044
222. Panyod S, Wu WK, Chang CT, Wada N, Ho HC, Lo YL et al (2024) Common dietary emulsifiers promote metabolic disorders and intestinal microbiota dysbiosis in mice. Commun Biol 7(1):749
223. Naimi S, Viennois E, Gewirtz AT, Chassaing B (2021) Direct impact of commonly used dietary emulsifiers on human gut microbiota. Microbiome 9(1):66
224. Gao Y, Kong Q, Zheng H, Xue C (2024) Tailored edible 3D porous scaffolds constructed by Pickering emulgel templating for cell-cultured fish meat. Food Hydrocoll 157:110453
225. Annabi N, Nichol JW, Zhong X, Ji C, Koshy S, Khademhosseini A et al (2010) Controlling the porosity and microarchitecture of hydrogels for tissue engineering. Tissue Eng Part B Rev 16(4):371–383
226. Hasanzadeh R, Azdast T, Mojaver M, Darvishi MM, Park CB (2022) Cost-effective and reproducible technologies for fabrication of tissue engineered scaffolds: the state-of-the-art and future perspectives. Polymer 244:124681

227. Costantini M, Barbetta A (2018) Chapter 6 – Gas foaming technologies for 3D scaffold engineering. In: Deng Y, Kuiper J (eds) Functional 3D tissue engineering scaffolds. Woodhead Publishing, pp 127–149. https://www.sciencedirect.com/science/article/pii/B9780081009796000069
228. Takei T, Aokawa R, Shigemitsu T, Kawakami K, Yoshida M (2015) Fabrication of uniformly cell-laden porous scaffolds using a gas-in-liquid templating technique. J Biosci Bioeng 120(5): 577–581
229. Stampella A, Rizzitelli G, Donati F, Mazzarino M, de la Torre X, Botrè F et al (2015) Human hepatoma cell lines on gas foaming templated alginate scaffolds for in vitro drug-drug interaction and metabolism studies. Toxicol In Vitro 30(1, Part B):331–340
230. Poursamar SA, Hatami J, Lehner AN, da Silva CL, Ferreira FC, Antunes APM (2015) Gelatin porous scaffolds fabricated using a modified gas foaming technique: characterisation and cytotoxicity assessment. Mater Sci Eng C 48:63–70
231. Mei JC, Wu AYK, Wu PC, Cheng NC, Tsai WB, Yu J (2014) Three-dimensional extracellular matrix scaffolds by microfluidic fabrication for long-term spontaneously contracted cardiomyocyte culture. Tissue Eng Part A 20(21–22):2931–2941
232. Tang Y, Lin S, Yin S, Jiang F, Zhou M, Yang G et al (2020) In situ gas foaming based on magnesium particle degradation: a novel approach to fabricate injectable macroporous hydrogels. Biomaterials 232:119727
233. Li X, Gou J, Santhanam AP, Maiti C, Ilegbusi OJ (2025) Tissue mimicking hydrogel foam materials with mechanical and radiological properties equivalent to human lung. Sci Rep 15(1):7471
234. Jin C, Zhu L, Zhang H (2025) Study on plant-based proteins and their complexes in improving food foam system: a review. Food Res Int 212:116514
235. Zink JI, Zeneli L, Windhab EJ (2023) Micro-foaming of plant protein based meat analogues for tailored textural properties. Curr Res Food Sci 7:100580
236. Su L, Jing L, Zeng S, Fu C, Huang D (2024) 3D porous edible scaffolds from rye secalin for cell-based pork fat tissue culturing. J Agric Food Chem 72(20):11587–11596
237. Radakisnin R, Majid MSA, Jamir MRM, Tahir MFM, Meng CE, Alshahrani HA (2021) Physical, thermal, and mechanical properties of highly porous polylactic acid/cellulose nanofibre scaffolds prepared by salt leaching technique. Nanotechnol Rev 10(1):1469–1483
238. Wulf A, Mendgaziev RI, Fakhrullin R, Vinokurov V, Volodkin D, Vikulina AS (2022) Porous alginate scaffolds designed by calcium carbonate leaching technique. Adv Funct Mater 32(14): 2109824
239. Bäckdahl H, Esguerra M, Delbro D, Risberg B, Gatenholm P (2008) Engineering microporosity in bacterial cellulose scaffolds. J Tissue Eng Regen Med 2(6):320–330
240. Miller JS, Stevens KR, Yang MT, Baker BM, Nguyen DHT, Cohen DM et al (2012) Rapid casting of patterned vascular networks for perfusable engineered three-dimensional tissues. Nat Mater 11(9):768–774
241. Wei X, Luo Y, Huang P (2019) 3D bioprinting of alginate scaffolds with controlled micropores by leaching of recrystallized salts. Polym Bull 76(12):6077–6088
242. Ben-Arye T, Shandalov Y, Ben-Shaul S, Landau S, Zagury Y, Ianovici I et al (2020) Textured soy protein scaffolds enable the generation of three-dimensional bovine skeletal muscle tissue for cell-based meat. Nat Food 1(4):210–20 [cited 2023 Jul 19]. https://www.nature.com/articles/s43016-020-0046-5
243. Lee M, Park S, Choi B, Kim J, Choi W, Jeong I et al (2022) Tailoring a gelatin/agar matrix for the synergistic effect with cells to produce high-quality cultured meat. ACS Appl Mater Interfaces 14(33):38235–38245
244. Guan X, Yan Q, Ma Z, Zhou J (2023) Production of mature myotubes in vitro improves the texture and protein quality of cultured pork. Food Funct 14(8):3576–3587
245. Ikuse M, Marchus CR, Schiele NR, Ganjyal GM (2024) Extruded plant protein two-dimensional scaffold structures support myoblast cell growth for potential applications in cultured meat. Food Res Int 195:114981

246. Gilbert T, Sellaro T, Badylak S (2006) Decellularization of tissues and organs. Biomaterials 27:3675. pii: S0142961206001682
247. Golebiowska AA, Intravaia JT, Sathe VM, Kumbar SG, Nukavarapu SP (2024) Decellularized extracellular matrix biomaterials for regenerative therapies: advances, challenges and clinical prospects. Bioactive Mater 32:98–123
248. Neishabouri A, Soltani Khaboushan A, Daghigh F, Kajbafzadeh AM, Majidi Zolbin M (2022) Decellularization in tissue engineering and regenerative medicine: evaluation, modification, and application methods. Front Bioeng Biotechnol 10:805299
249. Gershlak JR, Hernandez S, Fontana G, Perreault LR, Hansen K, Larson SA et al (2017) Crossing kingdoms: using decellularized plants as perfusable tissue engineering scaffolds. Biomaterials 125:13–22
250. Allan SJ, Ellis MJ, De Bank PA (2021) Decellularized grass as a sustainable scaffold for skeletal muscle tissue engineering. J Biomed Mater Res A 109(12):2471–2482
251. Fontana G, Gershlak J, Adamski M, Lee J, Matsumoto S, Le HD et al (2017) Biofunctionalized plants as diverse biomaterials for human cell culture. Adv Healthc Mater 6(8):1601225
252. Yun J, Robertson S, Kim C, Suzuki M, Murphy WL, Gopalan P (2023) Aligned skeletal muscle assembly on a biofunctionalized plant leaf scaffold. Acta Biomater 171:327–335
253. Liu Y, Gao A, Wang T, Zhang Y, Zhu G, Ling S et al (2025) Growing meat on autoclaved vegetables with biomimetic stiffness and micro-patterns. Nat Commun 16(1):161
254. Banavar A, Sarkarat R, Amirvaresi A, Li X, Nguyen C, Kaplan DL et al (2024) Decellularized banana leaves: eco-friendly scaffolds for cell-based seafood. Front Sustain Food Syst 8. https://www.frontiersin.org/journals/sustainable-food-systems/articles/10.3389/fsufs.2024.1341151
255. Hong TK, Do JT (2024) Generation of chicken contractile skeletal muscle structure using decellularized plant scaffolds. ACS Biomater Sci Eng 10(5):3500–3512
256. Ong S, Loo L, Pang M, Tan R, Teng Y, Lou X et al (2021) Decompartmentalisation as a simple color manipulation of plant-based marbling meat alternatives. Biomaterials 277:121107
257. Perreault LR, Thyden R, Kloster J, Jones JD, Nunes J, Patmanidis AA et al (2023) Repurposing agricultural waste as low-cost cultured meat scaffolds. Front Food Sci Technol 3. https://www.frontiersin.org/journals/food-science-and-technology/articles/10.3389/frfst.2023.1208298
258. Jones JD, Thyden R, Perreault LR, Varieur BM, Patmanidis AA, Daley L et al (2023) Decellularization: leveraging a tissue engineering technique for food production. ACS Biomater Sci Eng 9(5):2292–2300
259. Hickey R, Szereszewski K, Pelling AE, de Sousa Faria Tischer PC, Canto A, Salamun J et al (2024) Plant-derived aerogels, hydrogels, and foams, and methods and uses thereof. US 2024307592 A1
260. Piantino M, Muller Q, Nakadozono C, Yamada A, Matsusaki M (2025) Towards more realistic cultivated meat by rethinking bioengineering approaches. Trends Biotechnol 43(2):364–382
261. Ianovici I, Zagury Y, Redenski I, Lavon N, Levenberg S (2022) 3D-printable plant protein-enriched scaffolds for cultivated meat development. Biomaterials 284:121487
262. Kim WJ, Kim Y, Lu Y, Ovissipour R, Nitin N (2025) Evaluation of plant-based composite materials as 3D printed scaffolds for cell growth and proliferation in cultivated meat applications. Food Hydrocoll 160:110823
263. Sealy MP, Avegnon KLM, Garrett A, Delbreilh L, Bapat S, Malshe AP (2022) Understanding biomanufacturing of soy-based scaffolds for cell-cultured meat by vat polymerization. CIRP Ann 71(1):209–212
264. Jing L, Sun J, Liu H, Wang X, Huang D (2021) Using plant proteins to develop composite scaffolds for cell culture applications. Int J Bioprint 7(1):298
265. Wang J, Dai S, Xiang N, Zhang L, Zhong W, Shao P et al (2024) Cell-based meat scaffold based on a 3D-printed starch-based gel. J Agric Food Chem 72(34):19143–19154

266. Dutta SD, Ganguly K, Jeong MS, Patel DK, Patil TV, Cho SJ et al (2022) Bioengineered lab-grown meat-like constructs through 3D bioprinting of antioxidative protein hydrolysates. ACS Appl Mater Interfaces 14(30):34513–34526
267. Fima S, Gal I (2021) Cultured edible meat fabrication using bio-printing. WO2021007359A1
268. Guo X, Wang D, He B, Hu L, Jiang G (2024) 3D bioprinting of cultured meat: a promising avenue of meat production. Food Bioprocess Technol 17(7):1659–1680
269. Gurel M, Rathod N, Cabrera LY, Voyton S, Yeo M, Ozogul F et al (2024) A narrative review: 3D bioprinting of cultured muscle meat and seafood products and its potential for the food industry. Trends Food Sci Technol 152:104670
270. Huang Z, Niu R, Du Z, Xin Q, Lao J, Hu H et al (2025) Cold-temperature 3D bioprinting for cultured meat: a case of rainbow trout cell cultivation. Sustain Mater Technol 44:e01422
271. Ianovici I, Zagury Y, Afik N, Hendel M, Lavon N, Levenberg S (2024) Embedded three-dimensional printing of thick pea-protein-enriched constructs for large, customized structured cell-based meat production. Biofabrication. http://iopscience.iop.org/article/10.1088/1758-5090/ad628f
272. Jeong D, Seo JW, Lee HG, Jung WK, Park YH, Bae H (2022) Efficient myogenic/adipogenic transdifferentiation of bovine fibroblasts in a 3D bioprinting system for steak-type cultured meat production. Adv Sci 9(31):2202877
273. Koranne V, Jonas OLC, Mitra H, Bapat S, Ardekani AM, Sealy MP et al (2022) Exploring properties of edible hydrolyzed collagen for 3D food printing of scaffold for biomanufacturing cultivated meat. Procedia CIRP 110:186–191
274. Levenberg S, Ianovici I, Zagury Y, Lavon N, David S (2022) 3D-printable protein-enriched scaffolds. WO2022162662A1
275. Kleger N, Cihova M, Masania K, Studart AR, Löffler JF (2019) 3D printing of salt as a template for magnesium with structured porosity. Adv Mater 31(37):1903783
276. Guo CF, Zhang M, Bhandari B (2019) A comparative study between syringe-based and screw-based 3D food printers by computational simulation. Comput Electron Agric 162:397–404
277. Dai H, Wen L, Liang H, Li B, Li J (2025) Enhancing 3D food printing precision: development and interaction behavior of soy protein isolate-konjac glucomannan-xanthan gum composite ink based on hot-melt extrusion. Int J Biol Macromol 304:140854
278. Su L, Jing L, Zeng X, Chen T, Liu H, Kong Y et al (2023) 3D-printed prolamin scaffolds for cell based meat culture. Adv Mater 35(2):2207397
279. Handral HK, Tay SH, Chan WW, Choudhury D (2022) 3D printing of cultured meat products. Crit Rev Food Sci Nutr 62(1):272–281
280. Kang DH, Louis F, Liu H, Shimoda H, Nishiyama Y, Nozawa H et al (2021) Engineered whole cut meat-like tissue by the assembly of cell fibers using tendon-gel integrated bioprinting. Nat Commun 12(1):5059
281. Xu E, Niu R, Lao J, Zhang S, Li J, Zhu Y et al (2023) Tissue-like cultured fish fillets through a synthetic food pipeline. npj Sci Food 7(1):17
282. Shiwarski DJ, Hudson AR, Tashman JW, Feinberg AW (2021) Emergence of FRESH 3D printing as a platform for advanced tissue biofabrication. APL Bioeng 5(1):010904
283. Prendergast ME, Davidson MD, Burdick JA (2021) A biofabrication method to align cells within bioprinted photocrosslinkable and cell-degradable hydrogel constructs via embedded fibers. Biofabrication 13(4). https://doi.org/10.1088/1758-5090/ac25cc
284. Shapiro L, Cohen S (1997) Novel alginate sponges for cell culture and transplantation. Biomaterials 18(8):583–590
285. Yadavalli NS, Darya A, Sergiy M, Alexander T, Vladimir R, Brianna B, et al (2023) Multi-layer nanofiber scaffolds. US; US 2023/0174928 A1. https://lens.org/006-665-753-399-808
286. Yon JS (2023) Porous cell support containing plant protein and cultured meat prepared using the same. US; US 2023/0067465 A1. https://lens.org/102-279-030-297-175
287. Martens M-E, Enn R, Õun S, Veinla M, Toots KM, Järvekülg M (2022) Device and method for producing polymer fibers and its uses thereof. EP; EP 3954811 A1. https://lens.org/125-928-019-761-901

288. Morel A, Eglauf JS, Razza N, Das SK (2023) Edible microcarrier for the preparation of cultured meat and method of producing same. WO; WO 2023/104768 A1. https://lens.org/070-868-530-238-362
289. Takle S, Sachdeva NS, Srivastava A (2022) A scaffolding for cultivated meat and a process for making the same. WO; WO 2022/208525 A1. https://lens.org/175-385-212-168-275
290. Studart AR, Masania K, Kleger N, Fehlmann S, Cihova M (2023) Method for the additive manufacturing of casting molds. WO; WO 2023/001418 A1. https://lens.org/108-414-080-319-472
291. Jae LH, Ho KJ, Rim JH, Young KM, Hyun SS, Keun JT (2020) Scaffold for cell culture and manufacturing method thereof. KR; KR 102151203 B1. https://lens.org/130-533-070-500-146
292. Smart MCs (2024) X1 Microcarrier – Smart biomaterials & μ-solutions [cited 2025 Jul 31]. https://smartmcs.com.au/product/x1-microcarrier/
293. TruSpin Nanomaterials. Hyaluronic acid (HA) nanofibers [cited 2025 Jul 31]. https://truspin.com/products/ha

Adv Biochem Eng Biotechnol (2026) 193: 201–220
https://doi.org/10.1007/10_2025_290

Published online: 15 October 2025

From Muscle Fibers to Functional Foods: Bridging Conventional and Cultivated Approaches

Martin Krøyer Rasmussen, Rebekka Rinder Thøgersen, Jette Feveile Young, and Margrethe Therkildsen

Contents

Abstract This chapter explores the quality attributes of cultivated meat, drawing parallels and distinctions with conventional meat to support future product development. It examines key technical, sensory, and nutritional parameters, like texture, water-holding capacity, color, flavor, and nutritional composition. The chapter highlights how cultivated meat's quality is shaped by cellular composition, scaffold materials, and postharvest processes and how these factors influence consumer-relevant traits such as tenderness, juiciness, and taste. It also discusses the potential of cultivated meat to match or surpass conventional meat in nutritional value, including protein content, amino acid profile, and micronutrient composition. Contributions from five cultivated meat companies provide an industrial context.

M. K. Rasmussen (✉), R. R. Thøgersen, J. F. Young, and M. Therkildsen
Department of Food Science, Aarhus University, Aarhus, Denmark
e-mail: Martink.rasmussen@food.au.dk

Graphical Abstract

Keywords Amino acids, DIAAS, Flavor, Meat quality, Muscle cell, Protein, Texture

Abbreviations

DIAAS	Digestible Indispensable Amino Acid Score
FAA	Free amino acids
NMR	Nuclear magnetic resonance
PUFA	Polyunsaturated fatty acids
TPA	Texture profile analysis
WHC	Water-holding capacity

1 Introduction

The quality of conventional meat has been a focus for numerous studies. These studies have identified parameters like tenderness, juiciness, flavor, visual appearance, and nutritional value as key quality attributes for meat. Today, the underlying mechanism behind these quality attributes is to some extent understood and has added to the high quality of several meat products. For example, we have a good understanding of the significance of postmortem maturation of meat to improve quality. We know that during maturation, pH changes are significant for not only water-holding capacity (WHC) but also proteolytic activity adding to increased tenderness and flavor development. The events are essential for securing high-quality products for the consumer. Thus, quality is paramount and is what makes the consumer willing to buy the product, and this justifies why much research has been directed into controlling and improving these aspects of conventional meat. Although the aforementioned quality attributes are often singularly described, it must be acknowledged that they are often interrelated. For example, for the consumer, juicier meat is typically also deemed more tender.

With the prospering transformation of the meat production industry to also encompassing cultivated meat products, more effort must be directed into understanding the fundamental mechanism shaping the quality of cultivated meat. In this chapter, we explore, with an offset in knowledge obtained from conventional meat, the quality of cultivated meat as a food product (Fig. 1). Although research into this aspect is still at an infant stage, some parallels and differences between conventional meat and cultivated meat can be identified. It must be acknowledged that conventional meat consists of well-organized muscle fibers (including different fiber types), fat (adipocytes and inter- and intramuscular fat), and connective tissue, which each contributes to the quality of the final product. This mixture of cells and organization is not given for cultivated meat products, where it is often simpler. However,

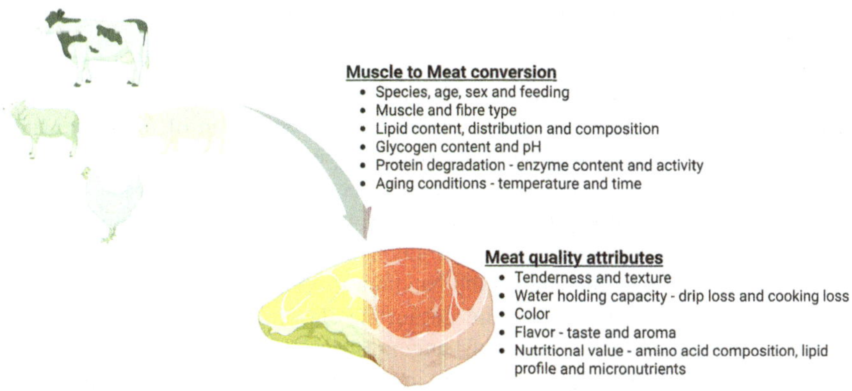

Fig. 1 Overview of physiological and biochemical attributes shaping the conversion of muscle into conventional meat and its final quality

standing on the edge of a new food era also provides the opportunity to reinvent meat into a more functional food and create new quality standards.

To bridge the gap between the theoretical understanding of cultivated meat quality and its practical relevance to industry, five companies were asked to provide insights into their business-level priorities. These companies represent a diverse range of cultivated product types, including minced beef, hybrid beef products (minced and whole cuts combined with plant proteins), chicken-based ingredients, fat-based ingredients, and integrated beef products containing both muscle and fat components.

Throughout this chapter, we incorporate both collective and individual perspectives on key quality attributes. These include aspects that the companies are currently addressing, anticipated focal points for their own product development, and broader views on which quality parameters will be critical for the cultivated meat sector moving forward.

While the quotes and examples are specific to each company, many of the insights reflect shared priorities across the industry. These contributions help contextualize theoretical quality frameworks within real-world commercial strategies and product development goals.

Industry Perspectives on Consumer Acceptance

All participating companies emphasized that consumer acceptance is the central focus in the development of cultivated meat products.

- "*Our primary quality focus is on delivering an organoleptic experience for consumers.*" (Aleph Farms – Neta Lavon)
- "*Consumers make their buying decisions mainly on taste, price, and convenience.*" (Re:Meat – Jacob Schaldemose Peterson)

2 Basic Knowledge of Meat Quality from Conventional Meat

The quality of conventional meat is basically defined as attributes related to the muscle per se, as well as attributes defined by the conversion of muscle to meat (aging). Some aspects of meat and sensory quality are settled during the production of conventional meat, where species, breed, feeding and growth rate, age, and specific muscle will impact the amount of fat, proteolytic enzyme activity in the tissue, glycogen content, fiber-type distribution, and amount and solubility of collagen. Other traits are dependent on the handling during the slaughter process, like stress prior to slaughter, temperature during the slaughter and rigor process, and stretch of the carcass as an effect of hanging method, and some traits are further affected by the final aging period and environmental conditions. All together, these factors will have an impact on the final quality of the meat, that is, the texture, water-holding capacity, juiciness, color, flavor, and nutritional value.

3 Quality of Cultivated Meat

With an offset in conventional meat, this section will highlight knowledge as well as expected quality attributes of cultivated meat and cultivated meat-containing products. For more information on meat quality in conventional meat, see, e.g., Pogorzelski et al. [1].

3.1 *Texture*

Texture of conventional meat is determined by the myofibrillar structure in the muscle fibers, the connective tissue surrounding the fibers and the fascicles (endomysium and perimysium, respectively), and the amount of fat in the tissue, which acts as a diluter of the other components as well as a lubricant during chewing. But the texture is not just given by these factors. The contribution from the myofibrils and the connective tissue depends on the activity of proteolytic enzymes in the tissue. Proteolytic activity is dependent on the postmortem pH profile in the tissue, which is mainly related to the glycogen level prior to slaughter and the slaughter and rigor process. In addition, texture is dependent on the degree of contraction of myofibrils, which is regulated both physically and biochemically. Proteolytic enzymes are involved in the degradation of myofibrillar and cytoskeletal proteins postmortem, and this degradation opens the myofibrillar structure and releases protein fragments, peptides, and amino acids, important for the texture and flavor development. The proteolytic enzymes are indigenous enzymes involved in the protein turnover in the living muscle, and thus the amount and activity are dependent on in vivo factors like growth rate, age, and muscle [2], which settles the potential for proteolysis postmortem. High growth rate at the time of slaughter is usually aligned with a larger protein turnover and thus also a larger protein degradation and proteolytic activity of enzymes [3–5]. The protein turnover decreases with age [5, 6], and there seems to be an effect of the fiber type in muscles in relation to protein turnover [7–9]. Whereas myofibrillar and cytoskeletal proteins are degraded postmortem, there are only minor changes in the proteins that constitute the perimysium and endomysium postmortem; thus, the connective tissue plays a major role in the experienced texture of conventional meat after cooking [10]. In cultivated meat, there is no contribution from perimysium and endomysium to the texture of the product. For products where the cultivated cells are used as ingredients or a minor part of the product, the texture will be largely dependent on the used scaffolds or formulations of the final product, whereas the postharvest processes in the myotubes will play a larger role for products of almost entirely cultivated nature. However, whether these processes can be manipulated as in live animals through the origin of the cells, the growth of the myotubes and the glycogen level at harvest are not established at this time. There is research indicating the same mechanisms in cultured meat as in conventional meat. First of all, the key proteolytic enzyme, the calpains (calpain 1-, 2-, 3- and calpastatin), is expressed during proliferation and differentiation [11–13], and the mRNA expression increases during differentiation [11] (Skrivergaard et al., unpublished data). Likewise, the expression of the caspase family increases during myogenesis [12], and some of these are known to be part of the apoptotic cell death which, in conventional meat, is related to the postmortem changes [14, 15]. Mehmood et al. [16] have further demonstrated active calpains at the time of harvest of differentiated bovine myotubes, and expression of both calpain-1 and -2 is in the myotubes postharvest where they autolyze with time as seen in conventional meat, upon activation. Thus, a similar proteolysis as in conventional meat is expected in cultivated cells postharvest, supported by research demonstrating degradation of the cytoskeletal proteins (e.g., desmin) which increases with time [16]. What is still to be evaluated is the optimal postharvest conditions, like the established knowledge about the optimal temperature conditions during the rigor process for conventional meat [17] and the optimal time for the maturation process of the myotube biomass. In the study by Mehmood et al. [16], it seems that the processes occur faster, i.e., within the first 24 h, as opposed to conventional meat where the aging period can be 1–21 days or more [18].

In conventional meat, the myofibrillar and muscle fiber structure is highly organized, which impacts the sensorial experience of texture [19]. However, in cultivated meat, we cannot expect the same alignment of myotubes, which seems to depend on the scaffold used [20]. This can have an impact on the texture experience, despite proteolytic processes as in conventional meat. In a study using texture profile analysis (TPA) on cultivated rabbit skeletal muscle proliferated in a gelatin scaffold, compared with conventional rabbit meat and grounded beef, the level of hardness was similar. However, the hardness of the cultivated meat product resembled mostly ground beef when the change in texture from fresh product to cooked product was compared [20], suggesting that the matrix in which the myofibers are embedded is important for the final texture. Likewise, Wang et al. [21] used the TPA method to quantify hardness of cultivated meat in a combined scaffold of collagen and yeast proteins and compared the texture with beef. The hardness of the products depended on the scaffold and the developmental stage of the cells (proliferation vs. differentiation), but in this comparison, the conventional beef was evaluated to be much harder compared with the cultivated meat. Unfortunately, both studies lack information about conventional meat, like the age of the animal and aging process, which makes it difficult to compare the texture of the different products. Importantly, the studies show that similar methods can be used for both products and that the measured texture traits are within the same scale [22, 23]. The research demonstrates that in most cases [21–23], cultivated meat does not have the same hardness and chewiness as conventional meat, which might be a drawback for the experience of mouthfeel, so from a texture perspective, postharvest maturation might not be important. On the other hand, postharvest proteolysis is desired in order to release peptides and amino acids for flavor development. The efforts regarding texture of products with high content of cultivated product should therefore be allocated to development of scaffolds that can resemble the cytoskeletal structure of meat and support an alignment of the myofibers to reflect conventional meat as much as possible. For products where the cultivated element is only used as an ingredient or as a minor component, the post-processing strategy is determining the structure and composition as modified by processing technologies [24] and enzymes [25].

Industry Perspectives on Texture
Texture is a key sensory attribute of meat, and several companies highlighted its importance – particularly mouthfeel, melting behavior, and product homogeneity.

- "*The tactile experience while cutting – ensuring the product remains intact – is important, but most critical are the taste, flavor, and texture during consumption.*" (Aleph Farms – Neta Lavon)
- "*CultiFat offers improved melting behaviour and shelf life compared to traditional plant-based fats, thanks to its intracellular production, which limits auto-oxidation and allows for better control of lipid composition.*" (Cultimate Foods – Jordi Morales-Dalmau)

3.2 *Water-Holding Capacity*

Water-holding capacity is an important quality parameter in conventional meat, which not only affects the weight of the product, but also adds to the sensory evaluation of the meat. Basically, most water can be found associated with the protein structure of actin and myosin inside the muscle fiber. Hence, changes to these protein structures and their interrelated architecture, caused by, e.g., proteolysis, sarcomere shortening, and basic chemical properties caused by pH changes, would impact the water-holding capacity of cultivated meat. In conventional meat, water holding properties have been evaluated using methods like drip loss over time or nuclear magnetic resonance spectroscopy. Unfortunately, none of these methods have been applied in order to evaluate water-holding capacity in cultivated meat. However, simple methods, like evaluating gross water content in the product, have been applied. In a study by Yen et al. [26] using bovine mesenchymal stem cells, it was demonstrated that the moisture content in two different prototypes (layer-by-layer or burger-like) of cultivated meat was similar to or reduced compared to beef patty. The moisture content was quantified by weighing the products before and after freeze-drying. To provide a more consumer-related estimate of the water-holding capacity, cooking loss was assessed, demonstrating comparable losses between the cultivated meat prototypes and beef patty. Likewise, it was demonstrated that scaffold-free cell sheets made of bovine myoblasts staked to produce a <3 mm thick structure had water content of approx. 88% compared to 60% in beef [27]. However, given the abovementioned information, it is still difficult to predict the water-holding capacity of cultivated meat, and more focused investigations using methods like nuclear magnetic resonance (NMR) spectroscopy [28] are needed.

From the detailed knowledge obtained from conventional meat, some predictions can be made for the water-holding capacity of cultivated meat. As the myofibrillar structure is important for water-holding capacity, it can be speculated that the content of myosin and actin is highly correlated with the capacity of the myocytes to hold water. It has been demonstrated that when satellite cells undergo myogenesis in vitro, the differentiated cells have significantly greater expression of myosin compared to cells in the proliferative stage [29]. This implies that the basic proteins for creating myofibrils are larger in differentiated cells compared to proliferating cells, and therefore better water-holding capacity could be expected in differentiated cells.

Although some features improving water holding inside the myofiber can be suggested, others compromise this. One major event is the very rapid increase in lactate postharvest in cultures of differentiated bovine satellite cells [16]. This increase suggests a corresponding decline in pH, which could affect the net charge of the proteins, causing lower WHC if the ultimate pH is approaching pH 5.1–5.3 (Fig. 2a). Likewise, rapid degradation of myofibrillar proteins has been observed postharvest [16], causing changes in water holding. Together, this will potentially affect the water-holding capacity negatively. Moreover, it could be speculated that the sarcomere length is greatly shortened postharvest, as the cells are detached from their growth surface and thereby have the potential to reshape and round up, resulting in shorter sarcomeres and less water-holding capacity because of less space (Fig. 2b).

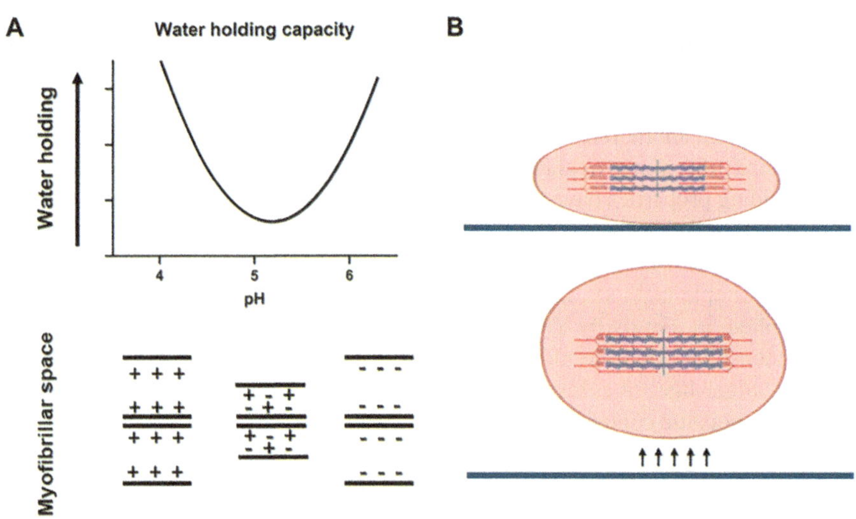

Fig. 2 Water-holding capacity is dependent on basic muscle structures. (**a**) pH of the meat dictates the electrostatic charge of the contractile proteins, myosin and actin, which make up the myofilaments. At the isoelectric point, the net charge is zero, resulting in less myofibrillar space, causing lower water-holding capacity. (**b**) When satellite cells are in suspension, they round up, possibly causing the sarcomere structure to shorten. This will potentially result in lowering the water-holding capacity

Industry Perspectives on Water-Holding Capacity

Although water-holding capacity was not explicitly highlighted by the companies, related attributes such as mouthfeel, cooking behavior, and humidity were mentioned as important indicators of product quality.

- *"Our blended product, which includes undifferentiated bovine adipogenic precursors, offers added value in terms of taste, mouthfeel, and cooking behavior compared to conventional plant-based alternatives."* (Mosa Meat – Mark Post)
- *"At Vital Meat, we have a business to business approach, so making sure the format of our product suits our clients is a continuous focus. Currently, our cultivated chicken is used by industrial partners in combination with plant-based ingredients. Its water content is well-suited to existing dry ingredient formulations and processing methods, which supports seamless integration into current manufacturing workflows."* (Vital Meat – Claude Rescan)

3.3 Color

The color of meat is an important quality trait and is probably the most important single factor for the average consumer to evaluate the quality of the meat in the buying situation. The color of conventional meat ranges from pale red (chicken) to deep red (beef). Thus, the color of a cultivated meat product must meet the expected color based on species origin. The color of meat is generally determined by the presence and state of color pigments, mainly myoglobin. The color of conventional meat is mainly dictated by the content of myoglobin and the state of the iron atom bound within. In its ferrous (Fe^{2+}) state, the meat will have a purple color, when no oxygen is present. However, upon exposure to oxygen, the color of the meat will change to red, while if the iron atom is oxidized to ferric (Fe^{3+}), the color will become brown. Thus, this implies that manipulation of the intracellular antioxidant capacity of the myofibers could foster color optimization. Likewise, the amount of myoglobin could also foster coloration of cultivated meat. Interestingly, the increased expression of myoglobin has been observed with differentiation of bovine satellite cells [29].

The state can be manipulated by, e.g., packaging strategies because it is linked to oxygen availability. For beef, the deep red color is often associated with a high-quality product. However, the color of bovine satellite cells in culture is very pale (Fig. 3).

In prototypes of cultivated meat, darker color was observed both before and after cooking compared to conventional beef patty [26]. For pure bovine satellite cells, however, more pale colors can be seen. Consequently, in a study comparing bovine satellite cells grown in fibrin hydrogels with conventional beef, not only were lower

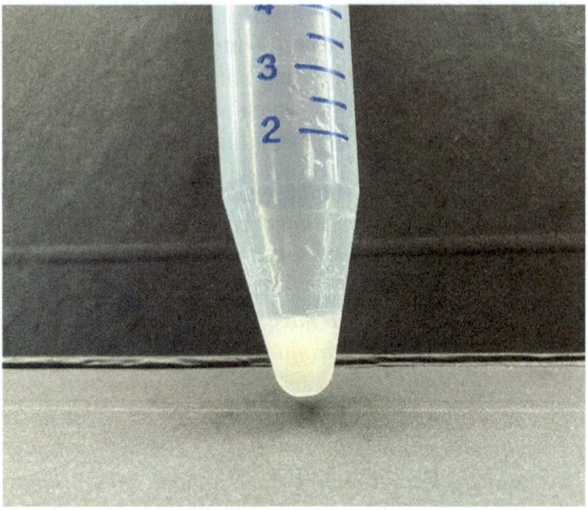

Fig. 3 Representative pictures of bovine satellite cells. Cultures of bovine satellite cells were cultured until clear myotube formation, trypsinized, and washed with PBS before pelleted by centrifugation

pigment levels observed but also large ΔE values (based on the CIE76 formula) [30]. If the cells were cultured with the addition of hemoglobin or myoglobin (both 3 mg/mL) for 9 days, the pigment content increased slightly and the ΔE values were reduced, but still far from that of conventional beef. However, cooking the sample diminished the differences.

Genetic engineering of cells offers the possibility to improve the coloration of the final product. In an attempt to increase nutritional quality, Stout et al. [31] produced transgenic bovine satellite cells containing various enzymes boosting the carotenoid biosynthesis pathway. This resulted in production of cell pellets with a more orange appearance compared to control samples. While orange color is not associated with high-quality meat products, this approach highlights the possibility of using genetic engineering to optimize coloration of satellite cell cultures to resemble that of conventional meat.

Another interesting aspect of cultivated meat is the prospect of using supporting structures like scaffolds to optimize coloration of the final product. In a study by Wang et al. [32], it was demonstrated that scaffolds based on proanthocyanidins and dialdehyde chitosan as cross-linkers of collagen could produce products with coloration comparable to beef.

Industry Perspectives on Visual Appearance: Color and Homogeneity
Appearance, including attributes such as color and visual uniformity, was mentioned by only a few companies. This is likely because some of the companies focus on producing ingredients or cultivated cells intended for use in hybrid products, where the final appearance is shaped by the complete formulation rather than the cultivated component alone.

(continued)

- "*Delivering an organoleptic experience for consumers encompasses various sensory attributes, including the product's appearance – such as color and homogeneity.*" (Aleph Farms – Neta Lavon)
- "*Appearance similar to regular meat.*" (Mosa Meat – Mark Post)

3.4 Flavor

Several compounds add to the sensory perception of flavor of conventional meat. These compounds range from well-defined single compounds, like inosine 5′--monophosphate, to groups of compounds like fatty acids [33, 34]. While the flavor of raw meat is important, more important are the flavor precursors and compounds produced during maturation and cooking. During heating of the meat, reducing sugars will react with the amino group of amino acids to create a mixture of flavor-rich compounds, a process known as the Maillard reaction. Hence, to make a true and applicable evaluation of the flavor attributes of cultivated meat, heating must be considered.

With Mark Post's presentation of the first burger made from cultivated meat in 2013, it became obvious that the taste rating of cultivated meat was low. Especially, it was identified that the taste, although meaty, missed some fatty taste. This is not surprising, given that the burger patty was made from pure muscle fibers. From conventional meat, it is known that, within the 3–7% fat percentage area in the uncooked meat, the content of fat is greatly correlated with palatability [35]. Hence, it could be expected that increasing the fat content of the cultivated meat product would increase the flavor and consumer acceptance. Research from our lab has shown that cultures of both bovine and porcine satellite cells take up fatty acids and store them inside the muscle fiber for several days (Abraham et al., unpublished data). However, this is not straightforward, as the fatty acids when added to the cell culture medium can introduce cytotoxic effects. Thus, the concentration of fatty acids added to the media is vital, as well as the type of fatty acids (saturated is more toxic compared to unsaturated). Moreover, the composition of the media before adding fatty acids also determines the cellular lipid uptake and accumulation. Satellite cells grown in no- or low-serum (2%) media have significantly higher lipid accumulation compared to cells grown in high serum (≥10%). Attempts to mimic the fatty acid composition of conventional meat demonstrated that cocktails of fatty acids matching the in vivo lipid distribution in beef did not introduce cytotoxicity of bovine satellite cells (Mehmood et al., unpublished data).

In the conversion of muscle to meat, time of maturation is paramount. This is especially true for beef. To what extent this is crucial for cultivated meat has only recently started to be elucidated. In a study from Mehmood et al. [16], it was demonstrated that differentiated bovine satellite cells expressed active forms of the protease calpain, which are heavily involved in maturation of beef. Moreover, the same study also demonstrated proteolytic degradation of structural proteins like desmin. Thus, the degradation of proteins, as well as the presence of sugars originating from glycogen, could potentially cause more substrate for Maillard reactions and by that improve taste. In an unpublished study from our lab, it was demonstrated that porcine myotubes aged for up to 24 h had increased concentrations of several amino acids (Abraham et al., unpublished data). Interestingly, some of these amino acids, like glutamine and glutamate, have been ascribed to umami taste [36].

Although the improvement of the flavor score of myocyte-based cell mass per se is an appealing approach, the option of mixing in flavor-adding compounds in the final stage of the production is relevant, and here the option of adding cultivated fat to the product could be advantageous. In an evaluation of the flavor of porcine cultivated fat, where gas chromatography-mass spectrometry (GC-MS) was coupled to descriptive sensory analysis and compared to conventional pork fat [37], only minor differences were observed between cultivated fat and fat samples obtained from pork. Furthermore, the flavor profile (obtained using GC-MS) of conventional meat and cultivated meat assembled of 2D structures of muscle tissue and adipose tissue has been demonstrated to be similar [22].

Along this line, there is also the possibility of using scaffold materials that contribute to the flavor of the final product. Using a mixture of gelatin and agar to coat a scaffold of textured vegetable protein, similar sensory scores (using an electronic tongue) were observed, between cultivated and conventional meat [23].

With the in vitro production of cell mass to create the final meat product also comes the opportunity to avoid off-flavors. One example is the off-flavor phenomenon known as boar taint caused by the microbial metabolism of tryptophan producing the odorous compound skatole and the testicular production of androsterone [38]. The production of both compounds is bypassed in cultivated meat production. Likewise, off-flavor caused by specific feeding ingredients is eliminated [39], but whether media components used for cultivated meat can impact flavor is not known.

Industry Perspectives on Taste and Flavor: The Top Priority Across the Industry

All participating companies identified taste and flavor as the most critical quality parameters – both in their current development efforts and in their long-term product strategies. These sensory attributes are consistently viewed as the primary drivers of consumer acceptance and market success for cultivated meat.

- "*Taste will always be N°1 in food! This is particularly relevant for cultivated meat: rather than imitating conventional chicken, our cultivated chicken inherently replicates its authentic taste.*" (Vital Meat – Claude Rescan)
- "*CultiFat acts as a flavouring agent and improves taste and extends aroma to increase the meaty and piggy flavour in plant-based meat formulations.*" (Cultimate Foods – Jordi Morales-Dalmau)

3.5 *Nutritional Quality of Cultivated Meat Products*

Conventional meat is rich in nutrients and has likely played an important role in human evolution. As part of a human diet, meat can contribute to important macro- and micronutrients including highly digestible proteins, iron, vitamin B12, zinc, and selenium. The nutritional content of meat depends on various factors such as species, feeding and cut [40]. Given the high nutritional value of conventional meat, the nutritional quality of cultivated meat is an important parameter to take into consideration. A previous study indicated that the perception of cultivated meat as healthy and nutritious was positively associated with willingness to consume cultivated meat [41]. If the consumption of cultivated meat is to substitute or partially substitute conventional meat consumption, matching the nutritional value of conventional meat is important. The production of cultivated meat might offer the potential of modifying the nutritional value of the final product, producing healthier products that meet the nutritional requirements of the consumer. As an example, Stout et al. [31] engineered the production of antioxidant phytonutrients (carotenoids) into immortalized mouse myoblasts and primary bovine muscle stem cells. It was shown that the resulting endogenous production of carotenoids was able to reduce lipid oxidation, thereby potentially enhancing nutrition and quality of cultivated meat products [31].

3.5.1 Protein Content and Amino Acid Composition

Protein is an important nutritional component of meat. Protein from conventional meat is characterized by high digestibility and contains all essential amino acids with no limiting amino acids [42]. The Digestible Indispensable Amino Acid Score (DIAAS) is used as a measure of protein digestibility. Proteins with DIAAS values ≥100 are classified as excellent-quality protein [43]. Pork, beef, and chicken have been reported to have DIAAS scores above 100 [44–46]. The protein content of meat is 22% on average, but it has been reported to vary from 12.3% up to 34.5% depending on species and production strategy [40]. Since protein is considered an important nutritional compound of conventional meat, the protein content and quality of cultivated meat are crucial.

Very few studies have reported the protein content of cultivated meat products. In addition, comparison of nutritional values between products is difficult due to the diversity of products in relation to cell type, cell density, scaffold material, etc. Kim et al. (2025) investigated the amino acid composition of cultivated meat, which was based on pig muscle stem cells cultured on cross-linked gelatin hydrogel scaffolds [47]. The cultivated meat had a higher moisture content compared to pork but lower compared to the scaffold alone; thus, the composition of the scaffold will affect the protein content of the final product. Differences in total amino acid composition were found for most of the investigated amino acids, where the content of most amino acids was lower in cultivated porcine cells compared to pork. The total amino acid content did not differ significantly between pork and cultivated porcine cells but was significantly lower in the scaffold alone. When amino acids were expressed as g/kg protein, the concentration of most of the amino acids was equal or higher in cultivated pork compared to pork, although some still had lower values [47]. It should be noticed that the values reported are related to total amino acid composition, not bioaccessible amino acids, reflecting digestion of the proteins by gastric and intestinal enzymes. Recently, Wang et al. (2024) were able to produce a cultivated meat product matching the total protein content of beef via a co-culture of bovine myocytes and adipocytes [32]. By regulating the bovine myoblast-to-adipocyte ratio, the authors were able to control the protein and fat content of the final product. Hence, when using a bovine myoblast-to-adipocyte ratio of 10:0, 9:1, and 8:2, the cultivated meat could match the protein content of beef. In addition, it was found that the cultivated meat had higher digestibility compared to beef [32]. For cultivated chicken made from cultured chicken fat and high moisture-extruded soy protein, it has also been possible to match the protein content of conventional chicken breast [48].

Various factors in the production process of cultivated meat are likely to affect the protein content of the final product. One factor affecting the protein content of the final product is the degree of differentiation of the muscle cells. The differentiation of muscle satellite cells ultimately resulting in the formation of mature muscle fibers has been deemed necessary for consumer acceptance in order to produce cultivated meat products resembling conventional meat with regard to functionality, taste, and texture [49]. Furuhashi and Takeuchi (2025) investigated the effects of cell differentiation and aging on intracellular free amino acids (FAA) in bovine muscle cells [50]. They found that differentiation decreased FAAs, whereas aging increased FAAs. The total content of protein per gram of dry cell weight was higher in myotubes compared to myoblasts but decreased with aging. In addition, it was found that cultured cells had a higher content of FAAs per dry weight compared to beef [50]. The degree of differentiation can be affected by the scaffold material used. Wang et al. (2024) determined the protein content of cultivated meat from bovine myoblasts using scaffolds with varying proportions of yeast and collagen protein for bovine myoblasts. A significantly lower protein content was found in the cultivated meat compared to beef, except for the cultivated meat grown on scaffolds with a yeast protein/collagen ratio of 3:7. Together with fusion index (a measure of myotube formation) determination, the results indicated that the addition of yeast protein in the scaffold promoted myogenic differentiation and muscle tissue formation [21]. It has also been shown that the stiffness of a scaffold can affect myogenesis and adipogenesis. Hence, Lee et al. (2024) regulated cell differentiation by controlling the stiffness of gelatin scaffolds by varying degrees of alginate cross-linking [22]. Using these scaffolds, primary bovine myoblasts and bovine adipose-derived mesenchymal stem cells were cultured to produce 2D blocks of muscle and fat, respectively. These blocks were combined to produce an assembled cultivated meat product. A lower protein content was found in the assembled cultivated meat product compared to beef; however, the results indicated that the nutritional value of the final product was affected by the degree of differentiation [22]. Together, these studies indicate that scaffold material can promote cell differentiation and thereby protein content, but the scaffold material itself can also provide valuable protein and amino acids. Liu et al. (2022) produced engineered porcine muscle satellite cell meatballs and analyzed nutritional value. Compared to a commercial product (Shizitou), dried porcine muscle satellite cell meatballs had five times higher protein content, which was likely due to high cell density and a gelatin scaffold rich in proteins [51]. Hence, the scaffold material has the potential to improve the protein content and amino acid composition of the cultivated meat product [52].

3.5.2 Lipid Content

The fat content and fatty acid profile of conventional meat vary and are dependent on species, feeding and breeding style, and specific cuts. Cooking can also affect fat content with considerable fat loss depending on the cooking method [40, 53]. Conventional meat can contain a relatively large proportion of saturated fatty acids but does also contain monounsaturated and polyunsaturated fatty acids (PUFAs) [42, 53]. The relatively high content of saturated fatty acids in conventional meat has received attention due to a possible association with cardiovascular disease and diabetes; however, this association is being increasingly questioned [54].

Different approaches to incorporate fat into cultivated meat products have been investigated, including assembly of 2D fat and muscle blocks and co-culturing of myocytes and adipocytes [22, 32]. Few studies report the fat content of cultivated meat products containing both muscle and fat. For cultured chicken, it has been possible to match the fat content of conventional chicken. The fat content of chicken strips made from cultured chicken fat and high moisture-extruded soy protein exceeded the fat content of chicken breast. At the same time, the cultured chicken product had a lower amount of saturated fat and cholesterol compared to chicken breast [48]. An assessment of cultured chicken meat based on data submitted to the US Food and Drug Administration on a cultured chicken meat product found higher fat content in cell-cultured chicken meat compared to conventional chicken [55]. Matching the fat content of cultivated meat to conventional beef and pork has been challenging. Two studies using bovine myoblasts found lower fat in cultivated meat compared to beef [21, 22], and a third of fat in porcine muscle satellite cell meatballs was found compared to a commercial product (Shizitou) [51]. Using a co-culturing approach combining myocytes and adipocytes, Wang et al. (2024) were able to produce a cultured meat product reaching the fat levels of conventional beef. Hence, when using a bovine myoblast-to-adipocyte ratio of 8:2 and 5:5, the fat content in the cultivated meat was significantly higher than the conventional beef and statistically similar to a ratio of 0:10. The results indicated that the fat content of cultivated meat products can be controlled by the ratio of myoblasts to adipocytes [32]. Being able to modify the nutritional content is a possible advantage of cultivated meat production, and other studies have investigated the possibility of regulating the fatty acid composition of cultivated meat. Thus, it has been shown that it is possible to control the fatty acid composition of cultivated fat by adjusting the culture media composition during in vitro culture. As an example, murine and porcine adipocytes were grown in 2D cultures, and after in vitro adipogenesis, the cells were aggregated into 3D tissues. Treatment with a lipid emulsion based on soybean oil during in vitro culture affected the intracellular adipocyte triacyl-glycerides in the adipocytes including an increase in PUFA levels. For the murine cells, the lipid addition resulted in the fatty acids more closely resembling native mouse fat [56]. In addition, Louis et al. (2023) cultured bovine adipose-derived stem cells and found that the fatty acid composition of the culture medium during adipogenesis could be used to control the fatty acid composition of the final product [57]. It has also been found that the phospholipid, phosphatidylcholine, can not only promote adipogenic differentiation and lipid accumulation in a sodium alginate-gelatin 3D culture system, but also increase the content of PUFA and lower total cholesterol of cultivated fat [58].

The scaffold material may further affect the final fat content of cultivated meat products. As previously mentioned, the stiffness of the scaffold can affect myogenesis and adipogenesis. Hence, Lee et al. (2024) demonstrated that they were able to regulate and optimize the degree of adipogenic differentiation of bovine adipose tissue-derived mesenchymal stem cells to produce 2D fat blocks. Combining these blocks with 2D blocks of cultivated muscle cells resulted in an assembled cultivated meat product. Despite the optimized cultivated fat blocks, the assembled product had lower total fat content compared to raw beef, but the study demonstrates the possibility of controlling the amount of lipids in the product via the scaffold [22]. Wang et al. (2024) investigated bovine myoblasts cultured on scaffolds with varying proportions of yeast protein and found a lower amount of fat in cultivated meat compared to conventional beef. The content of fat was reported to be markedly lower in the cultured meat product compared to beef. The fat content in the cultivated meat varied between the products made with different yeast protein and collagen content [21].

3.5.3 Vitamins and Minerals

Meat is a rich source of vitamins and minerals, including bioavailable vitamin B12, vitamin B6, riboflavin, and niacin in addition to zinc, selenium, and iron [53]. Red meat in particular can provide over two-thirds of the daily requirement of vitamin B12 in a 100 g serving [42]. A large part of the iron found in meat is heme-bound iron, which is more bioavailable than nonheme iron, which is the form present in vegetables. Therefore, meat is an important source of iron with beef having the highest heme-iron content followed by pork and chicken [53]. Cultivated meat prototypes have shown high mineral content values compared to conventional meat, which might be due to the elements being present in the culture media [52]. As an example, cultivated engineered pork meatballs contained relatively high levels of essential minerals including zinc, calcium, and iron [51]. In the assessment of cultivated chicken meat, using data submitted to the US Food and Drug Administration, it was reported to have a superior mineral content compared to conventional chicken, likely due to the presence of these elements in the culture media. The study also pointed out the importance of investigating bioavailability in future analyses [55]. Since most heme-iron in meat is present in myoglobin and to a lesser extent hemoglobin, increasing the amount of myoglobin in cultivated meat would probably be beneficial from a nutritional point of view [59]. Among the compounds found in meat, not all of them are directly produced in the muscle. Some of them might derive from digested animal feed and reach the muscle after modification by non-muscle organs [59]. As an example, a nutrient such as B12 will not be produced by myoblasts nor adipocytes, and therefore, it is likely that it needs to be added to the culture media and taken up by the cells [60]. However, whether vitamin uptake from the culture media will result in vitamin levels in the cultured meat corresponding to the levels in conventional meat still needs to be explored. Adding vitamins after culturing might be an alternative approach if addition via culture media is insufficient [59].

Industry Perspectives on Nutritional Value

While nutritional value is not yet the central focus in the current stage of cultivated meat product development, companies recognize its importance. Many view it as both a future area for innovation and a key indicator of overall product quality.

Some early-stage efforts have already incorporated nutritional targets:

- "*Nutritional value is a key consideration in Aleph Farms' product development. Our first product, Aleph Cut, contains at least 15% protein and less than 5% fat.*" (Aleph Farms – Neta Lavon)

Looking ahead, others emphasize the role nutrition will play in shaping the next generation of products:

- "*Nutritional composition and uptake will be important for future products.*" (Re:Meat – Jacob Schaldemose Peterson)

4 Future Perspectives on Cultivated Meat Quality

As the cultivated meat sector evolves, future developments will increasingly focus on refining product quality beyond basic functionality. A key distinction will be made between cultivated meat as a standalone ingredient and its role in hybrid products, where it is combined with plant-based or other components. This opens opportunities to optimize individual elements – such as flavor and nutrition – while designing them to work synergistically in final formulations.

Scaffold materials will play a more active role in quality enhancement, potentially serving as carriers for micronutrients like vitamin E or functional fats, contributing both to nutritional value and oxidative stability. Meanwhile, texture development will benefit from hybrid approaches, where plant-based matrices or novel structuring techniques can complement the mouthfeel and bite of cultivated muscle tissue.

A critical area of innovation lies in balancing production-stage optimization with postharvest processing. Minimizing downstream processing – while still achieving desired sensory and nutritional profiles – will be essential for sustainability, cost efficiency, and possibly consumer acceptance. This will require detailed characterization of postharvest processes and a clear understanding of their time and resource implications.

Product diversification will also expand, with differentiation based on muscle types, animal species, and developmental stages (e.g., calf vs. adult cattle), enabling tailored products for specific markets or consumer groups. This includes the potential to design targeted nutrition profiles for patient populations or other specialized dietary needs.

Another aim is understanding how cultivated cells interact with other food components – from fats and binders to flavor carriers – to form a cohesive, appealing final product. Achieving a balance between homogeneity and heterogeneity will be key: while uniformity ensures consistency, controlled heterogeneity may better mimic the complexity of conventional meat.

Finally, nutritional evaluation will become increasingly important, not only to meet regulatory standards but also to substantiate health claims and guide product development. As the industry scales, robust upscaling strategies and postharvest quality control will be vital to ensure that cultivated meat delivers on its promise of being a sustainable, nutritious, and affordable food of the future.

Acknowledgments The authors would like to express their sincere gratitude to the companies Vital Meat, Mosa Meat, Aleph Farms, Re:Meat, and Cultimate Foods for generously sharing valuable insights into the commercial perspectives on cultivated meat quality.

References

1. Pogorzelski G et al (2022) Towards an integration of pre- and post-slaughter factors affecting the eating quality of beef. Livest Sci 255:104795
2. Koohmaraie M (1996) Biochemical factors regulating the toughening and tenderization processes of meat. Meat Sci 43(S1):193–201
3. Kristensen L et al (2002) Dietary-induced changes of muscle growth rate in pigs: effects on in vivo and postmortem muscle proteolysis and meat quality. J Anim Sci 80(11):2862–2871
4. Therkildsen M et al (2004) Compensatory growth in slaughter pigs – in vitro muscle protein turnover at slaughter, circulating IGF-I, performance and carcass quality. Livest Prod Sci 88(1): 63–75
5. Maharjan P et al (2021) Effects of dietary energy levels on Pectoralis major mixed muscle protein turnover and body composition in two broiler lines housed in different grow-out environments. J Anim Physiol Anim Nutr (Berl) 105(3):535–548
6. Lobley GE et al (2000) The effects of breed and level of nutrition on whole-body and muscle protein metabolism in pure-bred Aberdeen angus and Charolais beef steers. Br J Nutr 84(3): 275–284
7. Garlick PJ et al (1989) Fiber-type composition of nine rat muscles. II. Relationship to protein turnover. Am J Phys 257(6 Pt 1):E828–E832
8. Therkildsen M et al (2002) Compensatory growth response in pigs, muscle protein turn-over and meat texture: effects of restriction/realimentation period. Anim Sci 75(3):367–377
9. Chang Y-M, Wei H-W (2005) The effects of dietary lysine deficiency on muscle protein turnover in postweanling pigs. Asian Australas J Anim Sci 18(9):1326–1335
10. Purslow PP (2005) Intramuscular connective tissue and its role in meat quality. Meat Sci 70(3): 435–447

11. Theil PK et al (2006) Changes in proteolytic enzyme mRNAs relevant for meat quality during myogenesis of primary porcine satellite cells. Meat Sci 73(2):335–343
12. Yang YB, Pandurangan M, Hwang I (2012) Changes in proteolytic enzymes mRNAs and proteins relevant for meat quality during myogenesis and hypoxia of primary bovine satellite cells. In Vitro Cell Dev Biol Anim 48(6):359–368
13. Wang W et al (2017) Changes in calpain and caspase gene expression at the mRNA level during bovine muscle satellite cell myogenesis and the correlation between the cell model and the muscle tissue. Russ J Bioorg Chem 43(3):270–277
14. Laville E et al (2009) Proteome changes during meat aging in tough and tender beef suggest the importance of apoptosis and protein solubility for beef aging and tenderization. J Agric Food Chem 57(22):10755–10764
15. Kemp CM, Wheeler TL (2011) Effects of manipulation of the caspase system on myofibrillar protein degradation in vitro. J Anim Sci 89(10):3262–3271
16. Mehmood W et al (2025) Postharvest changes in bovine satellite cells in vitro resemble postmortem conversion of muscle to meat. Food Chem 483:144292
17. Rosenvold K, Borup U, Therkildsen M (2010) Stepwise chilling: tender pork without compromising water-holding capacity. J Anim Sci 88(5):1830–1841
18. Joo ST et al (2023) Aging mechanism for improving the tenderness and taste characteristics of meat. J Anim Sci Technol 65(6):1151–1168
19. Listrat A et al (2016) How muscle structure and composition influence meat and flesh quality. ScientificWorldJournal 2016(1):3182746
20. MacQueen LA et al (2019) Muscle tissue engineering in fibrous gelatin: implications for meat analogs. npj Sci Food 3(1):20
21. Wang Y et al (2024) 3D edible scaffolds with yeast protein: a novel alternative protein scaffold for the production of high-quality cell-cultured meat. Int J Biol Macromol 259(Pt 1):129134
22. Lee M et al (2024) Cultured meat with enriched organoleptic properties by regulating cell differentiation. Nat Commun 15(1):77
23. Lee M et al (2022) Tailoring a gelatin/agar matrix for the synergistic effect with cells to produce high-quality cultured meat. ACS Appl Mater Interfaces 14(33):38235–38245
24. Lambert EG et al (2024) Enhancing the palatability of cultivated meat. Trends Biotechnol 42(9): 1112–1127
25. Karp SG et al (2024) Enzymes in the production of cultivated meat products. Syst Microbiol Biomanuf 5(1):57–68
26. Yen FC et al (2023) Cultured meat platform developed through the structuring of edible microcarrier-derived microtissues with oleogel-based fat substitute. Nat Commun 14(1):2942
27. Tanaka RI et al (2022) Production of scaffold-free cell-based meat using cell sheet technology. npj Sci Food 6(1):41
28. Bertram HC, Purslow PP, Andersen HJ (2002) Relationship between meat structure, water mobility, and distribution: a low-field nuclear magnetic resonance study. J Agric Food Chem 50(4):824–829
29. Skrivergaard S et al (2021) Bovine satellite cells isolated after 2 and 5 days of tissue storage maintain the proliferative and myogenic capacity needed for cultured meat production. Int J Mol Sci 22(16):8376
30. Simsa R et al (2019) Extracellular heme proteins influence bovine myosatellite cell proliferation and the color of cell-based meat. Foods 8(10):521
31. Stout AJ et al (2020) Engineering carotenoid production in mammalian cells for nutritionally enhanced cell-cultured foods. Metab Eng 62:126–137
32. Wang Y et al (2024) A rich-nutritious cultured meat via bovine myocytes and adipocytes co-culture: novel prospect for cultured meat production techniques. Food Chem 460(Pt 3): 140696
33. Calkins CR, Hodgen JM (2007) A fresh look at meat flavor. Meat Sci 77(1):63–80
34. To KV et al (2023) Inosine 5′- monophosphate derived umami taste intensity of beef determination by electrochemistry and chromatography. Meat Sci 206:109343

35. Miller RK (2004) Chemical and physical characteristics of meat/palatability. In: Jensen WK (ed) Encyclopedia of meat sciences. Elsevier, Oxford, pp 256–266
36. Diepeveen J, Moerdijk-Poortvliet TCW, van der Leij FR (2022) Molecular insights into human taste perception and umami tastants: a review. J Food Sci 87(4):1449–1465
37. Lew ET et al (2024) Chemical and sensory analyses of cultivated pork fat tissue as a flavor enhancer for meat alternatives. Sci Rep 14(1):17643
38. Rasmussen MK, Zamaratskaia G (2014) Regulation of porcine hepatic cytochrome p450 – implication for boar taint. Comput Struct Biotechnol J 11(19):106–112
39. Melton SL (1990) Effects of feeds on flavor of red meat: a review. J Anim Sci 68(12): 4421–4435
40. Pereira PM, Vicente AF (2013) Meat nutritional composition and nutritive role in the human diet. Meat Sci 93(3):586–592
41. Leung AK et al (2023) Higher well-being individuals are more receptive to cultivated meat: an investigation of their reasoning for consuming cultivated meat. Appetite 184:106496
42. Williams P (2007) Nutritional composition of red meat. Nutr Dietet 64(S4):S113–S119
43. Dietary protein quality evaluation in human nutrition. Report of an FAQ Expert Consultation. FAO Food Nutr Pap 2013;92:1–66
44. Herreman L et al (2020) Comprehensive overview of the quality of plant- and animal-sourced proteins based on the digestible indispensable amino acid score. Food Sci Nutr 8(10): 5379–5391
45. Ertl P, Knaus W, Zollitsch W (2016) An approach to including protein quality when assessing the net contribution of livestock to human food supply. Animal 10(11):1883–1889
46. Marinangeli CPF, House JD (2017) Potential impact of the digestible indispensable amino acid score as a measure of protein quality on dietary regulations and health. Nutr Rev 75(8):658–667
47. Kim M, Jung HY, Jo C (2025) Fundamental study on structural formation, amino acids and nucleotide-related compounds of cultivated meat from 3D-cultured pig muscle stem cells. Food Sci Biotechnol 34(2):457–469
48. Pasitka L et al (2023) Spontaneous immortalization of chicken fibroblasts generates stable, high-yield cell lines for serum-free production of cultured meat. Nat Food 4(1):35–50
49. Martins B et al (2024) Advances and challenges in cell biology for cultured meat. Annu Rev Anim Biosci 12:345–368
50. Furuhashi M, Takeuchi S (2025) The effects of differentiation and aging on free amino acid profiles in cultured bovine muscle tissue. Food Chem 488:144753
51. Liu Y et al (2022) Engineered meatballs via scalable skeletal muscle cell expansion and modular micro-tissue assembly using porous gelatin micro-carriers. Biomaterials 287:121615
52. Lim PY, Suntornnond R, Choudhury D (2025) The nutritional paradigm of cultivated meat: bridging science and sustainability. Trends Food Sci Technol 156:104838
53. Pereira PC, Vicente F (2022) Meat nutritive value and human health. In: Purslow P (ed) New aspects of meat quality. Woodhead Publishing, pp 561–577
54. Geiker NRW et al (2021) Meat and human health-current knowledge and research gaps. Foods 10(7):1556
55. Sikora D, Rzymski P (2024) Assessment of the potential nutritional value of cell-cultured chicken meat in light of European dietary recommendations. J Food Compos Anal 135:106663
56. Yuen Jr JSK et al (2023) Aggregating in vitro-grown adipocytes to produce macroscale cell-cultured fat tissue with tunable lipid compositions for food applications. elife 12:e82120
57. Louis F et al (2023) Mimicking Wagyu beef fat in cultured meat: progress in edible bovine adipose tissue production with controllable fatty acid composition. Mater Today Bio 21:100720
58. Liu SQ et al (2024) Production of sodium alginate-gelatin composite hydrogel-based 3D cultured fat with low cholesterol and high polyunsaturated fatty acids. Food Hydrocoll 154: 110156
59. Fraeye I et al (2020) Sensorial and nutritional aspects of cultured meat in comparison to traditional meat: much to be inferred. Front Nutr 7:35
60. Zidarič T et al (2020) Cultured meat: meat industry hand in hand with biomedical production methods. Food Eng Rev 12(4):498–519

Adv Biochem Eng Biotechnol (2026) 193: 221–272
https://doi.org/10.1007/10_2026_333

Published online: 30 June 2026

Safety Aspects of Cell Culture-Derived Food for Human Consumption

Fernando Rivero-Pino, Diana C. Rueda, Chee Seng Hee, Oana. N. Kubinyecz, and Hannah E. Lester

Contents

Abstract The emergence of cell culture-derived foods such as cultured meat and seafood could represent a transformative shift in the food production system. However, the novel nature of these products raises safety considerations that must be addressed prior to market entry. This chapter provides a comprehensive overview of the safety assessment framework for cell culture-derived foods intended for

---

F. Rivero-Pino, D. C. Rueda, C. S. Hee, O. N. Kubinyecz, and H. E. Lester (✉)
Atova Regulatory Consulting, SLU, Barcelona, Spain
e-mail: hannah@atovaconsulting.com

human consumption. It summarises the critical components of the risk assessment to be applied, including cell line selection, gene editing and genetic modification of cell lines, culture media composition, nutritional information, and toxicological considerations. Special attention is given to potential risks such as contamination, genetic instability, allergenicity, and unintended effects from novel ingredients or culture media components. The chapter also explores current research gaps that need to be considered to ensure safety and support innovation in the food sector.

Graphical Abstract

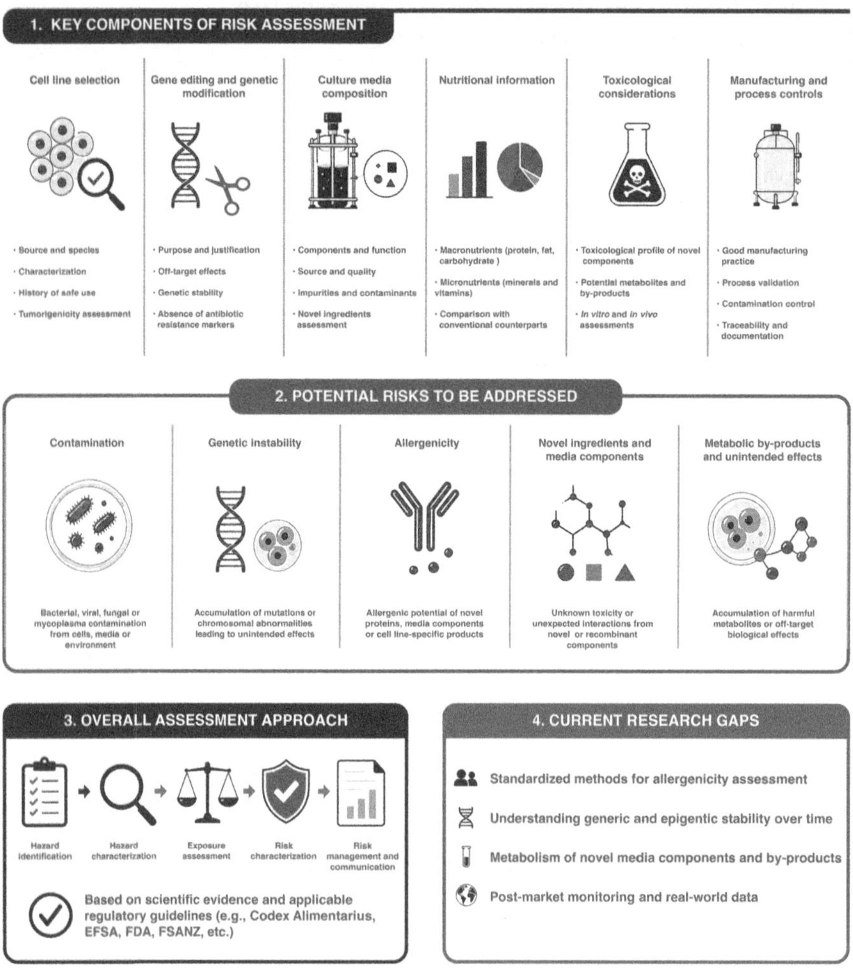

Keywords Cultivated meat, Risk assessment, Safety assessment

Abbreviations

ADME	Absorption, distribution, metabolism, and excretion
AI	Adequate intake
APAC-SCA	Asia-Pacific Society for Cellular Agriculture
BSA	Bovine serum albumin
CCP	Cell-cultivated product
CJD	Creutzfeldt–Jakob disease
CM	Cultivated meat
CMSI	Cultured Meat Safety Initiative
COA	Certificate of Analysis
COI	Cytochrome c oxidase subunit I
DIAAS	Digestible indispensable amino acid score
DNA	Deoxyribonucleic acid
EFSA	European Food Safety Authority
ELISA	Enzyme-linked immunosorbent assay
EMA	European Medicines Agency
ERV	Endogenous retroviral
EU	European Union
FAO	Food and Agriculture Organization (of the United Nations)
FBS	Foetal bovine serum
FSA	Food Standards Agency
FSANZ	Food Standards Australia New Zealand
GCCP	Good Cell Culture Practices
GFI	Good Food Institute
GLP	Good Laboratory Practice
GMO	Genetically modified organisms
GMP	Good Manufacturing Practices
HACCP	Hazard Analysis and Critical Control Points
HBGV(s)	Health-based guidance value(s)
IgE	Immunoglobulin E
INFOGEST	International consensus in vitro digestion method
LOD	Limit of detection
LOQ	Limit of quantification
MCB	Master cell bank
MFDS	Ministry of Food and Drug Safety
MOE	Margin of exposure
NAMs	New approach methodologies
NGTs	New genomic techniques
NHANES	National Health and Nutrition Examination Survey
NOAEL	No Observed Adverse Effect Level
OECD	Organisation for Economic Co-operation and Development
PBOs	Precision bred organisms
PCR	Polymerase chain reaction

PDCAAS	Protein digestibility-corrected amino acid score
PTMs	Post-translational modifications
qPCR	Quantitative polymerase chain reaction
QSAR	Quantitative structure-activity relationship
RDA	Recommended dietary allowance
RNA	Ribonucleic acid
SAMI	Safety-assessed media ingredients
SFA	Singapore Food Agency
STR	Short tandem repeats
TALENs	Transcription activator-like effector nucleases (TALENs)
TSE	Transmissible spongiform encephalopathies
TTC	Threshold of toxicological concern
UK	United Kingdom
UL	Upper intake level
WCB	Working cell bank
WGS	Whole-genome sequencing
WHO	World Health Organization
WoE	Weight of evidence
ZFS	Zinc finger nucleases

1 Introduction

1.1 Cultivated Foods and the Importance of Safety

Cell culture-derived foods, commonly referred to as cultured or cultivated foods, are produced from animal or plant cells grown in vitro under controlled conditions. These cells proliferate in sterile, nutrient-rich media within bioreactors, forming edible biomass or structured tissue [1, 2]. Among these products, cultivated meat (CM) and seafood have attracted commercial and regulatory interest. While the underlying production concept is well established, its implications for food safety assessment are more complex. Unlike conventional meat, CM introduces novel considerations, and ensuring its safety is critical for both regulatory approval and consumer acceptance.

Owing to its innovative production process and lack of a history of consumption, CM is subject to pre-market authorisation in jurisdictions with established regulatory frameworks. Central to this process is a comprehensive safety assessment conducted in accordance with internationally recognised risk assessment principles. Nevertheless, the primary responsibility for food safety remains with food business operators (FBOs), who must ensure that products placed on the market are safe for human consumption.

The safety framework for CM spans the entire production continuum. Regulatory considerations should be integrated from the earliest stages of research and development. This begins with cell sourcing and extends through process development, scale-up, and commercial manufacturing. Such an integrated approach enables product development, process control, and safety assessment to progress in parallel. Failure to address key safety or regulatory aspects early in development may compromise regulatory feasibility, necessitate substantial process modifications, and ultimately lead to delays and increased costs during authorisation.

Although established food safety principles provide a strong foundation, CM production presents distinct challenges that require additional consideration. Key aspects include the following:

- **Cell sourcing and traceability** – ensuring that the source material is well documented, ethically obtained, and free from transmissible agents
- **Cell line characterisation and stability** – confirming species identity, absence of adventitious agents, and long-term genetic and phenotypic stability
- **Genetic modification considerations** – evaluating the purpose, stability, and safety of any genetic alterations introduced into production cell lines
- **Culture media composition** – fully characterising media components, including recombinant growth factors, and assessing their presence, clearance, and fate in the final product
- **Production process controls** – demonstrating effective downstream processing, including removal or control of residue culture media components, recombinant growth factors, and processing aids, as well as ensuring the absence of unintended contaminants (e.g. microbial agents or extraneous cell types) and validating Hazard Analysis and Critical Control Point (HACCP)-based control measures
- **Compositional and nutritional assessment** – establishing the nutritional profile relative to conventional meat and identifying potential chemical, biological, or nutritional risks
- **Toxicological assessment** – determining whether existing data are sufficient or if additional studies are required
- **Allergenicity** – assessing potential allergenic risks, particularly in relation to novel proteins, residual media components, or altered expression profiles

These considerations can largely be addressed using established food safety assessment methodologies, although their application to CM often requires adaptation and case-specific approaches. Regulatory precedents from foods produced via microbial fermentation such as microbial biomass and recombinant proteins demonstrate how regulatory authorities evaluate products without a history of consumption. Many of the same principles are applicable to CM, particularly in areas such as compositional analysis, nutritional equivalence, allergenicity, and process validation.

Importantly, CM products have already been authorised in several jurisdictions, including Singapore, the United States, and Australia/New Zealand. Publicly available assessments from these regions provide valuable insight into how existing regulatory frameworks are being applied in practice. In parallel, dedicated guidance for the risk assessment of cell-cultured foods is beginning to emerge, as regulatory authorities and international organisations work to refine existing frameworks to address the specific characteristics of CM.

Building on this context, this chapter explores the safety considerations unique to CM, their interaction with regulatory requirements, and the areas in which both scientific understanding and regulatory approaches continue to evolve.

1.2 Regulatory Context

The regulatory landscape for cultivated foods varies considerably across jurisdictions, with direct implications for data requirements, submission complexity, and timelines for pre-market authorisation. Although regulatory systems share the common objective of ensuring food safety and consumer protection, they differ in terminology, procedural frameworks, and the authorities involved. These differences influence the efficiency with which companies can achieve market access and, consequently, have broader implications for innovation and international trade in this emerging sector [3, 4].

Even in jurisdictions where regulatory guidance is available, the safety assessment of CM is conducted on a case-by-case basis. This approach reflects the diversity of CM products and processes, requiring consideration of the specific characteristics of the final food, its source material, and the production process, including all relevant inputs and conditions. Applicants are therefore expected to provide a comprehensive and transparent description of these elements, supported by robust scientific data, generated using internationally recognised, validated methods to enable a thorough evaluation by the competent authorities.

This reliance on applicant-generated data can present practical challenges, particularly where regulatory expectations differ between jurisdictions. As a result, CM companies must anticipate and systematically address the unique safety considerations associated with CM, often tailoring their data packages to meet multiple regulatory frameworks. Greater international alignment in data requirements and assessment approaches would improve consistency, reduce duplication of effort, and enhance the overall efficiency of regulatory evaluation.

1.3 Current Risk Assessment Guidance for Cultivated Meat

Dedicated risk assessment guidance for cell-cultivated foods is still in its early stages, and no internationally harmonised framework is yet in place. Nonetheless, several authorities and international organisations have begun to adapt existing food safety principles to the specific challenges of these novel foods, publishing documents that guide both regulators and applicants on data requirements, methodological approaches, and areas of concern such as toxicology, allergenicity, nutritional adequacy, and production process integrity [5]. The extent and clarity of guidance vary considerably across jurisdictions, but each provides insight into how the unique safety aspects introduced in Sect. 1.1 can be integrated into regulatory evaluations.

Across different regions, guidance consistently converges on core safety themes: cell source, identity and stability, genetic modifications, cell banking and quality control, culture media safety, process conditions, and compositional, nutritional, and safety evaluation of the final product. However, differences in terminology, scope, and mandatory data requirements mean that applicants may still face duplication of effort when seeking approval in multiple regions. The development of internationally harmonised, science-based guidance would help drive regulatory convergence and reduce inefficiencies, but such alignment has not yet been achieved for any novel food and is likely to be a gradual and complex process. In the meantime, applicants are advised to structure dossiers around these shared principles wherever possible, while tailoring the presentation and depth of data to the specific expectations of each target jurisdiction.

1.3.1 Global Initiatives

In 2023, the Food and Agriculture Organization (FAO) published a landmark report on the food safety aspects of cell-based foods. It provides a structured overview of potential hazards and risks throughout the production process. Many hazards were considered manageable within existing frameworks, but the report highlighted areas unique to CM such as genetic stability of cell lines, use and clearance of growth factors, and residual scaffolds that require tailored risk assessment. For applicants, this report is a valuable reference when conducting early hazard mapping and designing a dossier that anticipates regulator concerns [6].

Ongoing Codex Alimentarius discussions have increasingly focused on the safety assessment of cell culture media components used in the production of cell-based foods. The relevant working document (CX/FA 26/56/13) recognises culture media as a critical input in the manufacturing process, with direct implications for product safety, composition, and regulatory evaluation.

The document highlights that cell culture media are complex mixtures designed to support cell growth and, in certain cases, differentiation, typically comprising nutrients (e.g. amino acids, carbohydrates, vitamins), growth factors, salts, and other functional components. Given their central role in upstream production, the safety of these constituents and their potential carryover into the final product must be systematically assessed.

A key principle outlined in the Codex work is the need for a structured and component-based approach to safety assessment. Media constituents may vary widely in origin and function, ranging from well-characterised food-grade substances to novel or highly purified compounds (e.g. recombinant proteins or growth factors). The document therefore proposes the development of a categorisation framework to group media components according to their level of familiarity and associated risk, with corresponding data requirements tailored to each category.

Particular attention is given to the potential for residual media components, their metabolites, or transformation products to remain in the final food. This necessitates evaluation of their fate during cell culture and downstream processing, including degradation, consumption by cells, or removal during purification steps. Where relevant, exposure assessment should consider both the identity and concentration of any residues present in the final product.

The document further emphasises that, while many media components have established safety profiles (e.g. commonly used nutrients), others may require additional toxicological, allergenicity, or compositional assessment, particularly where they are novel, produced via biotechnology, or used under conditions that differ from traditional applications. As such, the safety assessment of culture media is closely linked to broader considerations of process design, control measures, and product characterisation.

Importantly, Codex identifies several areas where further scientific input from FAO and World Health Organization (WHO) is required, including the development of harmonised risk assessment frameworks and the identification of data gaps. This reflects the current lack of internationally agreed guidelines specific to cell culture media, despite their recognised importance in the safety evaluation of cell-based foods.

1.3.2 European Union

While the European Union (EU) has not yet issued specific risk assessment guidance for cell-cultured foods, the European Food Safety Authority (EFSA) Scientific Guidance on Novel Foods was recently updated to include provisions for products derived from cell culture or tissue culture [7]. These updates highlight safety issues unique to CM that are not typically encountered in other novel or conventional foods. They cover traceability of donor animals, cell line development techniques, genetic and phenotypic stability of cell lines, pathogen risk management, cell banking and propagation practices, and safety of microbially derived growth factors in culture media, including evaluation of the source microorganism.

Applicants are expected to follow EFSA's established guidance for dossier preparation, with CM-specific information transparently mapped into each section. Practical approaches include documenting the source and health status of donor animals, applying validated assays to demonstrate sterility and genetic stability, and adhering to Good Cell Culture Practices (GCCP) [8]. Related EFSA guidance documents, such as microorganisms in the food chain [9], and established biotechnology precedents, such as European Medicines Agency (EMA) guidance on cell substrates for biotechnological products [10], can help address methodological gaps.

EFSA has also actively engaged with stakeholders on this topic. In May 2023, a Scientific Colloquium titled "Cell Culture-Derived Foods and Food Ingredients" was convened to discuss the emerging challenges and opportunities associated with cell culture-derived foods within the EU regulatory framework. From that Colloquium, an event report was published, in which some key considerations for FBOs were described [11].

1.3.3 United Kingdom (UK)

Following the UK's departure from the EU, the Food Standards Agency (FSA) has retained assimilated EU Novel Food Regulation (EU) 2015/2283, alongside relevant EFSA guidance, as the foundation for the safety assessment of novel foods, including CM. Within this framework, applicants are required to demonstrate that products are safe, not nutritionally disadvantageous, and not misleading to consumers.

To support the application of these principles to cell-cultivated products (CCPs), the FSA has undertaken targeted initiatives to address the specific challenges associated with this emerging category. Early efforts included the publication of a hazard identification report for CM [12], which outlines potential biological, chemical, and process-related risks across the production continuum. Building on this work, the FSA launched a 2-year CCP Regulatory Sandbox programme in March 2025. This initiative brings together risk assessors, policymakers, scientists, industry stakeholders, and selected CCP companies to develop practical, science-based guidance tailored to cultivated products. A key objective of the programme is to facilitate early engagement between regulators and applicants, thereby improving clarity around data expectations and supporting more efficient regulatory pathways.

As an initial output of the Sandbox programme, the FSA, in collaboration with Food Standards Scotland (FSS), has published two complementary guidance documents addressing the regulatory classification and safety assessment of CCPs. The first provides an overview of how CCPs are classified under UK legislation, clarifying that products derived from animal cells are considered products of animal origin rather than "meat" and are therefore subject to existing food safety and hygiene requirements [13]. The guidance emphasises that, despite the novelty of the production process, established food safety principles remain applicable. Food business operators must implement robust food safety management systems, including HACCP-based approaches, adapted to the specific risks associated with cell culture processes. This includes consideration of inputs, process controls, and potential hazards arising during cell proliferation, differentiation, and downstream processing.

The second guidance document specifically addresses the assessment of allergenicity and nutritional aspects of CCPs [14]. The guidance focuses on the scientific requirements necessary to demonstrate that CCPs are safe from both a nutritional and allergenicity perspective. In particular, it emphasises the need to assess whether the product may pose a risk of nutritional disadvantage compared to conventional foods and whether it introduces new or altered allergenic risks. These considerations form a core component of the novel food safety assessment and must be supported by robust, evidence-based data within the application dossier. Consistent with broader risk assessment principles, the document reinforces a case-by-case approach, reflecting the diversity of CCP products and production methods. Applicants are expected to provide detailed information on the characteristics of the final product, alongside a structured evaluation of its nutritional profile and potential allergenicity. This includes consideration of compositional differences, presence of novel or modified proteins, and any residual substances arising from the production process.

Further guidance documents are planned for publication during 2026.

1.3.4 Singapore

Singapore was the first country to issue regulatory guidance that explicitly addresses CM, through its *Requirements for the Safety Assessment of Novel Foods and Novel Food Ingredients* – introduced in 2019 and most recently updated in 2025 [15]. The guidance applies broadly to novel foods, including those produced by precision fermentation and biomass fermentation, but contains a dedicated section for CM. For CM, applicants must submit detailed information on the cell line (history, banking, genetic stability, and evidence of absence of adventitious agents) and provide a full list of inputs used in production, including all culture media components. Residuals of culture media components in the final product must be quantified using validated methods and their safety contextualised against literature data, existing evaluations, or toxicological reference values. To support this, the Singapore Food Agency (SFA) has developed a decision tree for the safety assessment of biologically derived culture media components (e.g. recombinant proteins, growth factors, enzymes), which is a valuable tool for applicants.

In addition, SFA draws specific attention to the risk of genetic drift in cell cultures, as genome instability could lead to the production of substances of food safety concern. To address this, applicants are expected to conduct a systematic literature review to identify hazards associated with the donor species and establish a list of substances for targeted analysis. Genetic stability may then be evaluated either through in silico genome screening against allergen/toxin databases or by comparing cells at the start and end of production using omics-based approaches (e.g. transcriptomics, proteomics, metabolomics).

1.3.5 Australia and New Zealand

In June 2025, Food Standards Australia New Zealand (FSANZ) implemented Standard 1.5.4 establishing explicit requirements for the production and labelling of CM. The standard reflects experience gained during the review of Vow's cell-cultured quail application and requires evidence of cell line identity, purity (absence of bacteria, viruses, fungi, and prions), donor traceability, and validated indicators of process control during biomass generation in bioreactors. It also specifies that production inputs must not be unsafe or unsuitable, reinforcing the importance of HACCP-based oversight across sourcing, proliferation, differentiation, and downstream processing.

1.3.6 South Korea

In February 2024, the Ministry of Food and Drug Safety (MFDS) implemented a temporary standard for food derived from cell culture [16]. Applicants must provide information on the source of the cell line and cell line development, substances used in manufacture, and assessments of human impact, allergenicity, and genetic stability. Unlike most other regulatory authorities, South Korea's MFDS mandates new toxicological studies, including genotoxicity and 90-day subchronic studies, reflecting a precautionary stance.

1.4 Definition of the Risk Assessment Process

In the context of food safety, risk assessment is a science-based process, typically conducted using a four-step methodology: i) hazard identification, ii) hazard characterisation, iii) exposure assessment, and iv) risk characterisation [17]. In line with FAO definitions, *hazards* are biological, chemical, or physical agents in food with the potential to cause adverse health effects, while *risks* reflect both their probability and severity [18]. This methodology also applies to the evaluation of CM, although the use of animal cell lines, complex culture media, bioreactor systems, and downstream processing introduces considerations not typically encountered in conventional foods.

1. **Hazard identification** requires the assessment of potential biological, chemical, and physical hazards associated with the production process. For CM, this may include microbial contaminants (e.g. mycoplasma and adventitious viral agents), consequences of cell line instability (e.g. genetic drift or unintended effects of immortalisation that could alter product composition), residual media components or processing aids, microbial toxins such as endotoxins, heavy metals, as well as extractables from equipment or particulates from scaffolds.
2. **Hazard characterisation** considers the nature and magnitude of any adverse effects arising from the identified hazards. This step includes toxicological profiling of residual substances (such as novel media components, metabolites, or cell debris), evaluation of the allergenicity of proteins intrinsic to or expressed by the cell line, and assessment of whether genetic or phenotypic changes in long-term cultures could lead to unintended proteins, metabolites, or nutritional imbalances.
3. **Exposure assessment** estimates consumer exposure to identified hazards, based on anticipated consumption patterns, intended food uses and serving sizes, and concentration of potential contaminants or residuals in the final product.
4. **Risk characterisation** integrates the previous phases to provide an overall assessment of the probability and severity of adverse food safety outcomes. This includes a clear account of uncertainties, such as limited long-term data, variability between production systems, or presence of novel ingredients.

2 Risk Assessment Process: Unique Factors in the Risk Assessment of Cultivated Meat

Many of the food safety challenges associated with CM can be assessed and managed using established food safety frameworks. Hazards such as microbial contamination, chemical residues, or physical foreign material are not unique to CM and can be controlled through good hygiene and manufacturing practices, validated analytical methods, and HACCP-based preventive controls. The use of established methodologies and recognised analytical techniques – spanning conventional microbiology, impurity testing, and toxicology – enables systematic hazard identification and targeted safety assessments. The general principles and methodologies typically used for food safety assessment therefore provide a solid foundation for evaluating CM [19, 20].

Certain factors, however, are unique to CM and require particular attention:

- **Cell sourcing and stability**, including confirmation of species identity, long-term genetic stability, and absence of adventitious agents.
- **Immortalisation approaches**, whether spontaneous or engineered, which must be fully characterised to exclude unintended consequences of regulatory concern.
- **Culture media composition**, especially novel growth factors or processing aids with limited food use history, which must be quantified and assessed in the harvested biomass.
- **Scaffolds and carriers**, where materials may contribute particulates or residuals to the final product.
- **Compositional and nutritional comparability**, ensuring that nutrient profiles and potential impurities fall within safe ranges relative to conventional foods.

Notably, the safety assessment of CM cannot follow a single, standardised protocol. Products differ in a range of fundamental aspects, which include the animal species and cell types employed, the cell line development and culture methods, the composition of culture media across different stages, the production conditions, and the specifications and properties of the final biomass (Fig. 1). This heterogeneity means that each assessment must be tailored to the specific characteristics of the product and its production system, while the remaining being anchored in established risk assessment principles.

A practical approach is to distinguish **upstream steps** (cell isolation, adaptation, and banking) from **downstream steps** (cell expansion, differentiation, harvest, washing, and storage) (Fig. 2). Notably, cell culture media are relevant to both, as different formulations are typically used at different stages. Hazards introduced upstream, such as cell culture aid media used during cell line establishment, generally pose a lower level of concern because their use is limited in duration and they are effectively removed or diluted through subsequent passaging. In contrast, downstream steps occur closest to the finished food and therefore carry greater safety significance and require more stringent assessment.

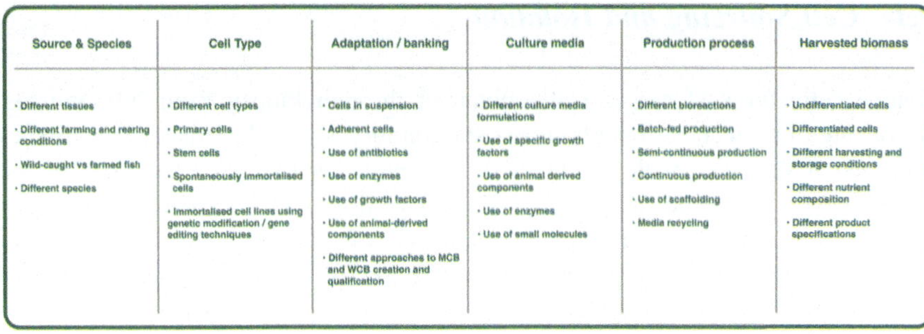

Fig. 1 Biological and technical aspects to be considered during the cultivation of cells for food production. *MCB* master cell bank, *WCB* working cell bank

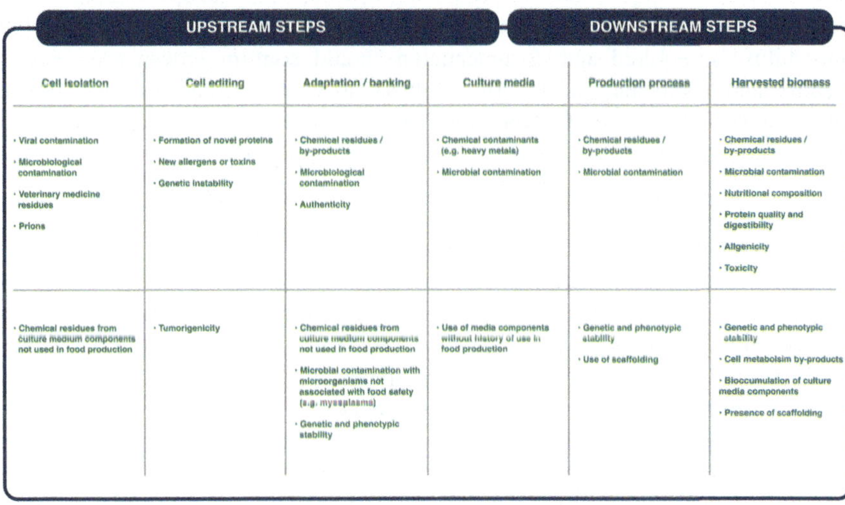

Fig. 2 Main hazards to be considered for upstream and downstream steps in the production process of cultivated meat

Taken together, these considerations highlight that CM safety assessments must be built on established risk assessment principles but adapted to the specific characteristics of the food product, its source, and its production system. In the absence of detailed guidance, regulators and risk assessors often rely on FBOs to provide a robust and well-documented assessment addressing both aspects.

The following sections therefore examine each critical element of the risk assessment in turn – spanning cell sourcing and banking, genetic manipulation, and culture media, as well as product-specific evaluations such as compositional analysis, nutritional considerations, intake and exposure assessment, ADME (absorption, distribution, metabolism and excretion), toxicological studies, and allergenicity – outlining the data that companies are expected to provide and the regulatory precedents available to guide such assessments.

2.1 Cell Sourcing and Isolation

Source cells for CM are typically obtained through biopsy from live animals, harvesting of stem cells from blastocyst-stage embryos, or cells isolated from tissues collected at slaughter, with the choice of the source depending on the intended cell lineage (e.g. myogenic, adipogenic) [20]. The origin and traceability of donor animals or embryos are critical elements of the safety assessment for CM. Regulatory evaluations consistently require the taxonomic classification of the source organism (family, genus, species, and subspecies when relevant) and documentation linking the donor tissue to its declared source species. This information supports traceability and aligns the cultivated product with its conventional counterpart for food safety and labelling purposes.

In addition to direct isolation, production cell lines may also be obtained from established sources, including commercially available lines. In such cases, full traceability is required and documentation should confirm provenance, species identity, and health status of the original donor, as well as provide evidence of cell line stability and absence of adventitious agents. The subsequent safety assessment should verify that this information is sufficient or supplement it if needed.

2.1.1 Veterinary Certification and Health Status

Biopsy donor animals should be demonstrably healthy and free from infectious disease, with supporting evidence in the form of veterinary health certificates signed by a licensed veterinarian or equivalent documentation. Such records are expected to cover the animal's health status, age, and sex and to include information on the farm of origin such as geographic location, rearing conditions, vaccination and feeding records, veterinary treatments, and presence of other animal species. In species where individual traceability is feasible (e.g. cattle), individual records are used, whereas for animals reared in groups (e.g. poultry or aquaculture species), flock- or batch-level certification is more common.

Where tissues are obtained at slaughter, pre-mortem health checks and post-mortem inspection certificates issued by an official veterinarian should confirm the absence of visible disease or contamination and the animal's suitability to enter the food chain. Records of medications or vaccinations, including antimicrobials, antiparasitics, or other prophylactic treatments, should be retained [21], as these inform the assessment of potential residues on the initial cell line and downstream products.

For wild-caught animals such as open-water fish, applicants are advised to provide as much information as possible. Ideally, animals should be destined for the food chain, and information regarding catch location, timing, and conditions, along with an inspection of overall health, is required.

2.1.2 Chemical Hazards at Isolation: Residues of Veterinary Drugs

The use of veterinary medicines in donor animals raises a legitimate question about residues at the point of cell isolation [4]. Where tissue is taken from animals entering the food chain (e.g. at slaughter), conventional withdrawal before slaughter and meat residue controls apply, and sampled tissues are therefore expected to meet legal residue limits at the time of collection. Even though veterinary drug residues from donor animals are recognised as a chemical hazard, no jurisdiction mandates a specific withdrawal period or fixed testing panel instead favouring risk-based documentation and analytical verification. For live-animal biopsy under veterinary oversight, records of recent treatments should be obtained.

Where relevant, applicants should justify why any residues present at isolation are unlikely to persist in the final cultivated biomass. In practice, this justification may reasonably rely on the expectation that repeated passaging, banking, and downstream processing dilute these residual substances to negligible levels [6].

2.1.3 Biological Hazards During Isolation and Early Culture

Potential biological hazards at the cell sourcing and isolation stage include the introduction of viruses, parasites, pathogenic bacteria and fungi, and prions into the cell line (Table 1) [22]. These hazards may be carried from donor animals or introduced during the biopsy and early storage steps and – if not controlled – could be propagated during subsequent culture [23]. A particular concern is the possible transmission of zoonotic infectious diseases and foodborne pathogens from the source animal. However, such hazards are also relevant in conventional meat production, and the overall risk of transmission is generally lower in CM, as cell culture systems operate in controlled environments that enable targeted monitoring and risk management from the outset.

Bacterial Hazards

Bacterial pathogens such as *Salmonella*, *Campylobacter*, *E. coli*, and *Listeria* are commonly associated with production animals and pose a risk during the cell isolation stage. Mycoplasma species are particularly of concern due to their persistent nature – these bacteria lack a cell wall, making them very difficult to detect and eliminate from cultures. Mycoplasmas can impair cell growth, alter metabolism, and even cause cell death, undermining the stability of the cell line [20, 24]. Antibiotics and antimycotics are commonly used in early cell culture to supress microbial contamination potentially introduced during tissue collection and transport. This can however mask low-level infections, particularly mycoplasma, which will persist in culture if not specifically monitored using dedicated polymerase chain reaction (PCR) testing [25].

Table 1 Summary of biological hazards during cell isolation

Hazard type	Description	Mitigation/control measures
Yeast and mould hazards	Pathogenic yeasts and moulds such as *Aspergillus* spp., *Penicillium* spp., and *Fusarium* spp. are ubiquitous and thrive in the moist conditions common to lab and processing environments	– Aseptic handling during cell sourcing and culture – Environmental monitoring – Sterility testing at each stage – Ensuring that if any raw materials are cereal based (common source), they are pathogen free
Bacterial hazards	Pathogenic bacteria such as *Salmonella*, *Campylobacter*, *E. coli*, and *Listeria* commonly associated with animals and their faeces. *Mycoplasma* species are also a concern in cell cultures	– Aseptic handling during cell sourcing and culture – Environmental monitoring – Sterility testing at each stage
Viral hazards	Species-specific viruses, including retroviruses, can be carried from the donor animal or animal-derived culture aids (e.g. BSA, FBS, trypsin) into the cell line. These viruses may persist and be expressed in vitro, potentially reducing cell viability and productivity	– Cell line procurement from animals free from viral diseases – Species-specific viral tests (e.g. PCR for retroviruses) – Avoidance of animal-derived raw materials (or procurement from reputable sources, with virus-free certificate)
Prions	Prions cause neurodegenerative diseases such as CJD in humans and BSE in cattle. They are typically found in high-risk tissues like the brain and spinal cord	– Exclusion of high-risk tissues – Sourcing from BSE-free herds – Certification of prion disease-free status

BSA bovine serum albumin, *BSE* bovine spongiform encephalopathy, *CJD* Creutzfeldt–Jakob disease, *FBS* foetal bovine serum, *PCR* polymerase chain reaction

Safety assessments should therefore provide information on sterility and aseptic measures applied during isolation and early culture, as well as results from microbiological and mycoplasma testing. Descriptions of monitoring practices (e.g. environmental controls, routine sterility assays, and validated PCR-based methods for mycoplasma) help to demonstrate that bacterial hazards are effectively managed.

Viral Hazards

Species-specific viruses represent a distinct hazard, as they may remain latent in the donor tissue and be carried into the cell culture. Endogenous retroviruses, for example, can integrate into the genomes of many animal species and may be expressed only in vitro [26]. While host tropism generally restricts viruses to their natural hosts, cross-species transmission is possible under certain conditions [27, 28].

Viral infections are detrimental to cell line maintenance, as lytic or persistent viral replication often reduces cellular viability and productivity [28]. In most cases, viral contaminants tend to be self-limiting within cell culture due to cytopathic effects. Nonetheless, latent viruses may remain undetected and still pose a risk if transferred to the final product [29].

The source animal itself may harbour latent or active viral infections. The UK FSA hazard identification report [23] flags this as a biological hazard at the biopsy/tissue collection stage, noting that viruses can be present in apparently healthy animals, warranting the need for testing the source animal to reduce the risk.

Adventitious viruses can alternatively be introduced in the cell culture from raw materials of animal origin, if not carefully monitored. For example, one of the most documented viral hazards is bovine viral diarrhoea virus (BVDV) detected in foetal bovine serum (FBS), together with other viruses like bovine herpesvirus 1, parainfluenza 3, bovine enterovirus type 4, and bacteriophages [30]. Even though the industry is increasingly moving toward serum-free media, which would eliminate this vector entirely, FBS is still widely used in early-stage isolation, and other animal-derived materials are regularly used in some production processes.

Another source of viral hazards is endogenous retroviral (ERV) sequences, present in many mammalian and avian genomes. While most are inactivated, some can produce replication-competent viruses in culture conditions, and it has been reported in the pharmaceutical industry that changes in the culture conditions can alter retrovirus output by altering cell metabolism [31]. For CM specifically, the relevance is that bovine, porcine, and avian cell lines are highly likely to contain ERV sequences. The FSA cell banking report [32] notes this as a concern requiring consideration but acknowledges that no cultivated meat-specific studies have quantified ERV expression in livestock cell lines used for food production.

To minimise viral contamination risk, sourcing healthy, certified pathogen-free animals is crucial. The safety assessment should include information on the health status of donor animals, veterinary certification or equivalent documentation, and results of molecular screening for relevant viruses at the isolation stage. This is usually then further monitored with targeted PCR assays for known viruses and techniques such as next-generation sequencing (NGS) or electron microscopy to capture unexpected agents at the banking stages.

Prions

Prions, infectious proteins associated with neurodegenerative diseases such as Creutzfeldt–Jakob disease (CJD) in humans, scrapie in sheep and goats, bovine spongiform encephalopathy (BSE) in cattle, and chronic wasting disease in cervids pose a unique risk in cell-based food systems [33]. Human transmission of variant CJD has been linked to the consumption of BSE-contaminated beef [34]. As infectivity is concentrated in high-risk tissues such as brain, spinal cord, blood, or lymphoid tissues, the relevance of prion hazards in CM depends primarily on the origin of the donor tissue [33]. Confirmation that high-risk tissues are excluded from biopsy material together with documentation verifying the prion-free status of donor animals is sufficient evidence that this hazard is being controlled.

However, when animal-derived components, such as bovine serum albumin (BSA) or FBS, are being used, the introduced risk needs to be assessed. Assessments usually include traceability documentation provided by the supplier, which includes certificate of origin and evidence of compliance with international transmissible spongiform encephalopathies (TSE) control standards to establish that prion risks are adequately managed.

Biological Hazard Risk Management

For pathogens to reach the final cultured meat product and consumers at levels of food safety concern, they would need to withstand initial antimicrobial treatments; propagate within the cell culture environment without disrupting cellular growth; remain undetected throughout cell line development, production, and downstream food processing; and ultimately persist through inspection and food preparation [6]. Therefore, there is a low likelihood that pathogens can become a risk to consumers. However, due to its unique production process and culture conditions, further studies to assess pathogen growth potential in cultivated meat production are warranted [35].

Because pathogens could be introduced in cell culture at multiple stages of production (e.g. operator risk, raw material), the safety assessment should include microbiological testing conducted at relevant points of the production process. Therefore, the recommended mitigation framework is a hybrid of HACCP, Good Manufacturing Practice (GMP), and GCCP, starting at the isolation step and spanning the whole production process [4, 36].

2.2 Cell Line Development and Engineering

2.2.1 Spontaneous Immortalisation

The finite proliferative capacity of primary animal cells, known as the Hayflick limit, necessitates the use of immortalisation strategies for large-scale cultivated meat production. Immortalisation can occur spontaneously, and selective culture can yield stable, proliferative lines without genetic modification, but it is a very rare event in agricultural species relevant for CM production. While avian fibroblasts were noted to have one of the highest rates of spontaneous immortalisation, it is still a very rare event [37]. Only recently was a spontaneously immortalised bovine fibroblast cell line demonstrated [38], and it was explicitly framed to be more difficult than avian cell lines. To note, fish cells are easier to immortalise compared to mammalian and even avian cells, and thus this bottleneck is removed in cultivated seafood [39]. For details on other cell line species differences, see the comprehensive review by Muller and Matsusaki [40], which compares the cell sources across species for cultivated meat and notes inter- and intra-species differences in proliferation potential and differentiation capacity.

From a regulatory perspective, the use of spontaneously immortalised lines may reduce the complexity of the safety assessment, as deliberate genetic modification is avoided and the cell line will therefore fall outside genetically modified organism (GMO) legislation in some jurisdictions. By contrast, immortalisation achieved through targeted genetic engineering (e.g. telomerase activation or inactivation of senescence-associated genes) is typically regulated as a GMO and introduces additional data requirements and specific safety considerations. These aspects are discussed further in Sect. 2.2.2 below. Due to the lower regulatory burden, the early approved cultivated meat products contained spontaneously immortalised cells (Table 2). Globally, as of April 2026, the only approved cultured meat product containing genetically engineered cells is UPSIDE Foods chicken.

2.2.2 Genetic Manipulation of Cell Lines

Gene editing tools such as CRISPR/Cas9, zinc finger nucleases (ZFNs), and transcription activator-like effector nucleases (TALENs) have emerged as powerful technologies for modifying cell lines used in CM production. These methods allow precise manipulation of the genome to introduce traits that enhance production efficiency and product quality. For example, CRISPR/Cas9 can be applied to immortalise cell lines by disrupting senescence-associated genes (e.g. p53, RB), thereby overcoming the Hayflick limit and enabling continuous proliferation. Beyond immortalisation, gene editing can optimise growth kinetics and cell behaviour, improve nutrient utilisation, and enhance metabolic pathways that influence protein, fat, or micronutrient composition. It also offers opportunities to reduce dependency on costly exogenous growth factors, for instance by engineering autocrine signalling pathways [41].

From a regulatory perspective, genome editing of production cell lines raises both opportunities and challenges. In the EU, edited cells fall under the definition of GMOs under Directive 2001/18/EC. This decision was made in 2018 by the European Court of Justice (CJEU Ruling Case C-528/16), the main argument being that DNA or RNA modifications introduced via new genome editing techniques do not occur naturally. Consequently, CM products developed from gene-edited cell lines will require authorisation under the EU GMO regulatory framework [42]. A recent study on new genomic techniques (NGTs) by the European Commission acknowledged however the limitations of the current framework and the potential need for a differentiated regulatory pathway [43].

As of April 2026, no CRISPR-edited food or feed has been submitted for authorisation in the EU. The requirements for foods produced using NGTs include a stringent safety assessment, with particular attention to the nature of the genetic modification, its source, and the characterisation of any unintended effects. Modifications must be stable, well characterised, and demonstrably free of harmful

Table 2 Summary of regulatory submissions and used immortalisation strategies

Company	Application number	Species	Cell type	Immortalisation strategy
USA				
GOOD Meat	CCC 000001	*Gallus gallus*	Chicken fibroblasts, DF-1 line	Spontaneous
UPSIDE Foods	CCC 000002	*Gallus gallus*	Chicken myoblasts	Mixed: spontaneous and engineered (telomerase induction)
Wildtype	CCC 000005	*Oncorhynchus kisutch*	Coho salmon mesenchymal cells	Spontaneous (selected/stabilised line)
Mission Barns	CCC 000008	*Sus scrofa domesticus*	Porcine adipocytes	Spontaneous
Believer Meats	CCC 000039	*Gallus gallus* domesticus	Chicken fibroblasts	Spontaneous
Australia and New Zealand				
Vow	A1269	*Coturnix japonica*	Quail fibroblasts	Spontaneous
PARIMA (Gourmey)	A1341[a]	*Anas platyrhynchos domesticus*	Duck embryonic stem cells	Embryonic stem cells (pluripotent with ability to self-renew indefinitely)
Singapore				
GOOD Meat	N/A	*Gallus gallus*	Chicken fibroblasts, DF-1 line	Spontaneous
Vow	N/A	*Coturnix japonica*	Quail fibroblasts	Spontaneous
Aleph Farms	N/A	*Bos taurus*	Bovine embryonic stem cells	Embryonic stem cells (pluripotent with ability to self-renew indefinitely)
PARIMA (Vital Meat)	N/A	*Gallus gallus*	Chicken embryonic stem cells	Embryonic stem cells (pluripotent with ability to self-renew indefinitely)
PARIMA (Gourmey)	N/A	*Anas platyrhynchos domesticus*	Duck embryonic stem cells	Embryonic stem cells (pluripotent with ability to self-renew indefinitely)

N/A not applicable
[a] Not yet approved, under risk assessment

downstream consequences, such as altered metabolism, unexpected protein expression, or creation of allergens or toxins. Markers such as antibiotic resistance genes, if any used in the cell line generation, would be considered red flags, and their absence needs to be demonstrated or their presence would require strong justification or mitigation.

In the UK, a distinct category of "precision bred organisms" (PBOs) has been introduced, exempting certain genome-edited organisms without foreign DNA from strict GMO rules, while still requiring novel food authorisation for products like CM. Despite this regulatory evolution, no genome-edited foods have yet been approved in the UK.

In the USA, there is a more flexible approach when it comes to gene-edited or bioengineered foods. Therefore, the USA is one of the main target jurisdictions for CM companies that use gene editing in their cell line development. However, so far, only UPSIDE Foods has obtained approval for their cultivated chicken cells, which they immortalise using gene editing (Table 2).

Nonetheless, regardless of the jurisdiction, safety assessments for genetically modified cultivated products must address several technical and safety aspects:

The Effect of the Gene Edits and Persistence of Editing Vectors

For instance, if a CRISPR/Cas9-based edit is used to change the cell line's phenotype, for example to enhance proliferation rate, the applicant must demonstrate the efficiency of the edit (for example by Sanger sequencing) and show phenotypic differences from non-edited lines. In addition, comparative "omics" analyses (proteomics, metabolomics, transcriptomics) can further characterise the new cell lines and identify unintended phenotypic consequences.

Rigorous screening for residual vector DNA, mRNA, or proteins used in the gene editing process is essential to demonstrate safety. Electroporation is often used for transient introduction of CRISPR components (guide RNAs and Cas9). While transient expression reduces the risk of genomic integration, companies must confirm this experimentally, proving the absence of residual Cas9 mRNA or protein [44, 45]. If other gene editing methods are used, which introduce foreign DNA in the cells (e.g. vectors), the absence of these DNA elements (genome integrated or cytoplasmic) needs to be demonstrated by whole-genome sequencing.

Stability and Safety of the Introduced Changes

Data needs to be provided across extended culture passages to evidence that the edit is stable and consistent across cell generations, especially throughout the intensive production processes where the cells need to proliferate extensively. This can be done either by Sanger sequencing at the edited sites or more comprehensively using whole-genome sequencing (WGS) or transcriptomics data. Stability studies are also usually recommended to demonstrate the consistency of the phenotype at various time points. In addition, in vitro and in silico allergenicity and toxicity assessments of any novel or overexpressed proteins provide further reassurance of product safety.

The Potential for Off-Target Effects

Although CRISPR is designed for precision, unintended edits can occur in other parts of the genome, called off-targets [46]. These may disrupt other genes or regulatory elements, introducing unpredictable risks. Therefore, comprehensive sequencing data and in silico off-target prediction tools are necessary to identify these [47] and, if found, to characterise them. Functional assays assessing protein expression and cellular behaviour, when compared to non-edited cells, must accompany this data to confirm that no harmful phenotypic changes have arisen.

Existing international guidance can help inform safety assessments. EFSA has published opinions on the applicability of existing GMO risk assessment frameworks to genome-edited organisms, noting that current principles (molecular characterisation, compositional analysis, toxicity, and allergenicity testing) remain valid but require case-by-case application [48]. FAO and WHO have also issued reports emphasising the need for transparency, whole-genome sequencing, and functional characterisation to evaluate potential unintended effects [49]. In addition, the Organisation for Economic Co-operation and Development (OECD) consensus documents on molecular characterisation and comparative assessment provide reference points for structuring data requirements [50].

In conclusion, while genome editing might be a promising alternative for optimising cultivated meat production, regulatory acceptance depends on robust scientific evidence. Applicants must deliver a thorough and robust safety narrative demonstrating the absence of unintended effects, stable phenotypes, and newly introduced hazards (e.g. toxins, allergens). Without such data, regulatory uncertainty will continue to pose a barrier to market entry.

2.3 Cell Line Characterisation and Qualification of Cell Banks

Once a cell line is isolated and developed (by gene editing for example), it must then be fully characterised. If it fulfils the desired requirements, it is expanded for the creation of a master cell bank (MCB) and further into working cell banks (WCBs). The MCB provides a long-term qualified source of cells to ensure that every batch of CM originates from the same starting material, thoroughly characterised to ensure safety. This standardisation supports consistent product quality and regulatory compliance.

2.3.1 Cell Line Characterisation

The quality, consistency, and safety of cultivated meat depend on the properties of the production cell line. Comprehensive characterisation prior to the creation of manufacturing cell banks ensures that only well-defined, stable, and contamination-free cells are preserved as the long-term source of production. These expectations apply whether a line is developed in-house or obtained commercially, in which case, supplier documentation should cover provenance, identity, stability, and adventitious-agent testing, and the manufacturer should verify or supplement it as needed.

Drawing on precedents from recent cultivated meat regulatory assessments (Table 3), the scope of characterisation typically includes the following:

- Provenance and history: source passage, adaptation, immortalisation strategy, GMO status (details in section above).
- Species identity and cross-species exclusion: verification by validated molecular methods; additional cross-species checks applied where justified.
- Cell type identity and function: lineage markers/functional assays; definition and trend monitoring of growth parameters (doubling time, viability, morphology).
- Phenotypic and genomic stability: periodic assessment (e.g. karyotyping, targeted variant panels, WGS, transcriptomics) at defined passages against preset criteria, with actions on deviation.
- Adventitious-agent control (cell line level): sterility, mycoplasma PCR, and targeted viral assays appropriate to species and inputs, using validated methods and clear pass/fail thresholds.

Once selected, the line proceeds to cell banking and bank-level qualification (Sect. 2.3.2). To avoid duplication, cell bank testing should reference the cell line characterisation results and focus on confirming continuity – that is, demonstrating that the banked cells retain the same identity, stability, and safety attributes as the characterised line and that no new adventitious agents or other risks have been introduced during the banking process.

Species and Cell Type Verification

Species authentication is central to both regulatory acceptance and consumer confidence. It demonstrates that the production line is derived from the intended species and excludes cross-contamination with non-target organisms that may introduce allergens or raise regulatory concerns. The scope of species excluded should follow a risk-based approach, considering other cell lines maintained in the same facility.

To support identity, authenticity, and traceability across the production process and after, analytical approaches such as DNA barcoding and DNA fingerprinting can be used to reliably link the harvested biomass and cultivated product to its source tissue, both at species and at individual levels (short tandem repeats (STRs)/microsatellite fingerprinting), respectively. These techniques are currently applied in conventional meat authentication and fraud detection, the most common one being the mitochondrial cytochrome c oxidase subunit I (COI) gene for species identification [51–53]. There are a range of methods available; however, DNA-based methods are the most robust option, as DNA survives processing, it does not change with cell state or differentiation, and it can uniquely identify the source animal. To ensure traceability, the ideal CM characterisation framework includes STR profiles (individual identity) established at key bioprocess checkpoints (e.g. MCB, WCB) and in the final product, providing an unbroken chain of identity. The suitability and

Table 3 Examples of cell line characterisation parameters from recent applications

Company	Application number	Species and cell type	Cell line characterisation
GOOD Meat	CCC 000001	*Gallus gallus* (chicken fibroblasts, DF-1 line)	Species identity confirmed by PCR; fibroblast markers characterised; monitoring of population doubling time and passage number; cytogenetic analysis of >100 cells (no aberrations); adventitious-agent testing (avian pathogens, *Mycoplasma*, *Salmonella*, etc.); suspension adaptation documented
UPSIDE Foods	CCC 000002	*Gallus gallus* (chicken myoblasts)	Karyotyping and cytogenetic analysis; monitoring of doubling time and cell morphology; stability testing across serial passages; species identity by COI barcoding; adventitious-agent and *Mycoplasma* screening
Wildtype	CCC 000005	*Oncorhynchus kisutch* (coho salmon mesenchymal cells)	Species identity confirmed by COI sequencing and gene expression profiling; phenotypic analysis of mesenchymal characteristics; stability data from master and working cell banks; documentation of growth parameters; adventitious-agent testing
Mission Barns	CCC 000008	*Sus scrofa domesticus* (porcine adipocytes)	Cell line stability evaluated through monitoring of growth characteristics and morphology; documentation of MCB establishment; testing for microbial and viral contaminants; phenotypic characterisation of adipogenic properties
Believer Meats	CCC 000039	*Gallus gallus domesticus* (chicken fibroblasts)	Species identity confirmed by PCR and qPCR; fibroblast identity verified via RNA sequencing; karyotyping (stable over >500 doublings); SNV analysis of TP53 gene; monitoring of doubling time (>800 days stable); differentiation potential (adipogenesis) confirmed; extensive adventitious-agent and *Mycoplasma* testing
Vow	A1269	*Coturnix japonica* (quail fibroblasts)	DNA barcoding (COI) for species identity; WGS for genetic stability (compared with quail genome, no instability of concern); phenotypic monitoring across production confirming fibroblast identity; pathogen screening (influenza A, Newcastle disease, *Mycoplasma*, endogenous retrovirus, *Chlamydophila*); allergenicity assessed by sequence homology and ELISA (all negative)

COI cytochrome c oxidase subunit I, *ELISA* enzyme-linked immunosorbent assay, *MCB* master cell bank, *PCR* polymerase chain reaction, *qPCR* quantitative polymerase chain reaction, *SNV* single-nucleotide variant, *TP53* tumour protein p53 gene, *WGS* whole-genome sequencing practicability of these techniques to identify markers that could differentiate conventional meat from CM are an emerging topic, since it is considered that the CM sector will not be exempt from the threats of food fraud.

Cell-type verification requires evidence that the cells retain characteristics of the intended lineage. Depending on the cell type, this may involve immunocytochemistry, expression of lineage-specific markers, or transcription factors. Functional differentiation assays may also be applied. Monitoring additional phenotypic traits such as doubling time, viability, and morphology helps ensure that the cells continue to represent the intended cell type and provides early indicators of phenotypic drift or senescence.

Phenotypic and Genomic Stability

Both phenotypic and genomic stability should be assessed at defined passage intervals to detect potential drifts or mutations. Typically, stability of cells from the MCB through to end of production is assessed.

Phenotypic drift (e.g. changes in morphology, metabolic activity, or differentiation potential) can compromise product yield or alter chemical and nutritional composition. Genomic instability may lead to accumulation of mutations affecting growth or, more critically, introduce safety concerns such as production of unintended metabolites or proteins. Monitoring should begin from early passages and continue throughout long-term culture, using a combination of approaches such as karyotyping, single-nucleotide polymorphism genotyping, whole-genome sequencing, transcriptomic, and proteomic and metabolic profiling. The approach will be dependent on the cell type and cell attributes; not all analyses are required, and the approach should be based on risk assessment and fully justified. Phenotypic stability can be assessed through morphological scoring, viability assays, and functional differentiation [54]. A combination of short-interval monitoring (e.g. every 10–20 passages) and periodic benchmarking provides both early detection and long-term assurance. Pre-defined acceptance criteria are essential, with deviations triggering corrective actions such as re-establishing cultures from early passage research and development cell banks.

For immortalised cell lines, a statement on the mechanism of immortalisation (e.g. spontaneous vs. induced) and its safety implications must be provided. This includes, where appropriate, tumorigenicity evaluation or evaluation of unintended traits (e.g. expression of new toxins/proteins). Non-GMO declarations should be included where relevant. The implications of genetic modifications and associated potential hazards have been discussed in Sect. 2.2.2.

Sterility and Adventitious-Agent Control

Sterility of the production cell line is fundamental for both product safety and cell culture performance. GCCP should be applied [8], covering aseptic handling, sterile reagents, and environmental controls. Antibiotics are best restricted to early stages to reduce the risk of residues in the final product. Safety assessments should include results of sterility testing for adventitious agents, including bacteria, yeasts, moulds, and viruses, with particular attention to *Mycoplasma*, for which routine monitoring using sensitive methods such as PCR or enzyme-linked immunosorbent assay (ELISA) is essential. Analytical methods should be validated and, when possible, carried out by accredited laboratories [36, 55].

As detailed in Sect. 2.1, special attention must be given to potential contaminants introduced via feeder layers (if applicable), serum, or animal-derived inputs. Risk assessment should therefore cover species-specific viruses and, where relevant, infectious prions. Verification of media components and supplier documentation is typically part of this evaluation.

2.3.2 Qualification of the Master Cell Bank (MCB)

The MCB is created by expanding the characterised cell line under controlled, sterile conditions; aliquoting them into multiple vials; and storing them under cryogenic conditions using cryoprotectants such as dimethyl sulfoxide. Secure, traceable storage with continuous temperature monitoring and redundancy systems is expected to safeguard reliability and enable consistent production of safe, high-quality cultivated products [56, 57]. It is important and good practice to store a few vials from the MCB externally, as a contingency plan.

The MCB is the key point at which the cell line undergoes comprehensive testing and documentation before manufacturing begins. By thoroughly characterising the MCB, applicants can demonstrate that the cells are safe: genetically stable, free from contamination, and phenotypically consistent.

Qualification of the MCB is essential before any downstream use. The number of vials tested should be statistically justified, and testing normally includes the following:

- **Post-thaw performance**: demonstration of viability, recovery, and consistent growth kinetics under production-relevant conditions.
- **Phenotypic stability across multiple generations post-thaw**: evaluation of morphology, doubling time, gene expression, and, where relevant, differentiation potential.
- **Identity and lineage**: confirmation of species and individual (e.g. DNA barcoding) and verification of cell type using lineage markers or functional assays. This stage can serve as the midpoint of a traceability assay: the starting point could involve cells shortly after isolation, while the endpoint should be cells taken from the harvested biomass. Together, these steps can demonstrate that cellular identity is preserved throughout the production process.
- **Genetic stability:** assessment of chromosomal integrity (e.g. karyotyping) or molecular approaches including quantitative PCR, digital PCR, and high-throughput sequencing.

- **Adventitious agents (risk-based panel)**: a documented panel covering bacteria, fungi, *Mycoplasma*, and viruses of concern. Panel design should reflect the line's species of origin, isolation/adaptation/banking procedures, storage, and materials used (sera, feeder cells, enzymes). Given the nature of the final product, inclusion of common foodborne hazards (e.g. *Salmonella*, *Listeria*, *E. coli*) is a common practice to provide additional assurance. Validated and sensitive methods such as PCR, culture, or targeted viral assays are typically used, with clear pass/fail criteria.

Defined acceptance criteria should be established for each of these parameters. Once qualified, the MCB serves as a reference point: any deviation observed in production cell banks or working cell stocks during production can be traced back to the original, qualified source, strengthening safety assurance and traceability.

Where comprehensive viral testing has already been undertaken during cell line characterisation after isolation, and no new risk factors have been introduced (e.g. no change of facility, raw materials, or process), a risk-based scientific justification may support reduced or non repeat viral testing at the MCB stage. In such cases:

- Limited confirmatory testing (e.g. *Mycoplasma* and one or two sentinel viral assays) may be appropriate to demonstrate continuity.
- Reliance on prior results may be acceptable where a documented risk assessment confirms no new potential for exposure.

Such approaches should be anchored in a formal change control and comparability framework. If any variable changes, the scope of retesting is expected to expand proportionally to the assessed risk.

2.3.3 Creation and Qualification of the Working Cell Bank (WCB)

The WCB is derived from a qualified MCB to provide the immediate source of cells for routine production. One or more vials of the MCB are thawed; expanded under controlled, aseptic conditions; and aliquoted into multiple cryovials that are cryopreserved and stored using the same procedures as for the MCB. Vials from the WCB are further used in the production process, without rapidly exhausting the original MCB. Thus, its main function is to link the MCB and the production process, providing a ready-to-use, standardised supply of cells that maintains the quality and characteristics already established.

Testing requirements for WCB qualification largely mirror those of the MCB, but the focus shifts to demonstrating continuity with the MCB rather than repeating the full qualification. In practice, this means:

- **Identity**: verification consistent with the MCB.
- **Adventitious agents**: sterility and *Mycoplasma* testing are always repeated; viral testing may be reduced to a confirmatory set if no new risks have been introduced, with a written justification. Any change in facility, raw materials, or procedures warrants proportionally broader testing.

- **Genetic/phenotypic stability**: full stability studies are not normally repeated; continuity is demonstrated through monitoring of morphology, doubling time, and growth performance.
- **Post-thaw performance**: viability, recovery, and growth kinetics assessed under production conditions.

Each WCB batch is released with a Certificate of Analysis (CoA) [56, 57]. This monitoring allows early detection of deviations that could compromise product safety or quality and provides a documented traceability record for regulatory compliance. Together, these measures ensure that the WCB functions as a reliable bridge between the qualified master bank and large-scale manufacture, maintaining consistency, safety, and regulatory confidence across production runs.

2.4 Culture Media

Cell culture media are critical in the production of cultivated meat and seafood, as they define the conditions under which cells grow, proliferate, and differentiate. In addition to supplying essential nutrients such as amino acids, vitamins, minerals, and energy sources, media formulations often include biologically active components that support cell viability and expansion. Because cells are directly exposed to these inputs throughout production, culture media form a key part of the safety assessment of cultivated meat, with evaluations addressing their composition, origin, and potential for any components to persist in the final product.

In the context of CM production, culture media are not regulated on their own. Across jurisdictions, they are treated as manufacturing inputs and are therefore evaluated as part of the overall CM safety assessment. This results in a process-based approach, where regulators consider not only the identity of media components, but also how they are used during production and whether they remain in the final biomass.

For the purposes of safety assessment, media are often considered in relation to their use at different stages of the process. Upstream media refer to early stages, typically up to and including cell banking, while downstream media are used during biomass production and differentiation, closer to harvest. This distinction does not necessarily align with technical terminology used in bioprocessing but provides a useful framework for evaluating potential food safety risks.

2.4.1 Upstream Media: Composition and Key Considerations

Upstream media are primarily used to support cell attachment, survival, and rapid proliferation during early expansion following isolation. In addition to basal nutrients, they may include more complex or less defined components such as FBS or other animal-derived materials, as well as functional additives such as growth factors and, in some cases, antibiotics [20].

These components can raise specific regulatory concerns. Animal-derived materials may introduce batch-to-batch variability or carry risks of adventitious agents, while some additives do not have a history of use in food. Antibiotics, although sometimes used during early stages, are generally discouraged due to concerns related to residues and antimicrobial resistance.

From a regulatory perspective, a key question is to understand what the cells have been exposed to throughout production including the early stages of isolation. Upstream media are therefore assessed based on the identity and function of their components, the levels used, and the extent to which they are removed, diluted, or metabolised during subsequent process steps. Repeated passaging and media exchanges are typically expected to reduce the presence of early-stage inputs, but this assumption should be supported by appropriate data or justification.

Proprietary media formulations, commonly used during early development, can complicate this assessment. Where composition is not fully disclosed by media developers, it may be difficult to determine whether higher-risk or novel components are present. Regulators generally accept confidential treatment of media formulations but still require sufficient information to allow a meaningful safety evaluation, including the nature of components, their functional roles, and their concentration ranges [58].

Decisions made at this stage can have significant implications for the regulatory fate of CM. The use of animal-derived materials, recombinant proteins, or substances without a history of use in food may introduce additional data requirements or complicate safety assessment. Conversely, selecting well-characterised, food-safe inputs can facilitate evaluation and reduce the likelihood of reformulation at later stages. In practice, post-development adjustments to media composition to address regulatory considerations are not uncommon and can result in delays, additional data generation, and overall increased costs.

2.4.2 Downstream Media: Composition and Regulatory Focus

Downstream media, used during biomass production and differentiation, are typically more chemically defined and rely primarily on well-characterised components such as amino acids, vitamins, minerals, and carbohydrates, most of which have a history of safe consumption.

At this stage, greater attention is given to components with biological activity, particularly recombinant growth factors, cytokines, and other signalling molecules involved in differentiation and tissue maturation. These substances are relevant due to their biological activity and their potential to persist in the final biomass, which could raise toxicological or allergenicity concerns.

In current practice, regulators often take a risk-proportionate approach to culture media components, whereby media components are considered according to their history of use in the food system and their potential for carry-over into the final product.

Biologically active components are assessed on a case-by-case basis. Key considerations include their origin, production system, and purity, including the presence of impurities such as host-cell proteins, residual DNA, or endotoxins. Their fate during processing, including potential inactivation and any residual activity in the final product, may also need to be addressed. Regulators typically expect sufficient information on identity, function, use levels, and evidence of clearance to support the safety assessment.

2.4.3 Fate of Media Components and Exposure Considerations

A consistent expectation across jurisdictions is that applicants demonstrate the fate of media components during production, as this underpins the assessment of potential dietary exposure. This includes evaluating whether substances are metabolised by the cells, incorporated into biomass, or removed during processing or remain as residues. Together, these considerations form the basis for estimating intake levels and determining whether residual levels are of safety concern.

Where residues may be present, their levels should be considered in the context of conventional foods or relevant safety thresholds and/or health-based guidance values (HBGVs). In some cases, comparative data may be used to show that levels are similar to, or lower than, those naturally present in conventional meat. For certain components, such as growth factors, additional considerations may include their stability and whether they retain biological activity after processing or cooking [59].

Analytical measurement is generally preferred where feasible, but may be complemented by process-based arguments, mass-balance calculations, or conservative exposure modelling, provided that these are scientifically justified. Where analytical data are generated, regulators expect the use of recognised methods, validated for use in the specific cell matrix with defined limits of detection and quantification to demonstrate that residual concentrations or activity are negligible [59]. This is technically demanding, as recombinant proteins are typically present at very low levels in complex matrices. Assay development is often case-specific, with challenges including low abundance, matrix interferences, and distinguishing biologically active from inactive forms.

2.4.4 Quality and Specifications of Media Inputs

Safety assessments also consider the quality and sourcing of media components. Materials used in production are expected to meet appropriate specifications and to be manufactured under suitable quality systems. While research-grade materials may be used during early development, regulatory submissions should demonstrate that inputs are of consistent quality and supported by appropriate documentation, such as certificates of analysis.

2.4.5 Harmonisation Initiatives

In parallel with evolving regulatory practice, several international and regional initiatives are working toward greater alignment in the safety assessment of cultivated foods, with emphasis on culture media and their components. These include industry-led efforts such as the Safety-Assessed Media Ingredients (SAMI) list [58, 60], as well as regional initiatives led by the Good Food Institute (GFI) and the APAC Society for Cellular Agriculture (APAC-SCA) to support regulatory convergence in the Asia Pacific region. At the multilateral level, work led by the FAO, including joint FAO/WHO expert consultations, has also contributed to the development of shared scientific principles for evaluating inputs used in cultivated food production [61]. While these efforts do not replace national regulatory requirements, they reflect a broader convergence toward risk-based, exposure-driven approaches.

Taken together, these considerations ensure that the assessment of culture media addresses not only their composition, but also their role within the production process, the origin and quality of their components, and the potential for any substances to remain in the final product.

2.5 *Production Process*

Cultivated meat may be produced through different process configurations depending on the cell type, product format, and manufacturing scale. Differences in bioreactor platform, operating mode, system closure, and use of scaffolds or microcarriers may affect not only process performance but also the conditions under which hazards arise and are controlled during manufacture.

From a regulatory perspective, the production process is a central component of the food safety assessment and should be considered within an overarching food safety plan, including HACCP, supported by appropriate hygienic and manufacturing controls. The key question is whether safety can be maintained across the entire manufacturing process, particularly during prolonged culture, transfer, harvest, and downstream handling. This section therefore considers production as a controlled manufacturing system, focusing on critical control points, process-related hazards, preventive measures, and process change.

2.5.1 Production Process and Hazard Analysis

For food safety purposes, cultivated meat production may be understood as a sequence of unit operations extending from cell expansion to proliferation, differentiation or maturation, harvest, separation, formulation, and final packing. The exact process may vary considerably, but its relevance lies in providing the basis for hazard analysis by identifying the stages at which hazards may be introduced, amplified, or removed. Parameters such as temperature, pH, dissolved oxygen, agitation, residence time, and system integrity are significant not only for process performance, but also because deviations may affect contamination risk, material carryover, and process/product consistency. Harvest is of particular importance, as it marks a transition from the controlled culture environment to downstream food processing operations that may be more open or only partially closed, where risks associated with transfer, washing, structuring, and handling may become more pronounced.

For regulatory purposes, the process description used in hazard analysis should be representative of intended industrial manufacture rather than limited to laboratory or early development conditions. Small-scale systems may not capture the transfer complexity, equipment interfaces, hold times, or handling steps encountered in commercial production, and these factors may materially affect the hazard profile of the process.

2.5.2 Hazard Identification and Evaluation

The principal hazards associated with production are biological, chemical, physical, and, in some cases, allergen-related. Biological hazards include bacterial or fungal contamination, *Mycoplasma*, adventitious agents, and cross-contamination between batches or cell lines, particularly where operations involve open handling, repeated transfers, or in-process sampling. Chemical hazards may arise from processing aids, cleaning agents, contact materials, scaffold constituents, microcarriers, or other substances introduced during manufacture and insufficiently removed thereafter. Physical hazards may include particles or fragments derived from equipment, tubing, filters, single-use components, or support materials. Although allergenicity is addressed elsewhere in this volume, the production process may still contribute to allergen-related risk through the use of allergenic materials or through cross-contact in shared manufacturing environments. Hazard analysis should therefore proceed stepwise across the process, with attention to where hazards enter the system, the conditions under which they may persist, and the measures by which they are prevented or controlled.

This evaluation should remain sensitive to scale and process modification. Changes in reactor design, operating conditions, material inputs, harvest methods, facility layout, or packaging interfaces may alter the nature or significance of identified hazards, and hazards characterised at laboratory or pilot scale may not fully reflect those encountered under commercial manufacturing conditions.

2.5.3 Control Measures, Monitoring, and Verification

Control of these hazards requires an integrated food safety management system based on preventive rather than solely reactive principles. A food safety plan, including HACCP, provides the most appropriate framework, supported by good manufacturing and hygienic practices adapted to both upstream culture operations and downstream food processing. Key elements include control of critical inputs, defined operating limits, environmental and in-process monitoring, sanitation procedures, traceability, deviation management, documented corrective actions, and verification. Sanitation and hygienic design are especially important because prolonged culture, complex equipment interfaces, and multiple transfers may create opportunities for contamination if zoning, personnel and material flow, cleaning, and equipment design are inadequate. Verification should extend beyond end-product testing to include review of process records, trend analysis, internal audit, and evidence that the process remained under control throughout manufacture.

Within this framework, change control is also critical. Scale-up and subsequent process modifications may require reassessment of hazards, review of established controls, and, where necessary, revalidation to demonstrate that safety is maintained under actual manufacturing conditions.

2.6 Compositional Data

2.6.1 Nutritional Profile

Nutritional characterisation is a core part of the safety assessment of cultivated food because it shows whether the product is nutritionally comparable to, or at least not nutritionally disadvantageous relative to, the food it is intended to replace. In many cases, the relevant comparator will be a conventional counterpart such as chicken, beef, or salmon. However, not all cultivated products are designed to replace conventional meat. Some are intended to enter the market as novel foods that expand consumer choice alongside conventional and plant-based products. In such cases, the key question may be nutritional adequacy for human consumption [62, 63] rather than strict equivalence to a conventional comparator.

The nutritional assessment should therefore reflect the nature and intended use of the product. Where the CM product is positioned as an analogue of conventional meat, comparison with the corresponding conventional food is highly relevant. Where it is marketed as a distinct food or as part of a formulated product, a broader assessment of nutritional adequacy may be more appropriate. In either case, the choice of comparator and the basis for concluding that the product raises no nutritional concern should be clearly justified.

Compositional analysis should be conducted on batches representative of the manufacturing scale. In most jurisdictions, at least three non-consecutive production batches are generally expected, while some authorities, including EFSA, request at least five independent batches. This matters because laboratory-scale or pilot-scale material may not reflect the composition of a commercially manufactured product. The tested batches should therefore represent the final process, including relevant downstream processing, and be sufficient to demonstrate consistency and process robustness.

Analytical testing should, where possible, use internationally recognised methods validated for their intended purpose by accredited laboratories. Where in-house methods are used, the method and its validation should be fully described. The analytical package should also report the limits of detection (LOD) and quantification (LOQ), together with the identity and certification status of the testing facility, so that regulators can assess the reliability and sensitivity of the data.

The compositional dataset should include the principal nutritional constituents of the product. These generally include proximate parameters such as protein, fat, carbohydrate, ash, fibre, and moisture; amino acid composition to assess protein quality; fatty acid composition, including trans-fat where relevant; and key vitamins and minerals such as vitamin D, vitamin B12, iron, zinc, and selenium. Additional constituents may also be relevant where they contribute to the nutritional value or identity of the food. The objective is not merely to provide a standard composition table, but to characterise the product sufficiently to support conclusions on nutritional quality and suitability for consumption.

For CM, the assessment should also address whether production inputs, including scaffolds, culture media components, enzymes, and growth factors, contribute residual nutrients, metabolites, or bioactive compounds to the harvested biomass. Although nutrients in the media are generally not intended to fortify the final product, it remains necessary to determine whether they are incorporated into the cells, transformed during culture, or largely removed during downstream processing. Similarly, metabolites, signalling molecules, or residual enzymes may alter the final composition and should be documented where relevant.

This is especially important where proprietary media formulations or novel supplements are used, particularly if they lack a history of safe food use. Risk assessors need sufficient information on what the cells were exposed to during manufacture and whether that exposure altered the composition of the final food in a nutritionally meaningful way. The assessment should therefore capture not only expected nutrients, but also any unintended nutritional contributions or losses arising from the production process.

Overall, nutritional profile should show that the cultivated product is compositionally well characterised, nutritionally appropriate for its intended use, and consistent across representative production batches. Low variability supports confidence in process robustness and product reliability, while clear analytical reporting allows regulators to judge whether observed differences are expected, acceptable, or potentially concerning.

2.6.2 Impurities

In addition to nutritional composition, cultivated meat must be assessed for impurities and potential contaminants that may arise from the production process, raw materials, or the final product itself. This assessment should follow a risk-based approach guided by the manufacturing process, composition of culture media and other inputs, and relevant scientific literature [58]. Because cultivated meat is produced under conditions distinct from those of conventional livestock production, the impurity profile may include both conventional food hazards and process-specific residues. These may include substances that are unique, or at least far more relevant, to cell culture systems, such as growth factors, signalling molecules, scaffold-derived substances, microcarrier-related residues, and cell culture-specific metabolites.

The analytical scope should therefore extend beyond routine food contaminant testing. Relevant parameters may include microbiological criteria, such as total viable count, yeasts and moulds, Enterobacteriaceae, *Salmonella* spp., *Listeria monocytogenes*, *Campylobacter*, and other organisms as appropriate to the product and process [36]. Chemical contaminants should also be considered, including heavy metals such as antimony, arsenic, mercury, lead, cadmium, chromium, nickel, and tin, as well as other elements where process materials or equipment indicate a potential source. Depending on the product and inputs used, additional contaminants may require evaluation, including mycotoxins, dioxins, polychlorinated biphenyls, pesticide residues, and toxins relevant to specific species or raw materials, such as marine toxins in fish-derived products.

Particular attention should be given to process-related residues and cellular by-products that are not typically considered in conventional meat assessment. These may include residual culture media components, growth factors, antibiotics, metabolites characteristic of the cell type, organic acids, and biogenic amines such as histamine or trimethylamine. Consideration should also be given to substances with the potential to persist or accumulate during production, particularly where their concentration may not decrease following media exchange or downstream processing.

Although not impurities in the strict sense, indicators of product deterioration may also be relevant where they affect safety or quality. For example, lipid oxidation markers such as peroxide value or anisidine value may be informative, particularly for products with higher fat content or during storage studies. Their inclusion should be justified by the nature of the product and its expected shelf life.

Cultivated meat also presents analytical challenges that are less common in conventional food testing. Some upstream inputs, including proprietary growth factors, novel scaffolds, hydrogels, or microcarrier systems, may lack validated analytical methods for detection and quantification in food matrices. This creates a practical challenge for developers, as new methods may need to be developed and validated for specific residues used in cultivated meat production, which can be technically demanding, time-consuming, and costly. Residual enzymes, signalling molecules, or cell culture-specific metabolites may likewise require tailored analytical approaches. Method development and validation can therefore be technically demanding, particularly where the relevant substances are novel, proprietary, or present at very low levels.

Additional attention may be needed for materials that are not ordinarily associated with meat production. Bioreactor materials, plastic-ware, scaffolds, hydrogels, and microcarriers may introduce residues, leachables, trace elements, or antinutritional factors that are not typically encountered in conventional meat. Antinutritional factors associated with certain scaffolding or structural materials may be particularly relevant where such materials remain in, or are intentionally incorporated into, the final product. Where such materials are used, the assessment should determine whether residues remain in the final product and whether they could affect safety or nutritional quality.

Regarding food contact materials, a full risk assessment must be performed on all materials used in production that come into contact with the cells, and details on the type of plastics, their regulatory status, and accompanying certificates of analysis/ declaration of compliance should be assembled.

The risk of nanoparticles and nanoplastics should also be assessed as part of the characterisation of the CM ingredient, considering the production process, culture media inputs, and food contact materials.

Overall, impurity assessment should demonstrate that both conventional contaminants and process-specific residues have been systematically identified, appropriately analysed, and controlled to acceptable levels. In this way, the assessment supports the conclusion that the final product is not only compositionally understood, but also safe for human consumption under its intended conditions of use.

2.7 Proposed Uses and Exposure

The estimation of dietary exposure to the components of the cultured food product is a critical part of its safety assessment. This includes exposure to both nutrients and potential impurities or residues arising from the production process. The approach should integrate compositional data, intended use levels, and food consumption data across different population groups.

Each geographical region typically has its own food consumption database and risk assessment tools to estimate dietary exposure. For example, Europe uses the EFSA Comprehensive European Food Consumption Database [64], while the United States relies on National Health and Nutrition Examination Survey (NHANES) data [65]. Other regions may use nationally developed dietary intake models or databases, such as FSANZ dietary modelling or the SFA using their own data which is publicly available [66]. These databases allow for estimation of both average and high (e.g. 90th percentile) intakes, stratified by age, gender, and other relevant population characteristics.

Nutritional exposure should focus on the components present in significant amounts in the CM product. These include protein, specific amino acids, fats, vitamins, and minerals. The inclusion rate of the product in foods should be factored in, depending on whether it is used as a whole piece (e.g. replacement) or as an ingredient in composite foods (e.g. fishcakes, ready meals). Parameters not relevant or present at negligible levels can be excluded from the assessment, with a justification (e.g. below the limit of detection or absent in all batches analysed). Exposure estimates should be compared against dietary reference values such as adequate intakes (AIs), recommended daily allowances (RDAs), or upper levels (ULs) when available. This helps to ensure that the inclusion of the product does not result in nutritional imbalances or excesses.

On the other hand, exposure to contaminants, such as heavy metals (e.g. mercury, cadmium, lead) or residual culture media components, must also be considered. Substances that were detected or expected based on the production process should be included. A reference point for substances of concern should be proposed (based on toxicological studies in literature), to derive an appropriate margin of exposure (MOE) [67], if no maximum levels are regulated for those substances of concern (e.g. heavy metals).

The overall exposure must be contextualised within the background diet [68]. The addition of the cultured food product should not result in total dietary exposure exceeding health-based guidance values or any toxicological reference point.

2.8 *Nutritional Information*

The nutritional value of cell-derived products depends not only on their proximate composition (e.g. protein, fat, moisture, and micronutrient content) and their estimated dietary exposure, as explained above, but also on the quality of the proteins they provide, if applicable.

Protein quality assessment involves evaluating amino acid profiles, digestibility, and bioavailability. Methods such as the digestible indispensable amino acid score (DIAAS) or the older protein digestibility-corrected amino acid score (PDCAAS) can help determine whether cell-derived proteins meet human nutritional requirements [69]. Different regions accept different protein quality assessment methods. Moreover, the presence of non-protein nitrogen may affect both nutritional and functional attributes of the protein [70, 71]. However, it is unlikely that the protein consumption of any population would depend entirely on CM products. If the protein in CM is lower in quality compared to conventional meat (for example, if certain essential amino acids are less abundant or not in the ideal proportions), it might still be acceptable as a dietary protein source, much like other lower-quality protein sources such as legumes or certain plant-based proteins [72]. The key is that protein quality can be improved through dietary diversity, where CM could be consumed alongside other protein-rich foods that compensate for any amino acid deficiencies. Recent studies have reported that the in vitro digestibility of cell-based foods is similar or even higher than the conventional counterparts [73, 74], using protein derived from cultured pork subjected to the standardised INFOGEST in vitro digestion protocol and comparing it with traditional proteins (e.g. pork, soy, casein, whey). The demonstrated high digestibility, amino acid release, and peptide generation support the nutritional adequacy of this type of novel food.

Regulators may also request information on factors that can reduce protein quality, such as the presence of antinutritional factors (e.g. protease inhibitors, lectins) introduced through culture media or scaffolding materials. Biogenic amines (e.g. histamine, tyramine) and purines, which can accumulate during cell culture, may also affect the nutritional and safety profile and should be quantified where relevant [75]. These purines and biogenic amines originate from both cellular metabolites, including nucleotides and amino acids naturally present in the cells, and from precursors supplied in the culture medium that support cell growth and metabolism.

Non-protein nitrogen compounds can influence both functional properties and the apparent protein content, and their contribution should be clearly distinguished. Overall, a robust nutritional assessment for cell-derived products integrates proximate composition, protein quality, digestibility, and potential inhibitors or secondary metabolites, ensuring that the product is safe and nutritionally adequate for human consumption.

Additionally, when assessing CM, some regulators are requesting the use of an appropriate comparator to evaluate safety and nutritional composition, such as the UK FSA. The comparator serves as a reference to contextualise the new product relative to conventional foods or established dietary sources. The main purpose of this comparison is to determine whether the cultivated product is equivalent in terms of macronutrients, micronutrients, bioavailability, and safety-relevant endpoints and to identify any differences that may require further risk assessment. However, selecting an appropriate comparator can be challenging. Many cultivated meat products are derived from a single cell type, such as myoblasts or adipocytes, which cannot fully replicate the complex structure and composition of conventional meat, which contains muscle fibres, connective tissue, fat, and other cell types. Consequently, direct compositional or functional comparisons may not fully reflect the properties of the final product, and limitations of the comparator should be clearly acknowledged.

Despite these challenges, comparison remains useful for risk assessment because it provides a reference point for evaluating potential differences in nutrient content, contaminants, or processing-related changes. When a single-cell product is not fully representative of the conventional counterpart, risk assessors may accept a broader approach, focusing on whether the cultivated product is nutritionally adequate and safe for its intended use rather than strictly equivalent to traditional meat.

2.9 *Absorption, Distribution, Metabolism, and Excretion (ADME)*

CM is a complex mixture composed primarily of proteins, lipids, moisture, and micronutrients, and its ADME is expected to follow patterns similar to conventional meat. Proteins are digested into amino acids and small peptides and lipids into fatty acids and monoacylglycerols, and carbohydrates and other nutrients are metabolised and absorbed according to their chemical properties. Therefore, for the major macronutrient components, ADME considerations are generally predictable and do not differ significantly from conventional foods.

However, the presence of residual culture media components, growth factors, enzymes, or other compounds without history of use in food used in the production process may warrant special consideration. Literature reviews and targeted analytical testing are recommended to assess whether such residues persist in the final product. If analytical testing confirms the absence of these components or if they are present at negligible levels, they are unlikely to affect ADME and do not represent a safety concern. Where residuals are detected, their chemical nature, potential metabolism, and possible systemic exposure should be evaluated. Regulators may request information on the digestibility and bioavailability of any novel compounds and their metabolic fate, including whether they are transformed during digestion, absorbed into systemic circulation, or excreted unchanged.

In vitro digestibility models, particularly standardised static or semi-dynamic gastrointestinal digestion systems such as INFOGEST, can provide relevant evidence to address ADME considerations for CM. These methods simulate oral, gastric, and intestinal digestion and allow characterisation of protein hydrolysis, lipid release, nutrient bioaccessibility, and stability of any residual process-derived compounds under gastrointestinal conditions. For conventional-like matrices such as cultivated meat, demonstration of digestion patterns may be sufficient to support expected absorption and metabolism of major nutrients, thereby reducing uncertainty regarding systemic exposure. In addition, in vitro digestion can help determine whether residual growth factors, media constituents, scaffold materials, or other novel inputs are degraded during digestion, remain bioaccessible, or generate transformation products requiring further assessment. Therefore, when appropriately designed and interpreted, in vitro digestibility studies may represent a proportionate and scientifically robust tool to address ADME questions without the routine need for animal testing.

Overall, ADME assessment for cultivated meat integrates knowledge of conventional macronutrient metabolism with targeted evaluation of any upstream process-derived residues. Demonstrating that the product is digested, absorbed, metabolised, and excreted similarly to conventional food supports both nutritional and safety assessments.

2.10 Toxicological Information

The need for toxicological studies for cultivated food products must be assessed based on a weight-of-evidence (WoE) approach, considering the cell source, production process, compositional data, and existing knowledge from conventional counterparts. This holistic evaluation can help determine whether specific studies, such as genotoxicity or subchronic toxicity testing, are required or can be waived [5, 76].

It must be noted that generally, the cell types used for cultivated products are typically derived from edible species with a known history of safe consumption or their potential to produce specific toxins is described (e.g. fish cells associated with the production of biogenic amines or marine biotoxins). In general, cells themselves are not inherently toxic. However, at the same time, some marine species may accumulate toxins such as ciguatoxins or tetrodotoxins, because these are acquired through the food chain in the wild, not synthesised endogenously. Cultivated cells grown in sterile, defined environments without environmental triggers or vectors are not expected to produce these compounds, and any compounds that could be harmful would generally have a cytotoxic effect, preventing cell survival and growth.

Potential safety concerns arise primarily from the culture media and other inputs. All components introduced during cell expansion, including media supplements, growth factors, or scaffolding materials, must undergo risk assessment to ensure that they do not introduce toxic compounds. Toxic or unsafe components should not be used at any stage, and any residual compound must be monitored in the final product. On top of this, maintaining genetic and phenotypic stability is critical. Cells are unlikely to acquire the ability to produce toxins if properly characterised and cultured under controlled conditions.

If an applicant uses a poorly characterised cell type, such as those from exotic or non-traditional species, additional safety concerns arise. These cells may produce unknown or potentially harmful metabolites, including toxins, due to uncharacterised metabolic pathways. Without a history of safe use, thorough evaluation of genomic stability, metabolite profiles, and potential toxin production is essential, and if the information collected is not sufficient, the toxicological studies' requirement might not be waived at all.

For example, EFSA recommends a tiered WoE approach for the safety assessment of novel foods, including cultivated meat. This approach integrates all available information in a stepwise manner to systematically identify and address potential hazards. Tier 1 relies on existing knowledge, literature, genotoxicity, and subchronic toxicity. Genotoxicity and subchronic toxicity are usually required when the novel food ingredient contains a substance or components of unknown toxicity, when there are no HBGVs for the components of a novel food, or when an uncharacterised fraction of the novel food may cause safety concerns despite thorough compositional analyses. If a subchronic study is not conducted, a well-reasoned justification should be provided. In the case of CM, all inputs are generally well defined and fully assessed, reducing the need for subchronic toxicity studies.

Tier 2 involves targeted experimental studies, such as reproductive, endocrine, and developmental toxicity, or neurotoxicity, and cardiovascular effects of immunotoxicity, to confirm or quantify hazards identified in Tier 1. Tier 3 is reserved for cases where uncertainty remains and may include in vivo studies of high complexity regarding the duration and the required number of animals to fully characterise potential risks (chronic toxicity, carcinogenicity, extended one-generation reproductive toxicity study). At all stages, the WoE approach emphasises the integration of multiple lines of evidence to support a scientifically justified risk assessment while prioritising resources and minimising unnecessary testing.

A literature review should be conducted to confirm that the species or cell type used is not inherently toxigenic. Multi-omics (WGS, transcriptomics, metagenomics, proteomics) provide robust and powerful tools for the characterisation of cell lines to identify genes or proteins linked to the biosynthesis of toxins, proteins, or other substances of concern [77]. Furthermore, integrated multi-omics can help evaluate whether the cellular metabolism could theoretically lead to undesirable compound formation under the cultivation conditions [78]. Assessing the genetic and phenotypic stability of the cell line throughout the production process is also critical. Methods such as karyotyping, whole-genome sequencing, multi-omics, and analysis of biomarkers can confirm that no transformation or unwanted shifts have occurred. Stability in morphology, growth rate, and viability over time also contributes to the assurance of safety and quality. Similarly, having complete compositional data in which no substances of concern are quantified or its consumption is under safe levels based on the proposed uses, animal toxicological studies might not be needed [79].

Recently, these authors [79] critically examined whether routine in vivo toxicological studies are scientifically justified for cultivated food products, given that these foods are manufactured under tightly controlled conditions and can be extensively characterised through compositional, molecular, microbiological, and exposure data. The authors argue that a tiered WoE framework provides a more scientifically robust and proportionate basis for safety assessment than default reliance on animal feeding studies. Under such an approach, potential risks are addressed through targeted evaluation of relevant hazards (e.g. media residues, scaffold materials, contaminants, novel expressed proteins, or allergenic changes), rather than testing the whole food matrix in animals.

The same publication [79] also refers to an unpublished 90-day repeated-dose oral toxicity study, performed in a third-party Good Laboratory Practice (GLP)-accredited laboratory, in which no toxicologically relevant or statistically significant adverse effects were observed at any tested dose. Accordingly, the No Observed Adverse Effect Level (NOAEL) was established at 5,000 mg/kg bw/day in both male and female rats. In any case, animal studies with CM should be interpreted in context: whole-food rodent studies often provide limited added value because complex food matrices can be difficult to administer at nutritionally balanced high doses, observed effects may reflect dietary imbalance rather than intrinsic toxicity, and any findings are rarely attributable to a specific constituent. In addition, rodent feeding models may have limited human relevance for products intended as part of a mixed human diet. Therefore, although animal studies may occasionally contribute supportive evidence in specific high-uncertainty scenarios, they should not be considered a default requirement for cultivated foods. More informative and human-relevant evidence is typically generated through analytical characterisation, comparative assessment versus conventional foods, exposure modelling, and targeted in vitro/in silico methods within a WoE framework.

However, core to all safety evaluations is whether the CM ingredient raises genotoxicity concerns. For that, a basic genotoxicity test battery may be proposed, following international guidelines [80]. Genotoxicity is an independent endpoint to be considered in the battery of toxicological studies. For CM products, the need for genotoxicity testing should be considered case by case rather than assumed universally. In many cases, where the final product is compositionally comparable to conventional meat and produced using well-characterised cell lines, food-grade media components, and controlled manufacturing processes, a full genotoxicity package may have limited added value. However, testing may be justified where novel inputs, processing aids, scaffold materials, novel growth factors, genetic modifications, unexpected metabolites, contaminants, or substantial compositional deviations from traditional meat are present. A key advantage of genotoxicity testing is that it provides an established and internationally accepted framework to identify DNA-reactive hazards early, thereby supporting regulatory confidence and risk management. It may be relevant when manufacturing introduces new substances or when process-related impurities cannot be excluded. However, limitations also exist for CM matrices. Standard in vitro assays were largely developed for chemicals or defined mixtures, whereas CM is a complex whole-food matrix rich in proteins, lipids, and other nutrients that may complicate extraction of fractions of concern (if applicable), dose preparation, bioavailability, cytotoxicity interpretation, or assay compatibility. Therefore, careful test item selection, justification of selection of relevant fractions, and demonstration of exposure (in in vivo tests if needed) relevance are essential. In some cases, a weight-of-evidence approach integrating compositional analysis, manufacturing controls, and targeted testing may be more scientifically appropriate than routine application of the full genotoxicity battery.

Finally, human studies are not mandatory prior to market entry unless the product is positioned for a sensitive population group or contains compounds at the upper safe intake limits. However, post-marketing surveillance may be recommended as a risk management tool, especially for first-in-market products [81]. Although this would set a precedent, so far, post-market monitoring has not been implemented for any other type of novel food and CM does not pose a greater risk compared to other foods or novel foods.

Additionally, FAO/WHO assessed the potential tumorigenicity of cell-based foods and concluded that it is not a safety concern. Animal cells used for cultivated meat may proliferate in controlled culture, but they lose this ability once removed from their growth environment and ingested. They cannot form tumours in consumers because they cannot survive, divide, or integrate into human tissues [19]. Importantly, cultivated meat does not pose a greater risk of tumorigenicity compared to conventional or other novel foods. Consequently, the main safety considerations are related to conventional hazards, such as microbial contamination, chemical residues, and quality of inputs and materials used during production. Rigorous process control and quality assurance remain essential to ensure the safety of these foods.

2.11 Allergenicity

Assessing the allergenic potential of cultivated food products is a critical component of safety evaluation, particularly due to the novelty of the production process and the use of animal cell culture technologies. Allergenicity can arise not only from the species used but also from the cell culture media components, post-translational modifications (PTMs), and processing aids employed throughout production [82–84].

The allergenicity assessment of CM products relies on a comprehensive WoE approach that integrates multiple methodologies to evaluate potential risks. This approach draws on existing scientific literature, proteomic analyses, and bioinformatics tools, which are particularly important when dealing with gene-edited or genetically modified cell lines. Literature reviews help identify known allergens and assess the potential for novel allergenic proteins introduced through genetic modifications.

The protein content and expression profile in the final product, including concentration of known allergens, should be quantified using validated immunoassays or proteomics techniques. Literature data on allergenic proteins in conventional species can support this evaluation, while in vitro tests using sera from allergic individuals may also be informative. A central aspect of the allergenicity assessment is the homology of protein sequences between the cultivated food and known allergens. Bioinformatic analysis, such as sequence alignment using databases like AllergenOnline, can be used to assess homology with known allergens. According to Codex guidelines, sequences sharing more than 35% identity over an 80-amino-acid window or having an exact match of 8 contiguous amino acids with a known allergen may trigger further investigation [85].

Proteomic profiling allows the detection and characterisation of both known and new proteins, providing insight into changes in protein composition resulting from gene edits or use of recombinant growth factors produced using various platforms. Bioinformatics analyses further support this assessment by predicting allergenic potential based on protein sequences and structural features, including their likelihood to bind IgE antibodies. Integrating these lines of evidence helps regulators and developers understand the allergenic risks of CM, ensuring that any modifications do not introduce unintended hazards. Importantly, the entire production process must be considered. For example, if recombinant growth factors produced in barley are used in the culture medium, it is crucial to assess the presence of residual barley proteins or gluten in the final product. This may require specific testing (e.g. ELISA for gluten) to exclude unintended allergenic residues [86].

Beyond primary sequence similarity, PTMs in the growth factors used in the production process, such as glycosylation, can influence allergenicity. For instance, glycan structures can act as cross-reactive carbohydrate determinants, which may elicit immune responses in sensitised individuals. If the growth factors used in the cultivated food product show altered glycosylation patterns compared to its wild-type counterpart, further evaluation of the immunogenic potential may be warranted

[87]. Analytical methods such as mass spectrometry and glycan profiling can be used to detect such changes.

In summary, allergenicity should be evaluated using a tiered WoE combining literature, bioinformatics, analytical data, and knowledge of the production process. Where possible, mitigation strategies should be implemented to reduce the presence of known allergens or undeclared cross-reactive compounds, ensuring consumer safety and regulatory compliance.

3 Research Gaps to Facilitate Approval Process

A major challenge in the safety assessment of CM lies in the analytical evaluation of components used during production, such as culture media ingredients or processing aids such as those protecting cells from shear stress. Often, these components are novel or complex, and internationally recognised analytical methods may not be available or validated for these specific matrices. This lack of validated methods complicates accurate identification, quantification, and monitoring of potentially hazardous substances, impurities, or residuals in the final product, raising concerns for food safety regulators and producers alike. To overcome this challenge, two main approaches can be pursued. First, avoiding the use of components that do not have a history of safe consumption in food and lack validated analytical methods can minimise uncertainty and facilitate regulatory approval. However, this may limit innovation or efficiency in cell culture processes. Alternatively, developing new analytical methods tailored to the unique matrices of CM is a proactive solution. This involves adapting or creating sensitive, specific, and robust analytical techniques, followed by thorough validation following international guidelines. Collaboration between academia, industry, and regulatory bodies is crucial to accelerate method development and harmonise standards, ensuring reliable safety assessment of novel components in CM production.

The Cultured Meat Safety Initiative (CMSI) aims to advance the safety assessment of cultivated meat and seafood. CMSI brings together experts in food safety, toxicology, cell biology, and biomanufacturing to identify safety priorities and develop standardised assessment frameworks. A key output is the Safety Assessed Media Ingredient (SAMI) List, which evaluates common media components such as vitamins, minerals, amino acids, and growth factors, providing a harmonised reference to streamline regulatory review. By focusing on defined inputs and evidence-based risk assessment, CMSI supports the safe integration of CM into the global food supply [58, 60, 88].

Also, new approach methodologies (NAMs), including in vitro tests, quantitative structure-activity relationships (QSAR) models, and read-across techniques [89], could offer valuable tools to support the safety assessment of cultivated meat. While not yet full substitutes for animal studies, NAMs help build a WoE by predicting potential hazards and estimating safe exposure levels, especially for novel culture media components with limited data. The use of existing data through approaches like QSAR and thresholds of toxicological concern (TTCs) to reduce animal testing should be promoted. TTC sets exposure limits for substances with limited toxicity data based on chemical structure and known toxicological thresholds, providing a practical safety benchmark [90]. By integrating NAMs early, developers can prioritise testing and simplify safety evaluations, ultimately reducing reliance on animal studies while ensuring product safety [91].

Scaffolds used in CM production, along with potential impurities from manufacturing (e.g. media components with unknown toxicity and degradation profiles), may introduce unknown substances, metabolites, or degradation products whose toxicity or allergenic potential is not well characterised. This uncertainty poses safety challenges, as these components could accumulate and cause adverse reactions. To address this, the TTC approach mentioned above could be applied, while in parallel, research should prioritise the use of well-characterised, safe materials for scaffolds to reduce risks from unknown substances [92, 93].

Multi-omics-based approaches are increasingly being adopted as valuable tools to support the safety assessment of CM. These techniques enable detailed characterisation of production cell lines and final products across multiple biological levels. For example, whole-genome sequencing can be used to confirm species identity and assess genomic stability; transcriptomics can provide insight into gene expression patterns and cellular responses to culture conditions; proteomics enables the identification and quantification of proteins, including the detection of novel or unintended proteins; and metabolomics can be used to profile metabolic activity and identify potential accumulation of undesirable metabolites [77].

Recent work by Mathieu et al. [78] and Kusters et al. [77] demonstrates how multi-omics approaches can be integrated to capture complex molecular interactions across genes, transcripts, proteins, and metabolites, providing a system-level understanding of cellular behaviour of CM. Such integrated datasets, including interactome models, offer a powerful means to assess genome stability, monitor batch-to-batch consistency, and detect unintended cellular changes, including the potential emergence of novel allergens. Despite their potential, the application of omics technologies in regulatory safety assessment remains at an early stage. There is a clear need for internationally recognised methodologies, standardised protocols, and guidance on data interpretation to ensure that omics-derived evidence can be applied consistently and reliably within the context of CM safety evaluation. Notably, similar approaches have already been implemented in the risk assessment of microorganisms used in the food chain, where whole-genome sequencing is used to support strain characterisation and detect virulence factors or antimicrobial resistance genes [7, 9]. These precedents provide a valuable foundation for the future integration of omics-based methods into the safety assessment of CM.

Consistency in CM production is critical to guarantee product safety, quality, and regulatory compliance. Variability can arise from differences in cell lines, culture conditions, media composition, and scaffold materials, potentially affecting the final product's composition and safety profile. To overcome this challenge, implementing robust process control and quality assurance systems is essential. This includes standardised protocols for cell line authentication, stringent monitoring of culture parameters, and validated analytical methods to assess key product attributes. Automated bioreactor systems and real-time sensors can help maintain controlled, reproducible conditions.

Using biomarkers for identification is a promising approach to ensure the traceability and authenticity of CM products. Biomarker-specific molecular signatures such as unique proteins, DNA sequences, or metabolites can confirm the cell line origin and verify production consistency. This helps prevent contamination, mislabelling, or adulteration, which are critical for safety and regulatory compliance [21]. Developing reliable biomarkers requires thorough characterisation of cell lines and their culture products.

4 Future Perspectives

The safety assessment of CM is an evolving field that requires continuous adaptation of risk assessment frameworks to address unique challenges. Traditional risk assessment pillars such as identity, composition, toxicological studies, and nutritional evaluation remain fundamental but must be enhanced with innovative approaches tailored to the complexities of CM production [25, 88].

Establishing clear cell identity through robust biomarkers is essential for traceability and consistency. Advances in genomics and proteomics will enable precise characterisation of cell lines and monitoring of production batches, reducing risks of contamination or misidentification. Comprehensive compositional analysis, supported by validated analytical methods, is critical to detect impurities, scaffold residues, and media components [58]. Where validated methods do not exist, collaborative efforts to develop and standardise novel techniques will be pivotal.

Integrating multi-omics approaches, encompassing transcriptomics, proteomics, and metabolomics, will be pivotal for optimising bioprocesses and ensuring product quality and safety. By adopting such integrative strategies, CM companies can achieve a system-level understanding of genomic and phenotypic attributes, cellular metabolism, and fundamental characterisation of the cells. These approaches provide robust scientific evidence for safety assessments.

Toxicological evaluation faces particular hurdles due to novel components and limited in vivo data. The integration of NAMs, including in vitro assays and computational models, promises to enhance hazard identification while reducing reliance on animal testing. Approaches such as TTC and read-across methodologies can fill data gaps for unknown substances like scaffold materials and unusual processing aids. Importantly, risk assessors must remain open and adaptive to these emerging methods, incorporating them into regulatory frameworks in a flexible, science-based manner to ensure comprehensive safety evaluations.

Nutrition assessments will benefit from deeper understanding of the bioavailability and functionality of nutrients in CM, ensuring that it meets or exceeds conventional standards. Additionally, addressing process consistency through advanced bioprocess controls and quality-by-design principles will be key to producing safe, uniform products at scale.

Future research should prioritise culture media safety assessments, safe scaffold materials, integration and validation of NAMs, and validated analytical methods to streamline regulatory acceptance. Cross-sector collaboration between academia, industry, and regulators will accelerate innovation, harmonise standards, and build consumer trust. Ultimately, a multidisciplinary, flexible risk assessment approach that embraces emerging technologies and addresses identified challenges will be essential to unlock the full potential of CM as a safe and sustainable food source.

Declaration of Competing Interest

The authors declare that they have no known competing financial interests or personal relationships that could have appeared to influence the work reported in this chapter.
Data Availability No data was used for the research described in the chapter.

Disclaimer Atova Regulatory Consulting is an independent scientific and regulatory advisory firm providing consulting services in the field of novel foods and ingredient safety. The authors declare that this manuscript was prepared in a personal capacity and not in the course of normal professional activities; accordingly, the views expressed herein do not represent those of the company. This work was not commissioned, and no specific financial interests or product endorsements influenced its content. The regulatory interpretations presented reflect the authors' understanding at the time of writing and are intended for academic and informational purposes only. They do not constitute legally binding advice, and readers are strongly encouraged to consult original legislation, official guidance documents, and competent authorities when making regulatory or compliance decisions.

References

1. Chriki S, Hallman W, Hocquette JF et al (2025) Food culture and cell-culture: technical, ethical and social frontiers. NPJ Sci Food 9:49
2. Lanzoni D, Rebucci R, Formici G et al (2024) Cultured meat in the European Union: legislative context and food safety issues. Curr. Res. Food Sci. 8:100722
3. Johnson H, Monaco A (2025) Global developments in the regulation of cultivated meat: a comparative study of the EU, Singapore, US and Australia and New Zealand. Rev Eur Comp Int Environ Law 34:496–511. https://doi.org/10.1111/reel.70007
4. Zandonadi RP, Ramos MC, Elias FTS, Guimarães NS (2025) Global insights into cultured meat: uncovering production processes, potential hazards, regulatory frameworks, and key challenges – a scoping review. Foods 14:129
5. Gu Y, Li X, Chan ECY (2023) Risk assessment of cultured meat. Trends Food Sci Technol 138:491
6. Food and Agriculture Organization of the United Nations, World Health Organization (2023) Food safety aspects of cell-based food. World Health Organization. https://www.who.int/publications/i/item/9789240070943
7. EFSA Panel on Nutrition, Novel Foods and Food Allergens (NDA), Turck D, Bohn T et al (2024) Guidance on the scientific requirements for an application for authorisation of a novel food in the context of Regulation (EU) 2015/2283. EFSA J 22(9):e8961. https://doi.org/10.2903/j.efsa.2024.8961
8. Pamies D, Leist M, Coecke S et al (2022) Guidance document on good cell and tissue culture practice 2.0 (GCCP 2.0). ALTEX 39. https://doi.org/10.14573/altex.2111011

9. Bennekou SH, Allende A, Bearth A et al (2025) Guidance on the characterisation of microorganisms in support of the risk assessment of products used in the food chain. EFSA J 23. https://doi.org/10.2903/j.efsa.2025.9705
10. European Medicines Agency (1998) Quality of biotechnological products: derivation and characterisation of cell substrates used for production of biotechnological/biological products. ICH Harmonised Tripartite Guideline. Dev Biol Stand 93.
11. Afonso AL, Gelbmann W, Germini A et al (2024) EFSA scientific colloquium 27: cell culture-derived foods and food ingredients. EFSA Support Publ 21. https://doi.org/10.2903/sp.efsa.2024.EN-8664
12. Smith-Uchotski R, Wanjiru P, Willis R, Adkin A (2022) Hazard identification: identification of hazards in cultured animal cells. Food Standards Agency. https://www.food.gov.uk/sites/default/files/media/document/Cultured%20meat%20hazard%20identification%20final_0.pdf
13. UK FSA (2026) Cell-cultivated products: guidance on classification and HACCP principles
14. UK FSA (2026) Introduction to supplementary guidance to applicants for assessment of Cell Cultivated Products (CCP) in food: Allergenicity & Nutrition
15. Singapore Food Agency (2025) Requirements for the safety assessment of novel foods and novel food ingredients
16. Ministry of Food and Drug Safety (2024) Guide to writing temporary standards and specifications for cell cultured food raw materials
17. Lammerding AM, Fazil A (2000) Hazard identification and exposure assessment for microbial food safety risk assessment. International J Food Microbiol 58:147
18. FAO (1999) Principles and guidelines for the conduct of microbiological risk assessment, vol 63. FAO, Agriculture and Consumer Protection
19. FAO/WHO (2023) Food safety aspects of cell-based food
20. Ong KJ, Johnston J, Datar I et al (2021) Food safety considerations and research priorities for the cultured meat and seafood industry. Compr Rev Food Sci Food Saf 20. https://doi.org/10.1111/1541-4337.12853
21. Mariano E, Lee DY, Yun SH et al (2023) Checkmeat: a review on the applicability of conventional meat authentication techniques to cultured meat. Food Sci Anim Resour 43:1055
22. Sogore T, Guo M, Sun N et al (2024) Microbiological and chemical hazards in cultured meat and methods for their detection. Compr Rev Food Sci Food Saf 23:e13392
23. UK FSA (2023) Hazard identification: identification of hazards in meat products manufactured from cultured animal cells
24. Food and Agriculture Organization of the United Nations/World. Health Organization (2014) Multicriteria-based ranking for risk management of food-borne parasites. Microbiological Risk Assessment Series No. 23. FAO. https://openknowledge.fao.org/items/3cf295a5-19b4-46af-8e24-100c168d7dc5
25. Manning L (2024) Responsible innovation: mitigating the food safety aspects of cultured meat production. J Food Sci 89:4638–4659
26. Denner J, Tönjes RR (2012) Infection barriers to successful xenotransplantation focusing on porcine endogenous retroviruses. Clin Microbiol Rev 25:318
27. Cho M, Son HS (2019) Prediction of cross-species infection propensities of viruses with receptor similarity. Infect Genet Evol 73. https://doi.org/10.1016/j.meegid.2019.04.016
28. Parrish CR, Holmes EC, Morens DM et al (2008) Cross-species virus transmission and the emergence of new epidemic diseases. Microbiol Mol Biol Rev 72. https://doi.org/10.1128/mmbr.00004-08
29. Aladhadh M (2023) A review of modern methods for the detection of foodborne pathogens. Microorganisms 11:1111
30. Zabal O, Kobrak A, Lager I et al (2000) Contamination of bovine fetal serum with bovine viral diarrhea virus. Rev Argent Microbiol 32:27
31. Brorson K, De Wit C, Hamilton E et al (2002) Impact of cell culture process changes on endogenous retrovirus expression. Biotechnol Bioeng 80. https://doi.org/10.1002/bit.10366

32. Strauss B, Kleinsorge B, Lotfian P (2025) Best practice for cell banking techniques used in cell cultivated products. FSA Res Evid. https://doi.org/10.46756/001c.144239
33. Gough KC, Maddison BC (2010) Prion transmission: prion excretion and occurrence in the environment. Prion 4:275
34. Chen CC, Wang YH, Wu KY (2013) Consumption of bovine spongiform encephalopathy (BSE) contaminated beef and the risk of variant Creutzfeldt-Jakob disease. Risk Anal 33. https://doi.org/10.1111/risa.12079
35. Romero R, Quill E (2025) How do microbial safety issues associated with meat apply to cell cultivated products? FSA Res Evid. https://doi.org/10.46756/001c.144237
36. Powell DJ, Li D, Smith B, Chen WN (2025) Cultivated meat microbiological safety considerations and practices. Compr Rev Food Sci Food Saf 24:e70077
37. Pasitka L, Cohen M, Ehrlich A et al (2023) Spontaneous immortalization of chicken fibroblasts generates stable, high-yield cell lines for serum-free production of cultured meat. Nat Food 4 (1):35–50. https://doi.org/10.1038/s43016-022-00658-w
38. Pasitka L, Cohen M, Regenbaum S et al (2025) Spontaneous immortalization of bovine fibroblasts following long-term expansion offers a non-transformed cell source for cultivated beef. Nat Food 6(11):1079–1094. https://doi.org/10.1038/s43016-025-01255-3
39. Xue T, Zheng H, Zhao Y et al (2024) A spontaneously immortalized muscle stem cell line (EfMS) from brown-marbled grouper for cell-cultured fish meat production. Commun Biol 7. https://doi.org/10.1038/s42003-024-07400-1
40. Muller Q, Matsusaki M (2026) Comparative analysis of cultivated meat cell sources and cell type usage across species: functional roles and engineering potential. Trends Food Sci Technol 169:105524
41. Riquelme-Guzmán C, Stout AJ, Kaplan DL, Flack JE (2024) Unlocking the potential of cultivated meat through cell line engineering. iScience 27:110877. https://doi.org/10.1016/j.isci.2024.110877
42. Van Der Meer P, Angenon G, Bergmans H et al (2023) The status under EU law of organisms developed through novel genomic techniques. Eur J Risk Regul 14. https://doi.org/10.1017/err.2020.105
43. European Commission (2021) Study on the status of new genomic techniques under Union law and in light of the Court of Justice ruling in Case C-528/16. Commission Staff Working Document SWD(2021)
44. Chiang TWW, Le Sage C, Larrieu D et al (2016) CRISPR-Cas9D10A nickase-based genotypic and phenotypic screening to enhance genome editing. Sci Rep 6. https://doi.org/10.1038/srep24356
45. Wu WH, Ma XM, Huang JQ et al (2022) CRISPR/Cas9 (D10A) nickase-mediated Hb CS gene editing and genetically modified fibroblast identification. Bioengineered 13. https://doi.org/10.1080/21655979.2022.2069940
46. Kimberland ML, Hou W, Alfonso-Pecchio A et al (2018) Strategies for controlling CRISPR/Cas9 off-target effects and biological variations in mammalian genome editing experiments. J Biotechnol 284:91
47. Bao XR, Pan Y, Lee CM et al (2021) Tools for experimental and computational analyses of off-target editing by programmable nucleases. Nat Protoc 16:10
48. EFSA GMO Panel, Casacuberta J, Barro F, Braeuning A et al (2025) New developments in biotechnology applied to animals: an assessment of the adequacy and sufficiency of current EFSA guidance for animal risk assessment. EFSA J 23. https://doi.org/10.2903/j.efsa.2025.9566
49. FAO/WHO (2023) Gene editing and food safety – technical considerations and potential relevance to the work of Codex Alimentarius
50. OECD (2025) Consensus documents: harmonisation of regulatory oversight in biotechnology
51. DawanS J, Ahn J (2022) Application of DNA barcoding for ensuring food safety and quality. Food Sci Biotechnol 31:1

52. Hu J, Wei H, Jiang Y et al (2025) DNA barcoding in meat authentication: principles, applications, and future perspectives. Foods 14:3522
53. Zia Q, Alawami M, Mokhtar NFK et al (2020) Current analytical methods for porcine identification in meat and meat products. Food Chem 324:126664
54. Naraoka Y, Mabuchi Y, Kiuchi M et al (2024) Quality control of stem cell-based cultured meat according to specific differentiation abilities. Cells 13. https://doi.org/10.3390/cells13020135
55. Habowski K, Sant'Ana AS (2024) Microbiology of cultivated meat: what do we know and what we still need to know? Trends Food Sci Technol 154:104759. https://doi.org/10.1016/j.tifs.2024.104759
56. Nausch H, Buyel JF (2021) Cryopreservation of plant cell cultures – diverse practices and protocols. New Biotechnol 62. https://doi.org/10.1016/j.nbt.2021.02.002
57. Zehorai E, Maor-Shoshani A, Molotski N et al (2024) From fertilised oocyte to cultivated meat – harnessing bovine embryonic stem cells in the cultivated meat industry. Reprod Fertil Dev 36:124
58. Ong K, Kukk K, Powell D, et al (2025) Development of a safety-assessed media ingredient (SAMI) framework for streamlined safety assessment of cultivated meat and seafood products
59. Huang T, Ko C, Paes D et al (2024) A review on the safety of growth factors commonly used in cultivated meat production. Compr Rev Food Sci Food Saf 23:e13350
60. Wonfor RE, Ong KJ, Ng W et al (2026) Standardising culture medium safety testing for cultivated meat: outputs from a workshop and case study. Foods 15:783. https://doi.org/10.3390/foods15040783
61. van der Heijden M (2026) Regulatory frameworks for cell-based food and precision fermentation-derived products: an overview of legal issues and solutions. FAO Legal Papers No. 114. FAO, Rome. https://openknowledge.fao.org/items/3cf295a5-19b4-46af-8e24-100c168d7dc5
62. Lim PY, Suntornnond R, Choudhury D (2025) The nutritional paradigm of cultivated meat: bridging science and sustainability. Trends Food Sci Technol 156:104838. https://doi.org/10.1016/j.tifs.2024.104838
63. Sikora D, Rzymski P (2024) Assessment of the potential nutritional value of cell-cultured chicken meat in light of European dietary recommendations. J Food Compos Anal 135:106663. https://doi.org/10.1016/j.jfca.2024.106663
64. EFSA (2011) Use of the EFSA Comprehensive European Food Consumption Database in exposure assessment. EFSA J 9. https://doi.org/10.2903/j.efsa.2011.2097
65. Akinbami LJ, Chen TC, Davy O et al (2022) National Health and Nutrition Examination Survey, 2017–March 2020 prepandemic file: sample design, estimation, and analytic guidelines. Vital Health Stat 2. https://doi.org/10.15620/cdc:115434
66. Lim GS, Er JC, Bhaskaran K et al (2024) Singapore's total diet study (2021–2023): study design, methodology, and relevance to ensuring food safety. Foods 13. https://doi.org/10.3390/foods13040511
67. EFSA (2012) Statement on the applicability of the Margin of Exposure approach for the safety assessment of impurities which are both genotoxic and carcinogenic in substances added to food/feed. EFSA J 10. https://doi.org/10.2903/j.efsa.2012.2578
68. Jin J, Den Besten HMW, Rietjens IMCM, Widjaja-Van Den Ende F (2025) Chemical and microbiological hazards arising from new plant-based foods, including precision fermentation-produced food ingredients. Annu Rev Food Sci Technol 11:32. https://doi.org/10.1146/annurev-food-111523. www.annualreviews.org
69. FAO, IAEA (2024) Development of a protein database and the way forward for reviewing protein requirements
70. Li X, Zhang D, Ren C et al (2021) Effects of protein posttranslational modifications on meat quality: a review. Compr Rev Food Sci Food Saf 20:289
71. Venkatesan M, Semper C, Skrivergaard S et al (2022) Recombinant production of growth factors for application in cell culture. iScience 25. https://doi.org/10.1016/j.isci.2022.105054
72. Calvez J, Azzout-Marniche D, Tomé D (2024) Protein quality, nutrition and health. Front Nutr 11:1406618

73. Chen Z, Liu H, Tang X et al (2026) Analysis of nutritional properties and in vitro digestibility characteristics of cultivated biomass protein. Food Chem 513:148846. https://doi.org/10.1016/j.foodchem.2026.148846
74. Xie Y, Ding S, Wang J et al (2026) Cultured meat protein has significantly higher digestibility and more potential bioactive peptides than traditional animal and plant proteins. Food Chem 502. https://doi.org/10.1016/j.foodchem.2025.147659
75. Park H, Cho I, Heo S et al (2025) Metabolomic insights of cultured meat compared to conventional meat. Sci Rep 15. https://doi.org/10.1038/s41598-025-00719-7
76. Samarasiri M, Chai KF, Chen WN (2023) Forward-looking risk assessment framework for novel foods. Food Hum 1. https://doi.org/10.1016/j.foohum.2023.06.020
77. Kusters R, Mathieu T, Kamgang Nzekoue F et al (2026) Multi-omics characterization of duck embryonic stem cells for cultivated meat. bioRxiv. https://doi.org/10.64898/2026.04.27.720974
78. Mathieu T, Légaré S, Nzekoue AF et al (2025) Integrative multi-omics modelling for cultivated meat production, quality, and safety. Trends Food Sci Technol 166:105364
79. Ong KJ, Rivero-Pino F, Lim C et al (2025) Scientific opinion on animal toxicological studies for the safety assessment of cultivated food products. Preprints (Basel). https://doi.org/10.20944/preprints202511.1087.v1
80. Ververis E, Ackerl R, Azzollini D et al (2020) Novel foods in the European Union: scientific requirements and challenges of the risk assessment process by the European Food Safety Authority. Food Res Int 137:109515. https://doi.org/10.1016/j.foodres.2020.109515
81. Amanor-Boadu V (2004) Post-market surveillance model for potential human health effects of novel foods. Food Policy 29. https://doi.org/10.1016/j.foodpol.2004.09.001
82. Fernandez A, Mills ENC, Koning F, Moreno FJ (2021) Allergenicity assessment of novel food proteins: what should be improved? Trends Biotechnol 39:4–8. https://doi.org/10.1016/j.tibtech.2020.05.011
83. Mills ENC, Orsenigo F, Salgado D et al (2024) Novel strategies for predicting allergenicity: development of a ranking method and screening tools to assess the allergy risk of innovative proteins. EFSA Support Publ 21. https://doi.org/10.2903/sp.efsa.2024.en-8840
84. Rahaman T, Vasiljevic T, Ramchandran L (2016) Effect of processing on conformational changes of food proteins related to allergenicity. Trends Food Sci Technol 49:24–34. https://doi.org/10.1016/j.tifs.2016.01.001
85. WHO/FAO (2009) Principles and methods for the risk assessment of chemicals in food. Int J Environ Stud 68:251
86. Aiassa E, Higgins JPT, Frampton GK et al (2015) Applicability and feasibility of systematic review for performing evidence-based risk assessment in food and feed safety. Crit Rev Food Sci Nutr 55. https://doi.org/10.1080/10408398.2013.769933
87. Hale RC, Morais D, Chou J, Stowell SR (2024) The role of glycosylation in clinical allergy and immunology. J Allergy Clin Immunol 153:55
88. Ong KJ, Tejeda-Saldana Y, Duffy B et al (2023) Cultured meat safety research priorities: regulatory and governmental perspectives. Foods 12. https://doi.org/10.3390/foods12142645
89. Roy K, Banerjee A (2024) Read-across and quantitative structure–activity relationships (QSAR) for making predictions and data gap-filling
90. Kroes R, Kleiner J, Renwick A (2005) The threshold of toxicological concern concept in risk assessment. Toxicol Sci 86. https://doi.org/10.1093/toxsci/kfi169
91. Cattaneo I, Astuto MC, Binaglia M et al (2023) Implementing new approach methodologies (NAMs) in food safety assessments: strategic objectives and actions taken by the European Food Safety Authority. Trends Food Sci Technol 133:277–290. https://doi.org/10.1016/j.tifs.2023.02.006
92. Bomkamp C, Skaalure SC, Fernando GF et al (2022) Scaffolding biomaterials for 3D cultivated meat: prospects and challenges. Adv Sci 9:2102908
93. Levi S, Yen FC, Baruch L, Machluf M (2022) Scaffolding technologies for the engineering of cultured meat: towards a safe, sustainable, and scalable production. Trends Food Sci Technol 126:13

Adv Biochem Eng Biotechnol (2026) 193: 273–294
https://doi.org/10.1007/10_2025_293

Published online: 15 October 2025

Environmental Sustainability of Cultivated Meat

Hanna L. Tuomisto and Toni Ryynänen

Contents

Abstract Cultivated meat is developed as an alternative to livestock meat in reducing the negative externalities of the animal-based food systems. Prospective life cycle assessment studies have been used to estimate the potential environmental impacts of cultivated meat production. Results indicate that cultivated meat

H. L. Tuomisto (✉)
Department of Agricultural Sciences, University of Helsinki and Natural Resources Institute Finland, Helsinki, Finland
e-mail: hanna.tuomisto@helsinki.fi

T. Ryynänen
Ruralia Institute, University of Helsinki, Mikkeli, Finland

production has relatively high industrial energy demand, but lower land use requirements compared to livestock meat production. The climate impacts of cultivated meat depend on the source of energy used. Due to the low land use requirements, cultivated meat could have benefits to biodiversity by reducing the pressure to convert forests and natural habitats to agricultural land. Cultivated meat production could also reduce water use and emissions to waterbodies. The environmental impacts of different cultivated meat production processes are wide depending on the process design and sources of inputs. Life cycle assessment studies can guide the development of cultivated meat production processes towards sustainable options.

Graphical Abstract

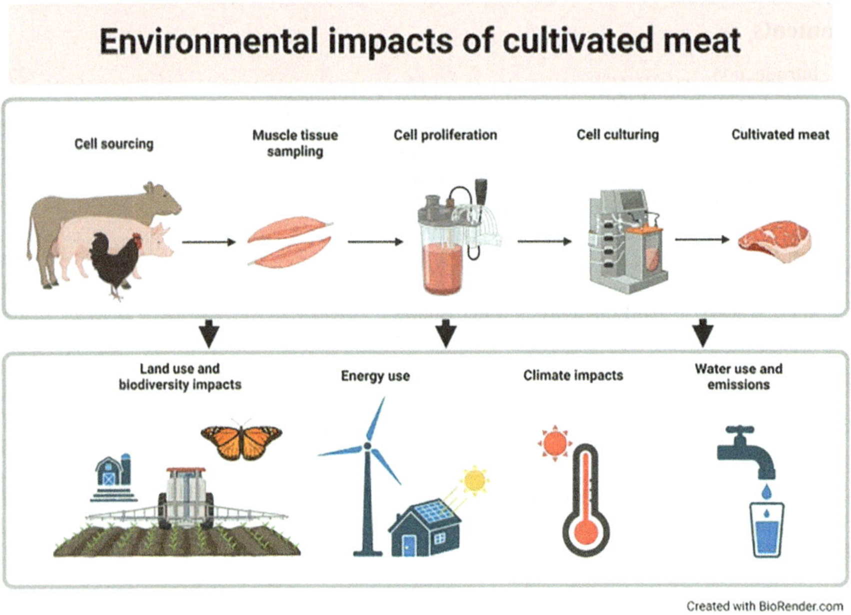

Keywords Biodiversity, Carbon footprint, Climate change, Energy use, Land use, Life cycle assessment, Water use

Abbreviations

ALCA	Attributional life cycle assessment
CHO	Chinese hamster ovary
CLCA	Consequential life cycle assessment
FBS	Foetal bovine serum
FU	Functional unit

GHG	Greenhouse gas
HOB	Hydrogen-oxidising bacteria
LCA	Life cycle assessment
LCI	Life cycle inventory
LCIA	Life cycle inventory assessment

1 Introduction

Agriculture and food production systems are fundamental in the provision of human life. However, these systems correspond to over 80% of the global biodiversity loss, major land use changes and nutrient losses [1]. Food systems produce approximately a third of anthropogenic greenhouse gas (GHG) emissions [2]. Livestock production is causing 60% of these food system-related emissions [3]. The consumption of animal-based products has been estimated to increase simultaneously with the expanding global population and cumulating wealth [4]. Meat consumption is also justified with the four Ns: meat eating is claimed to be necessary, natural, normal and nice [5]. These trends will increase supply and demand of food products from livestock food systems and therefore directly cause more negative environmental impacts from the food systems.

Major socio-technical and system-level transitions or even transformations [6] are needed in food production and diets to reduce the current environmental impacts. Switching animal-based diets to plant-based diets is shown to reduce the environmental impacts: transition to vegan diets globally could lead to 80% reduction of GHG emissions from food production systems compared to status quo scenario in 2050 [7]. Similarly, achieving the set climate targets and limiting the increase in atmospheric temperature to 1.5° above preindustrial levels need major restructuring of food supply and demand systems [8]. Consumer acceptance of vegan diet is currently rather low: approximately 1–4% of population in the European countries follow a vegan diet [9]. Plant-based meat alternatives are widely available on the markets globally, and consumer interest in the products is steadily increasing [10]. However, the market share of the alternative protein sources is small compared to preferred livestock products. Cultivated meats have the potential to improve the sustainability of food production systems: it offers appealing, ethical and environmentally sustainable analogue – not just an alternative – to livestock meat without a need to change culturally and socially meaningful eating habits and practices.

This chapter addresses the environmental impacts of livestock meat and cultivated meat production by reviewing recent environmental life cycle assessment (LCA) studies. An overview of the LCA methods and environmental challenges associated with livestock production is followed by a review of the LCA studies considering environmental impacts of cultivated meat. The chapter concludes by presenting a summary of the current scientific knowledge on the environmental impacts of cultivated meat production.

2 How Environmental Impacts of Cultivated Meat Are Studied?

2.1 Background Information and the Analysis Process Design

Environmental impacts of food products are estimated with LCA methods [11, 12] standardised by the International Organization for Standardisation [13]. LCA studies are typically attributional LCAs (ALCAs) that calculate the emissions and resource use along the life cycle of a product and use allocation methods to share the impacts between the co-products [12]. Consequential LCA (CLCA) is a less used method that assesses the marginal impacts of changes in a product system or production of an additional product [14]. CLCA quantifies the impacts of changes in all product systems affected rather than quantifying the impacts of the production of inputs needed for the product system. CLCA does not use allocation methods, but the system considers all co-products. Most of the available LCA studies of cultivated meat use the ALCA approach.

LCA studies follow four steps: (1) goal and scope definition, (2) inventory analysis, (3) impact assessment and (4) interpretation. The goal and scope definition details the purpose of the study and sets the functional unit (FU), system boundaries, data requirements and allocation processes. The data for a study is collected during an inventory analysis step, whereas an impact assessment step is used to modify the data to the environmental impact results. The assessment of the robustness of the analysis, conclusions and recommendations are presented in the interpretation step.

2.2 Setting Goals, Functional Units and System Boundaries

The first step is to define the goal of the LCA study and the intended application, such as policy design, hotspot analysis, product development, eco-design, marketing or research. Studies can target a specific group such as consumers, policymakers, researchers or business stakeholders. The aim of the study outlines the quality of data required from each step of the life cycle. The cultivated meat LCAs published are prospective studies aiming to estimate the potential impacts of future food technologies and therefore are based on a set of assumptions rather than factual data needed for studies assessing existing processes.

The environmental impacts are allocated according to a functional unit (FU). The FU answers to what, how much, how well and how long: the FU of a cultivated meat product can be 1 kg unpackaged cultivated meat (with X% protein content) at the factory gate. Complex indexes considering the nutritional content of the foods such as the health-promoting and health-reducing nutrients have been developed in addition to using simple FU measures as weight [15]. Although nutritional indexes are useful in comparing foods with differing nutritional profiles, their use as FU is not simple as diets consist of several food items: the nutritional content of the whole diet is a more valuable information than the nutrients in single food items. The environmental impacts of protein-rich foods are often compared using the protein content as an FU [12].

Setting the system boundaries for an LCA study determinates the included and excluded processes. It is not necessary to include packaging, retail and consumption steps if the study aims to compare different food production methods. Often used assessment types are a cradle-to-gate assessment and a cradle-to-grave assessment. The extant cultivated meat LCA system boundaries include processes from extraction of natural resources up to the harvesting of the cultured cell biomass from the bioreactors. These studies exclude the product processing or packaging. The production of bioreactors is included in some of the cultivated meat LCA studies, but other industrial infrastructure such as buildings are excluded. The limitation of the studied cultivated meat system boundaries to the cradle-to-gate approach is justified when compared to livestock meat: most of the environmental impacts of livestock production originate from the agricultural activities before the products leave the farm [12].

2.3 Inventory Analysis and Co-product Allocation

The inventory analysis step includes the collection of the material and energy input and emission flow data within the defined system boundaries. The data quality requirements are lower for a screening analysis aiming to understand the hotspots of impacts in the product system and to estimate the general impacts than for a more detailed LCA study. Higher quality of data is often required for analysing core processes such as cell culturing than for upstream processes such as production of inputs. LCA databases have available life cycle inventory (LCI) data for a wide range of products. Activity data for the core processes considering the material and energy input requirements and outputs are collected first, and the existing databases are used as a data source for the material and emission flows of the input production processes.

Several methods are used in ALCAs for allocating the impacts of the studied production process between the co-products. Creating sub-processes should be avoided, but not often possible in the food product assessments. A system expansion can also be used, where the impacts of alternative ways of producing the co-products are calculated. In the exemplary case of dairy production, meat from dairy cows is assigned similar impacts to producing meat in beef herd systems. System expansion is associated with the CLCA approach and only rarely used in ALCA studies, whereas allocation methods based on the dry or wet mass, energy content or economic value of the co-products are used [16]. Different allocation methods have a major impact on the results and can change the conclusions. Kyttä [17] showed that when system expansion was used for comparing the climate impacts of cultivated meat production with livestock and plant-based systems, the relative differences were different compared to using standard allocation methods. When the system expansion was used, cultivated meat had lower impacts compared to dairy, beef and pork systems. When economic allocation was used, cultivated meat had higher impact than pork and beef from dairy system, but lower impact compared to beef from beef cattle.

Lactate and ammonia are produced as cultivated meat co-products. Some of the impacts of the whole cultured meat production process can be allocated to these co-products if they are used as inputs in other product systems. A study compared alternative allocation methods that included product substitution that assumed lactate to replace lactic acid production and dry and wet mass-based allocations [18]. The substitution-based allocation resulted in 5–48% of the impacts of the whole production process being allocated to lactate depending on the impact category. Dry mass allocation resulted in 85% whereas wet mass allocation 63% of the impacts being allocated to lactate. Economic allocation is often the most reasonable choice but needs reliable price information.

2.4 *Assessing the Life Cycle Impacts and Interpreting the Results*

The life cycle impact assessment (LCIA) phase converts the LCI data to selected environmental impacts. The results of different LCA studies are challenging to be directly compared due to several LCIA method options. Some LCIA methods consist of up to 20 different environmental impact categories ranging from climate change impacts to eutrophication potential. Often, only the most relevant environmental impact categories are selected and analysed using an LCA software.

The first step of LCIA is characterised by grouping and characterisation of the resource and emission flows to form the environmental impact results for each impact category. In addition, normalisation and weighting steps may be used for aggregating the results to a single score in cases when a study involves several impact categories. The weighting factors aim to describe the relative importance of different environmental impact categories, for instance, based on the opinions of stakeholders and experts or based on the distance to targets [19].

The last interpretation step evaluates the robustness of the results based on the data quality. Monte Carlo method is commonly used for the uncertainty assessment in LCA studies, whereas sensitivity analyses are used for testing how set assumptions or uncertainties in the data impact the results. Contribution assessments are used to identify hotspots showing the impacts of each life cycle stage. The interpretation step includes limitations of the study analysis, whereas conclusions and recommendations are drawn from reflecting the results with the uncertainty assessments and limitations.

3 Baseline: Environmental Impacts of Conventional Livestock Production

3.1 Climate Impact

Approximately 20% of the global anthropogenic GHG emissions originate from livestock production, when direct and indirect emissions are considered [2, 3]. The direct atmospheric emissions consist of methane emissions from the ruminants' digestion and manure management processes and nitrous oxide emission from manure and use of nitrogen fertilisers. Indirect emissions consist of the production of agricultural inputs such as fertilisers, pesticides and fuels and transportation of livestock and fuel used at the farm. In addition, indirect emissions include land use change such as carbon emissions from deforestation caused by expansion of agricultural land area. The direct global GHG emissions from crop and livestock production account for approximately 11%, and the indirect emissions are approximately 14% of anthropogenic GHG emissions [4]. Over 80% of the GHG emissions of livestock products are allocated to land use change and agricultural practices. The processing, transporting, packaging and retailing have only a small overall contribution (Fig. 1).

Majority of all GHG emissions of global livestock production are related to bovine and sheep cattle production [3]. Beef has approximately six times higher emissions than pork meat and eight times higher than poultry meat per unit of

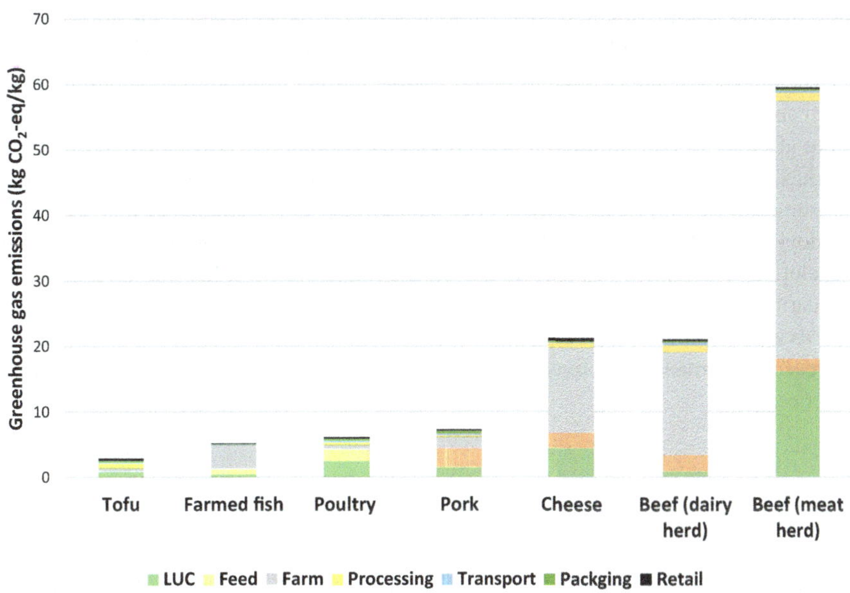

Fig. 1 Greenhouse gas emissions of livestock products, fish and tofu by life cycle stages. *LUC* land use change. (Adapted from Poore and Nemecek [12])

product [12]. This is explained by the methane emissions from ruminants and their greater feed requirements. The production of 1 kg of poultry needs approximately 3 kg, pork 6–7 kg and beef 25 kg of feed [20].

3.2 Land Use and Biodiversity Impacts

Livestock feed production covers a third of the world's total land area and 80% of all agricultural land area [4]. Livestock production is one of the main causes of global land use change accounting for 67% of agriculture-derived deforestation [12]. The deforestation is mainly driven by increasing production of soybean feed and beef for export. Global demand for feed and consumption of livestock meat has direct and indirect consequences to the deforestation of tropical forests, which support two-thirds of the world's biodiversity although covering only 10% of the Earth's land area [21]. Approximately 10 million ha are globally deforested per year, which may lead to major loss of biodiversity [22].

Food production contributes to farmland biodiversity and agrobiodiversity losses via changes in agricultural practices [23]. Decreasing livestock grazing practices have been associated with decline of wild species such as butterflies and beetles dependent on extensive grazing [24]. Increased use of fertilisers and pesticides, reduced number of crop varieties and general intensification of agriculture have all contributed to the decline of farmland biodiversity. In addition, reduced mosaic of farmland landscapes due to favouring of larger fields and shrinking non-crop habitats such as field boundaries, set-aside lands, wildflower meadows and agroforestry has contributed to increased number of endangered wild plant and animal species.

The land use changes such as deforestation and intensification of agriculture are often caused by the demand to produce more and cheaper food for the growing and more affluent global population in increasingly competitive and global agricultural product markets. The gross domestic product per capita tends to correlate with the consumption of livestock products per capita [25]. It is anticipated that livestock product demand will potentially increase in currently low meat-consuming and developing economies [4]. Global agricultural land area would need to expand on an alarming rate by 2050 to an area that equals approximately the size of all current forests and savannah on the Earth if anticipated consumption trends will hold and the increasing demand of livestock products is met with current agricultural practices [4].

3.3 Water Use and Emissions to Water Bodies

The environmental assessment methods produce differing results in estimating the contribution of livestock production to global water use. Simpler methods calculate only blue water usage including freshwater extraction from rivers, reservoirs, lakes and aquifers. In addition to blue water, the more complex water footprint method also includes the usage of green water or rainwater and grey water describing the amount of water needed to dilute wastewater to meet the regionally specific water quality standards. Livestock production contributes approximately 30% of all anthropogenic freshwater use globally [26]. Similarly, blue water use is approximately 8% [27].

Food products from ruminants have relatively high blue, green and grey water usage or the water footprint due to the larger agricultural land area requirements for feed production. The green water footprint is therefore significantly higher compared to monogastric animal meat production such as pork or poultry. The differences between livestock products tend to be low when only blue water usage is assessed. Studies reporting only blue water footprint results may include caveats: some water footprint methods use country-specific water scarcity factors in assessing the potential impact of livestock water use on the availability of water resources in the specific locations [28]. Water use in water-scare areas has more severe consequences than using the same amount in a water-rich area.

Livestock production contributes to the pollution of water bodies due to leaching of nitrogen and phosphorus from fields and manure-handling systems [7]. Regionally centralised livestock production units and expanded farm sizes create additional challenges. Traditionally, smaller and versatile farm units could use produced manure fully for fertilising the fields and producing the needed feed on site. The concentration of individual agricultural production sectors to certain regions has created a mismatch: excessive manure cannot be efficiently used as fertiliser nearby fields, whereas crop production in different regions depends on industrially produced fertilisers. Manure is therefore spread to the fields excessively or managed in unprotected systems outdoors. The industrial livestock farms import feed from global supply chains. Excessive nutrients leach to water bodies causing eutrophication, which increases growth of water plants and algae disturbing the aquatic ecosystems and causing negative impacts on aquatic species. The eutrophication impact of producing meat from ruminants is higher compared to pork, poultry and plant-based proteins [12].

4 Environmental Impacts of Cultivated Meat Production

4.1 LCA Studies on Cultivated Meat

The environmental impacts of cultivated meat production have been studied only with prospective LCA (Table 1). The studies have assumed several factors in modelling the impacts of cultivated meat production as large-scale industrial production units do not exist and data from real-world applications are largely unavailable or missing. The extant LCA studies involve uncertainties due to these assumptions. It is uncertain whether the assumed production amounts of cultivated meat are realistic or achievable with modelled production system designs and

Table 1 Comparison of the climate impact and land use of cultivated meat based on published life cycle assessments

	Carbon footprint (kg CO_2 eq/kg cultivated meat)	Land use (m^2/kg cultivated meat)	Details
Tuomisto and Teixeira de Mattos [29]	1.9–2.2	0.2	Cyanobacteria hydrolysate as a main source of nutrients in the culture medium, stirred-tank bioreactor
Smetana et al. [30]	23.9–24.6	0.4–0.8	Data mainly from Tuomisto et al. (2011), but cyanobacteria produced in bioreactors instead of open ponds
Mattick et al. [31]	3.0–25.5	1.5–9.5	Chinese hamster ovarian cells, standard serum-free culture medium, stirred-tank bioreactors
Tuomisto et al. [18]	4.9–25.2	1.8–6.9	C2C12 cells, standard culture medium with and without FBS, hollow-fibre bioreactors
Kim et al. [32]	15.4	0.08	Primary bovine cells, standard culture medium, data from a company
Sinke et al. [33]	2.9–14.3	2.5	Aggregated data from companies (several cell types and culture medium ingredients)
Risner et al. [34]	12–1,508		Used data from techno-economic assessments. The worst-case scenarios include energy-intensive purification steps for the medium ingredients

whether assumptions considering the needed ingredients of the culture medium, design or types of the bioreactors, duration of the cultivation process and nutritional content of cultivated meat align with the actual future production processes.

The first study that assessed the potential environmental impacts of cultivated meat [29] is a screening analysis providing estimates of the environmental impacts at the level of magnitudes. The LCA study is based on a hypothetical production scenario using cyanobacterial hydrolysate as the main source of amino acids and sugars in culture medium. The cyanobacteria were assumed to be produced in large-scale commercial production units using open ponds. Other ingredients of the culture medium such as vitamins, minerals and growth factors were excluded from the analysis. The modelling used a stirred-tank bioreactor, the bioreactor energy use requirements were simplified and assumed production locations were selected from three regions: Spain, Thailand and California in the USA. The seminal study had the lowest land use estimates compared to the other LCA studies conducted later due to the use of cyanobacteria as the main source of nutrients (Table 1). Also, low climate impacts resulted partly due to the usage of cyanobacterium-based culture medium and potentially underestimated energy requirements for bioreactor processes. Bioreactor cleaning processes were not included in the study.

Smetana et al. [30] drew from the seminal study [29] by modelling the production of cyanobacteria in bioreactors, which increased the energy usage and climate impacts. However, the land use requirements increased only marginally (Table 1).

Another study assessed the environmental impacts of cultivated meat production in the USA [31] using Chinese hamster ovarian (CHO) cells cultivated in a stirred-tank bioreactor and serum-free culture medium. The medium contained standard basal medium (synthetic amino acids, vitamins and minerals) and soybean hydrolysate. The study extended Tuomisto's and Teixeira de Mattos' [29] study by including the cleaning of the bioreactors and energy requirements of the production facilities. The results reported higher environmental impacts for cultivated meat production compared to the previous study [29]. However, the uncertainty ranges of the results were extensive (Table 1).

Yet another study presents an LCA based on data collected from start-up companies producing cultivated meat supplemented with modelling and data from literature [33]. The company data included various cultivated meat cell types and culture media options. The study estimated the potential ranges of the environmental impacts rather than presenting impacts of specific cultivated meat product type. The study included the future scenarios of improved cultivated meat and livestock production systems reaching 2050. The study reported lower climate and land use impacts from the previous study [31].

A recent study assessed the environmental impacts of cultivated meat production in the UK using hollow-fibre bioreactors [18]. The study focused on different cultivated meat process designs and their potential environmental impacts. The baseline scenario used immortalised murine myoblast cell line C2C12 grown in standard culture medium containing FBS and using standard average UK electricity mix for the production process. The alternative scenarios included serum-free culture medium, CHO cells, renewable energy use and various cell metabolism assumptions. The climate and land use impacts aligned with the uncertainty ranges presented in Mattick et al.'s [31] study.

Kim et al. [32] estimated the environmental impacts of cultivated meat beef patty produced in the USA. The data was obtained from a start-up company and, therefore, was not published in detail. The carbon footprint result of cultivated meat cells was in the upper end of the finding of Sinke et al. [33] and in the middle of the range found by Tuomisto et al. [18], whereas the land use result was the lowest of all studies.

Risner et al. [34] estimated the carbon footprint of cultivated meat by using data from techno-economic assessments. They also included a scenario where pharma-grade culture medium ingredients were assumed to be used. They assumed highly energy-intensive process for the purification process of the culture medium ingredients. This increased the energy consumption and carbon footprint of cultivated meat substantially (Table 1). However, this study highlights the importance of using feed- or food-grade ingredients, instead of pharma-grade ingredient in cultivated meat production.

4.2 Issues Affecting the Environmental Impacts of Cultivated Meat

4.2.1 Culture Medium

The published LCA studies show that the production of the culture medium has a major contribution to the environmental impacts of cultivated meat production. The volume of culture medium needed per kg cultured meat in the LCA studies had a wide variation ranging from 21 to 200 L. The medium requirements depend on the cell type, medium type, cultivation time and possible reuse of the nutrients of the spent medium. The published LCA studies assumed relatively short differentiation and maturation periods in cultivated meat production [18, 31, 33]. Muscle tissue may need more cell-culturing time especially when whole-meat cuts are cultivated, which also increases the medium consumption. However, also cells with short differentiation time or even undifferentiated cells can be used as ingredients in processed products (e.g. sausages). For instance, Mattick et al. [31] and Tuomisto et al. [18] assumed 7-day differentiation period in the baseline cases, which means that undifferentiated cells would require less medium. A shorter cultivation time would reduce the environmental impacts of cultivated meat production due to lower impacts related to the medium production and lower energy use requirements for the bioreactor processes. The data published in the supplementary information by Tuomisto et al. [18] shows that in the baseline scenario, 58% of the total medium and 61% of the bioreactor energy inputs were spent during the differentiation phase. This would mean that the environmental impacts of cultivated meat production would be only around 40% of those reported by Tuomisto et al. [18]. However, currently, it is not known what the protein content and quality would be in the undifferentiated cells.

The environmental impacts also depend on how well the nutritional content of the medium is optimised to meet the requirements of the cells. Tuomisto et al. [18] showed that adjusting the amino acid content of the medium to the cell requirements reduced the environmental impacts of cultivated meat over 20% compared to using a standard culture medium. The total nutrient use conditions depend also on factors such as recycling potential of unused nutrients in the cultivation process or utilisation of the leftover nutrients as ingredients for producing fertilisers or other suitable products. Advanced filtration solutions can be used in removing the surplus nutrients from the outflow medium and circulating them in cultivated meat production process or collecting, storing and selling unused nutrients to other industrial sectors. Part of the environmental impacts of used and recycled nutrition ingredients can be allocated to the co-products reducing the environmental impact of cultivated meat production. Also, environmental impacts of waste management processes decrease when recycling of nutrients is possible.

The sources for the nutrients used in the culture medium have a major effect on the environmental impacts. The culture medium often used in mammalian cell cultures consists of two main components: foetal bovine serum (FBS) and basal medium, which includes glucose, synthetic amino acids, vitamins and minerals. This culture medium is expensive and contains animal-based ingredients challenging its usability in large-scale cultivated meat production. Alternative culture media are being developed using non-animal-based growth factors derived from microalgae, yeast, fungi or plant extracts that would perform equally well or better in cultivated meat production (e.g. [35]). Trinidad et al. [36] estimated environmental impacts of common growth factors used in serum-free media and found that the differences in the impacts of different growth factors were high. They estimated that the contribution of the growth factor production to the carbon footprint of cultivated meat production would vary between 0.4% and 23% depending on the type of growth factor used.

The synthetic production processes of amino acids, vitamins and minerals needed for cultivated meat production are energy intensive and require relatively high amount of glucose sourced from agricultural crops [31]. Therefore, alternative sources for these ingredients are actively searched and developed within the industry. Recycled nutrients from agricultural and food-processing side streams are considered promising in composing culture medium for cultivated meat production [37].

4.2.2 Bioprocess Design

The bioprocess design has a major impact on the environmental implications of cultivated meat production. The energy use requirements for the functions of the bioreactor (e.g. temperature control, pumping, mixing, aeration), cleaning of the bioreactor, type of scaffold materials used, downstream processing of the cells and wastewater treatment are key factors affecting the level of environmental impacts. As cultured meat production is still at the development stage and data from larger scale bioreactors are not available, the published LCA studies are based on models assuming either stirred-tank or hollow-fibre bioreactors. The process designs included in the studies are not necessarily fully optimised to minimise the energy use.

The environmental impacts of cultivated meat production are also highly dependent on the type of energy sources used for the production processes [18]. As most of the climate impact of cultivated meat production is related to energy use, the carbon footprint can be lowered by using renewable or low-emission energy sources.

4.3 Impacts to Land Use and Biodiversity

Land use in cultivated meat production is mainly associated with production of ingredients to culture medium. Most of the land use is related to agricultural production of maize or sugar cane for the source of glucose that is directly used in the culture medium and indirectly in the production of synthetic amino acids,

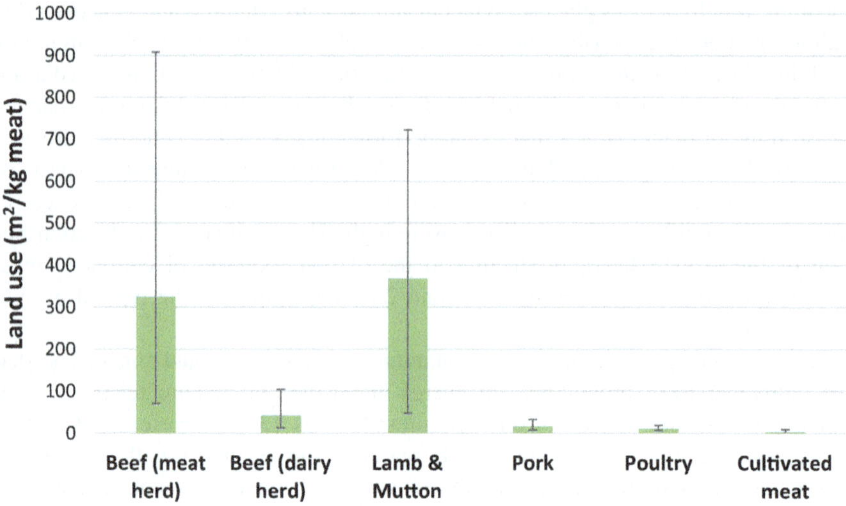

Fig. 2 Land use of cultivated meat (as reported in the LCA studies in Table 1) and livestock meat (based on global means, error bars represent the 5th and 95th percentiles, source of data for livestock products [12])

vitamins and growth factors. Cultivated meat is estimated to have lower land area requirements compared to conventionally produced beef but not necessarily to poultry production (Fig. 2).

The land use requirements of cultivated meat production potentially decrease in cases when novel and sustainable sources of feedstock ingredients are discovered. A potential source of nutrients for cultivated meat production is hydrogen-oxidising bacterium (HOB), which sources carbon from carbon dioxide and hydrogen from water molecule. HOB has minimal land area requirements: agricultural land is not needed in the production, and only indirect land use is allocated for the energy and other input material production [38, 39].

Potential biodiversity benefits of cultivated meat could be realised via reduced land use requirements. Especially pressure on land use changes and deforestation could decline [40]. Alternative use for lands released from agriculture would define the potential biodiversity advantages. Biodiversity impacts would be only neutral in cases where the freed land area would be converted to intensive and industrial usage such as forestry, bioenergy crop production or material crop production. The potential biodiversity benefits of cultivated meat production are not self-evident: expanding protected and nature conservation areas and use of wildlife-promoting land management practices could be advanced with appropriate policy measures.

In addition, grazing livestock provides ecosystem services and maintains farmland biodiversity as some of the endangered wild plants and insects depend on extensive grazing systems [41]. Grazing livestock provides landscape maintenance by keeping up high-nature-value farmland areas such as ecological and cultural benefits provided by semi-natural grasslands [42]. However, positive farmland biodiversity impacts provided by extensive grazing livestock production systems can be reached with a fraction of current livestock and farm animal numbers.

The biodiversity impacts depend also on what type of sources are used for the nutrients used in culture medium and how they are produced. Monocropping-based intensive production of crops for the sources of glucose and amino acids would have negative impacts on biodiversity. Cropping systems based on agroecological farming practices would provide higher biodiversity benefits. The type of crop has also biodiversity impacts: grass-clover leys and legume crops (e.g. beans and peas) are an important part of the crop rotations used in agroecological farming systems. Grass and legume crops can also be used as a source of glucose and amino acids for culture medium used in cultivated meat production. Therefore, agroecological farming systems could be optimal for providing nutrients for cultivated meat production.

4.4 Impacts of Energy Use

Direct energy use of cultivated meat production comprises pumping, mixing, aeration, temperature control, cleaning and drying processes. Indirect energy use consists of production of ingredients used in culture medium and industrial infrastructure materials. The total energy requirements depend on the culture medium ingredients and bioreactor design: the production processes of synthetic culture medium ingredients are energy intensive [18], and the energy requirements of the bioreactor processes consist of the level of heat integration or the use of heat exchangers, insulation and possibility to utilise cold air and fresh- or seawater for cooling.

Cultivated meat production is estimated to require higher amounts of industrial energy such as electricity, heat or gas than even beef production [18, 31]. Livestock produces energy from feed and uses it for maintaining the bodily functions. The energy in feed is replaced with industrial energy in cultivated meat production processes. However, cultivated meat required only a third of the total energy (industrial and feed combined) needed for beef production (Fig. 3).

Food production with cellular agriculture technologies will increase industrial energy consumption and would therefore require investments to sustainable energy production. However, a need for producing sustainable energy is a concern not only in food production but also in other sectors participating in the green energy transition. Conventional meat production uses efficient and natural energy production process: livestock feed with plants that convert direct sunlight into energy without a need to build complex industrial infrastructure and processes for energy conversion and supply. Livestock grazing systems especially benefit from taking advantage of these natural energy-generating processes eliminating energy usage for fertilising, harvesting, processing and transporting feed. Future food manufacturing in bioreactors such as cultivated meat production needs to be optimised for energy efficiency as sustainable industrial energy production is limited and increasingly competitive.

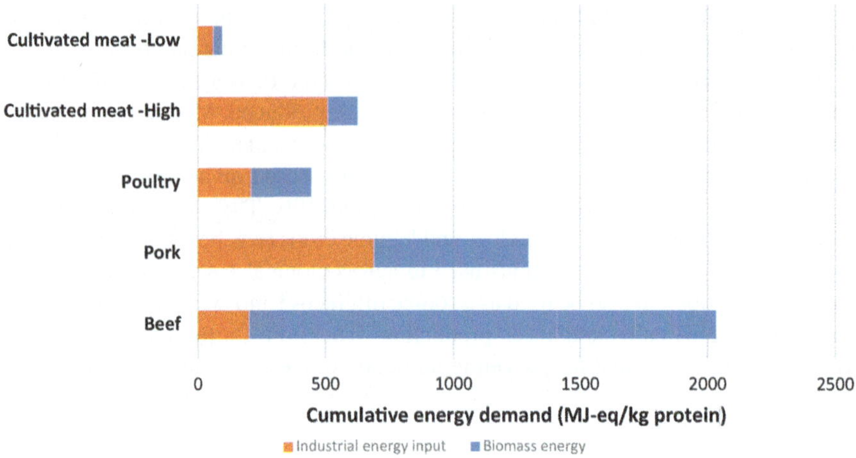

Fig. 3 Cumulative energy demand of cultivated meat (low and high indicate the lowest and highest impacts reported by Tuomisto et al. [18]) compared to livestock meat (data from Agribalyse [43] database). Industrial energy input includes electricity, oil and gas, whilst biomass energy includes energy content of feed

4.5 Climate Impact

The most GHG emissions of cultivated meat production originate from industrial energy consumption. Therefore, the energy source dictates most of the climate impact of the cultivated meat products. Fossil fuel usage in cultivated meat production leads often to higher climate impact in comparison with pork and poultry. However, GHG emissions from conventional beef production are nevertheless higher (Fig. 4) [18, 31]. Cultivated meat production processes that use clean or renewable energy sources substantially reduce GHG emissions and consequently climate impacts [18, 33].

In addition, production of culture medium ingredients used in cultivated meat production causes climate impacts. The ingredients sourced from agriculture or that require agricultural inputs in the processes such as production of synthetic amino acids cause GHG emissions from the agriculture. These include nitrous oxide emissions from the use of nitrogen fertilisers and fuel usage in planting, weeding, harvesting, drying, transporting and storing culture medium input materials from agriculture.

As cultivated meat production requires often less land area compared to conventional livestock production, it provides indirect climate benefits through reduced need to convert forests to agricultural land. In some areas, agricultural land previously used for feed production could potentially be afforested increasing the carbon storages in soil and vegetation. However, it is still uncertain how much large-scale cultivated meat production will need land area as such systems do not exist yet.

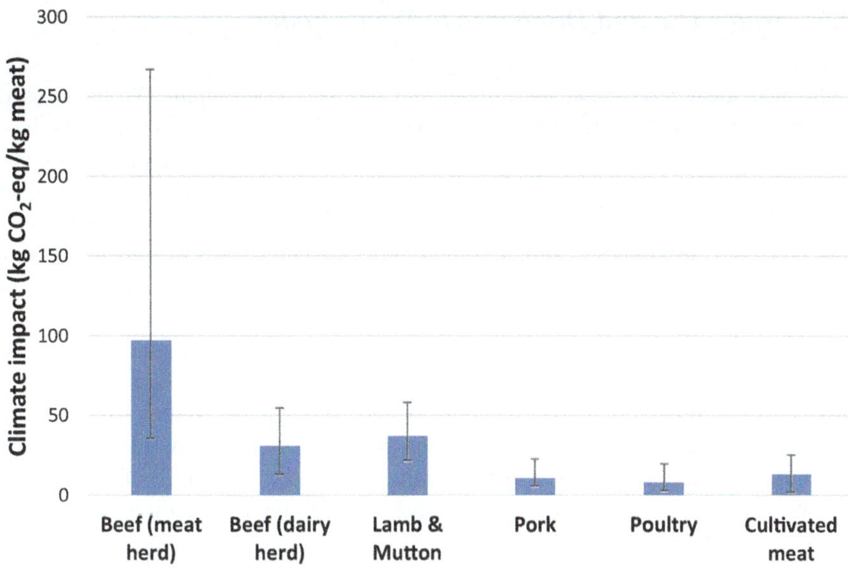

Fig. 4 Climate impact of cultivated meat (as reported in majority of the LCA studies in Table 1) and livestock meat (based on global means, error bars represent the 5th and 95th percentiles, source of data for livestock products [12])

Climate impacts are reported by converting all GHG emissions to carbon dioxide equivalents (CO_2 eq). The resulting figure represents the climate impacts of the emitted gases comparable to carbon dioxide during a 100-year time frame. Nitrous oxide is 265 and methane 28 times stronger GHGs than carbon dioxide. It is criticised that the commonly used methods for reporting climate impacts unwarrantably emphasise methane emissions although methane has an atmospheric lifetime of only 12 years whilst carbon dioxide remains for longer time periods in atmosphere. Lynch and Pierrehumbert [44] conducted a study using a 1000-year time frame for comparing the climate impacts of cultivated meat with beef. They concluded that in the longer-term analysis, some of the cultivated meat systems produced higher climate impacts than selected beef production systems.

Methane has higher climate impact during shorter periods because it traps more heat per mass unit than carbon dioxide: 84 times in a 20-year period and 28 times during a 100-year period. Therefore, focusing on rapid reduction of especially methane emissions has faster impacts on climate change mitigation. In addition, methane cannot be removed from atmosphere during its lifetime by currently known methods, whilst carbon dioxide is consumed by growing vegetation and absorbed in water bodies. The anthropogenic GHG emissions generated by 2030 will determine the seriousness of climate change [45]. In this context, switching from the 100-year time frame to the 20-year time frame would emphasise the short-term impacts of methane emissions.

4.6 Water Use and Discharge to Water Bodies

Cultivated meat production uses direct and indirect water inputs. Water is used directly in preparing culture medium, controlling temperature and cleaning bioreactors. Indirect water use considers production of inputs: culture medium ingredients, materials and energy. The water use of cultivated meat production depends on several factors such as the type of the culture medium ingredients and their production processes, used energy sources along the cultivated meat value chain, options for recycling water used in culture medium and whether fresh- or seawater is used in the cooling processes.

Water use between cultivated meat and livestock meat production depends on the methods used in calculating the relative differences. The water use of livestock meat production is relatively higher (Tuomisto and Teixeira de Mattos 2011) when calculated using water footprint approach that considers green, blue and grey water than using methods that include only blue water usage [18]. The green or rainwater use is significantly higher as livestock production requires agricultural land for feed production.

Cultivated meat production facilities are assumed to discharge only minor amounts of wastewater to water bodies due to closed-loop production design and recycling of valuable ingredients. Ingredients used for culture medium production determine potential emissions to water bodies. Culture medium ingredients originating from agricultural production such as nitrogen and phosphorus tend to leach to water bodies as agricultural fields are open systems in natural environments where plants use only a part of the applied nutrients. Cultivated meat production is anticipated to require substantially less agricultural inputs than livestock meat production leading to decreasing eutrophication impact. A study estimated that the eutrophication impact of cultivated meat could be in parity with poultry production, significantly lower (−70%) than pork production and just a fraction (−97%) compared to beef production in the USA [31].

Agricultural production and livestock farms produce other water emissions such as pesticide residues, antibiotics and pathogens from manure that impact the quality of surrounding water bodies. Cultivated meat production systems will completely avoid or drastically reduce these emissions to water bodies due to closed-loop production system design and assuming that antibiotics are not needed in cultivated meat production. However, total avoidance of livestock-based emissions to water bodies is not anticipated due to the need to maintain cell-donating farm animals. Regardless, livestock-based emissions in cultivated meat production would be rather negligible: number of livestock needed for cultivated meat production would be only infinitesimal compared to current amount of production animals globally. Cultivated meat production could be independent from rearing livestock or cell-donating animals if the development of immortalised cell lines proves to be successful in the future.

5 Conclusion

The LCA studies conducted on a product level indicate that cultivated meat production has a potential to reduce environmental impacts compared to pork and beef production systems. Environmental impacts of cultivated meat production depend mainly on bioreactor energy use and ingredients needed for the culture medium production. Large-scale cultivated meat production units are not yet in operation. Therefore, environmental impact assessments are anticipatory and based on a set of assumptions. Technological development and innovations in bioreactor and production process design, energy efficiency and culture media ingredients have the potential to reduce environmental impacts. Each cultivated meat production process has different levels of impacts depending on the process design and origin of inputs. Thus, the variation in the environmental impacts of different cultivated meat production processes is wide. More LCA studies are needed to estimate the impacts of different cell types, bioreactors and media formulations.

Anticipatory product-level assessments provide baseline information about environmental impacts of a cultivated meat product. However, real-world data from state-of-the-art cultivated meat production processes are needed to verify the results from anticipatory studies that are mainly based on sets of assumptions. These data are currently hard to obtain and only available in start-up companies developing the novel products of cellular agriculture. In addition, global-level assessments are needed to better understand the total impacts of cultivated meat production and its implications to the food systems. Major changes in land use dynamics caused by cultivated meat production call for global earth system impact modelling [46]. Several studies have assessed the potential climate benefits of switching to more sustainable vegan and vegetarian diets that originate from reforesting land previously used for livestock production [47, 48]. Similar global assessments could be used in modelling various rates of the cultivated meat product usage, in different parts of the globe and at several timescales. The environmental assessments should also model the potential development of energy technologies: cultivated meat production needs clean and sustainably produced energy in achieving the climate benefits.

The pressing environmental challenges and global targets set to 2030 (e.g. United Nations Sustainable Development Goals and climate targets) call for quick and impactful environmental measures in the food production systems. However, rather slow upscaling of cultivated meat production systems and associated challenges with technologies and culture medium composition indicate that products of cellular agriculture may not provide feasible solutions in time. Development of novel food products such as cultivated meat is needed as a part of the grand food system transformation [6]. However, environmentally beneficial changes in all parts of the current food systems need to be implemented: urgent reduction of food loss and waste, switching to plant-based diets, developing the sustainability of current agricultural practices, stopping fossil fuel usage and deforestation and tackling the farmland biodiversity loss. Alternative protein sources including algae, fungi and insects, novel food innovations such as cell-cultured microbial cells and recombinant proteins need to be also considered in improving the sustainability of global food systems before 2030.

References

1. Campbell BM, Beare DJ, Bennett EM, Hall-Spencer JM, Ingram JSI, Jaramillo F, Ortiz R, Ramankutty N, Sayer JA, Shindell D (2017) Agriculture production as a major driver of the earth system exceeding planetary boundaries. Ecol Soc 22(4):8
2. Crippa M, Solazzo E, Guizzardi D, Monforti-Ferrario F, Tubiello FN, Leip A (2021) Food systems are responsible for a third of global anthropogenic GHG emissions. Nat Food 2:198–209
3. Xu X, Sharma P, Shu S, Lin T-S, Ciais P, Tubiello FN, Smith P, Campbell N, Jain AK (2021) Global greenhouse gas emissions from animal-based foods are twice those of plant-based foods. Nat Food 2:724–732
4. WRI (2019) Creating a sustainable food future. A menu of solutions to feed nearly 10 billion people by 2050. World Resources Report. World Resources Institute, Washington
5. Piazza J, Ruby MB, Loughnan S, Luong M, Kulik J, Watkins HM, Seigerman M (2015) Rationalizing meat consumption. The 4Ns. Appetite 91:114–128
6. Moritz J, McPartlin M, Tuomisto HL, Ryynänen T (2023) A multi-level perspective of potential transition pathways towards cultured meat: Finnish and German political stakeholder perceptions. Res Policy 52(9):104866
7. Willett W, Rockström J, Loken B, Springmann M, Lang T, Vermeulen S, Garnett T, Tilman D, Declerck F, Wood A (2019) Food in the anthropocene: the EAT–Lancet Commission on healthy diets from sustainable food systems. Lancet 393:447–492
8. Clark MA, Domingo NGG, Colgan K, Thakrar SK, Tilman D, Lynch J, Azevedo IL, Hill JD (2020) Global food system emissions could preclude achieving the 1.5° and 2°C climate change targets. Science 370(6517):705–708
9. Statista (2023a) Share of vegans in European countries. https://www.statista.com/forecasts/1256518/share-of-vegans-in-european-countries. Accessed 17 Oct 2023
10. Statista (2023b) Meat substitutes – worldwide. https://www.statista.com/outlook/cmo/food/meat/meat-substitutes/worldwide#volume. Accessed 17 Oct 2023
11. Clark M, Springmann M, Rayner M, Scarborough P, Hill J, Tilman D, Macdiarmid JI, Fanzo J, Bandy L, Harrington RA (2022) Estimating the environmental impacts of 57,000 food products. Proc Natl Acad Sci 119:e2120584119
12. Poore J, Nemecek T (2018) Reducing food's environmental impacts through producers and consumers. Science 360:987–992
13. ISO (2006) ISO14044; Environmental management – life cycle assessment – requirements and guidelines. International Organization for Standardization, Geneva
14. Smetana S, Schmitt E, Mathys A (2019) Sustainable use of Hermetia illucens insect biomass for feed and food: attributional and consequential life cycle assessment. Resour Conserv Recycl 144:285–296
15. McAuliffe GA, Takahashi T, Lee MR (2020) Applications of nutritional functional units in commodity-level life cycle assessment (LCA) of agri-food systems. Int J Life Cycle Assess 25: 208–221
16. Kyttä V, Roitto M, Astaptsev A, Saarinen M, Tuomisto HL (2022) Review and expert survey of allocation methods used in life cycle assessment of milk and beef. Int J Life Cycle Assess 27: 191–204
17. Kyttä V (2022) Circular bioeconomy in life cycle assessment – addressing multifunctionality of agriculture. Faculty of Agriculture and Forestry, University of Helsinki, Helsinki, p 60

18. Tuomisto HL, Allan SJ, Ellis MJ (2022) Prospective life cycle assessment of a bioprocess design for cultured meat production in hollow fiber bioreactors. Sci Total Environ 851:158051
19. Tuomisto HL, Hodge ID, Riordan P, MacDonald DW (2012) Exploring a safe operating approach to weighting in life cycle impact assessment – a case study of organic, conventional and integrated farming systems. J Clean Prod 37:147–153
20. Alexander P, Brown C, Arneth A, Finnigan J, Rounsevell MD (2016) Human appropriation of land for food: the role of diet. Glob Environ Change 41:88–98
21. Giam X (2017) Global biodiversity loss from tropical deforestation. Proc Natl Acad Sci 114: 5775–5777
22. FAO and UNEP (2020) The state of the world's forests 2020. Forests, biodiversity and people. FAO and UNEP, Rome
23. Dudley N, Alexander S (2017) Agriculture and biodiversity: a review. Biodiversity 18:45–49
24. Filazzola A, Brown C, Dettlaff MA, Batbaatar A, Grenke J, Bao T, Peetoom Heida I, Cahill Jr JF (2020) The effects of livestock grazing on biodiversity are multi-trophic: a meta-analysis. Ecol Lett 23:1298
25. Tilman D, Clark M (2014) Global diets link environmental sustainability and human health. Nature 515:518–522
26. Mekonnen MM, Hoekstra AY (2012) A global assessment of the water footprint of farm animal products. Ecosystems 15:401–415
27. FAO (2006) Livestock's long shadow. FAO, Rome
28. Boulay A-M, Benini L, Sala S (2019) Marginal and non-marginal approaches in characterization: how context and scale affect the selection of an adequate characterization model. The AWARE model example. Int J Life Cycle Assess 25:1–13
29. Tuomisto HL, Teixeira de Mattos MJ (2011) Environmental impacts of cultured meat production. Environ Sci Technol 45(14):6117–6123
30. Smetana S, Ristic D, Pleissner D, Tuomisto HL, Parniakov O, Heinz V (2023) Meat substitutes: resource demands and environmental footprints. Resour Conserv Recycl 190:106831
31. Mattick CS, Landis AE, Allenby BR, Genovese NJ (2015) Anticipatory life cycle analysis of in vitro biomass cultivation for cultured meat production in the United States. Environ Sci Technol 49:11941–11949
32. Kim S, Beier A, Schreyer HB, Bakshi BR (2022) Environmental life cycle assessment of a novel cultivated meat burger patty in the United States. Sustainability 14(23):16133
33. Sinke P, Swartz E, Sanctorum H, van der Giesen C, Odegard I (2023) Ex-ante life cycle assessment of commercial-scale cultivated meat production in 2030. Int J Life Cycle Assess 28:234–254
34. Risner D, Negulescu P, Kim Y, Nguyen C, Siegel JB, Spang ES (2024) Environmental impacts of cultured meat: a cradle-to-gate life cycle assessment. ACS Food Sci Technol 5(1):61–74
35. Skrivergaard S, Young JF, Sahebekhtiari N, Semper C, Venkatesan M, Savchenko A, Stogios PJ, Therkildsen M, Rasmussen MK (2023) A simple and robust serum-free media for the proliferation of muscle cells. Food Res Int 172:113194
36. Trinidad KR, Ashizawa R, Nikkhah A, Semper C, Casolaro C, Kaplan DL, Savchenko A, Blackstone NT (2023) Environmental life cycle assessment of recombinant growth factor production for cultivated meat applications. J Clean Prod 419:138153
37. Upcraft T, Tu W-C, Johnson R, Finnigan T, Van Hung N, Hallett J, Guo M (2021) Protein from renewable resources: mycoprotein production from agricultural residues. Green Chem 23 (14):5150–5165
38. Järviö N, Maljanen N-L, Kobayashi Y, Ryynänen T, Tuomisto HL (2021a) An attributional life cycle assessment of microbial protein production: a case study on using hydrogen-oxidizing bacteria. Sci Total Environ 776:145764
39. Sillman J, Uusitalo V, Ruuskanen V, Ojala L, Kahiluoto H, Soukka R, Ahola J (2020) A life cycle environmental sustainability analysis of microbial protein production via power-to-food approaches. Int J Life Cycle Assess 25:2190–2203

40. Machovina B, Feeley KJ, Ripple WJ (2015) Biodiversity conservation: the key is reducing meat consumption. Sci Total Environ 536:419–431
41. Alkemade R, Reid RS, van den Berg M, de Leeuw J, Jeuken M (2013) Assessing the impacts of livestock production on biodiversity in rangeland ecosystems. Proc Natl Acad Sci USA 110 (52):20900–20905
42. Bengtsson J, Bullock J, Egoh B, Everson C, Everson T, O'Connor T, O'Farrell P, Smith H, Lindborg R (2019) Grasslands—more important for ecosystem services than you might think. Ecosphere 10(2):e02582
43. Agribalyse (2022) AGRIBALYSE LCA database. https://doc.agribalyse.fr/documentation-en/
44. Lynch J, Pierrehumbert R (2019) Climate impacts of cultured meat and beef cattle. Front Sustain Food Syst 3(5) Original Research. https://doi.org/10.3389/fsufs.2019.00005
45. IPCC (2018) Global Warming of 1.5 °C - an IPCC special report on the impacts of global warming of 1.5 °C above preindustrial levels and related global greenhouse gas emission pathways, in the context of strengthening the global response to the threat of climate change, sustainable development, and efforts to eradicate poverty. Intergovernmental Panel on Climate Change, Incheon. http://www.ipcc.ch/report/sr15/
46. Humpenoder F, Bodirsky BL, Weindl I, Lotze-Campen H, Linder T, Popp A (2022) Projected environmental benefits of replacing beef with microbial protein. Nature 605:90–96
47. Hayek MN, Harwatt H, Ripple WJ, Mueller ND (2021) The carbon opportunity cost of animal-sourced food production on land. Nat Sustain 4(1):21–24
48. Sun Z, Scherer L, Tukker A, Spawn-Lee SA, Bruckner M, Gibbs HK, Behrens P (2022) Dietary change in highincome nations alone can lead to substantial double climate dividend. Nat Food. https://doi.org/10.1038/s43016-021-00431-5

Adv Biochem Eng Biotechnol (2026) 193: 295–316
https://doi.org/10.1007/10_2026_332

Published online: 16 June 2026

Consumer and Social Perspectives to Cultivated Meat

Liisa Lähteenmäki and Angelica Rossi

Contents

L. Lähteenmäki (✉)
Department of Management, MAPP Centre, Aarhus University, Aarhus, Denmark
e-mail: liisal@mgmt.au.dk

A. Rossi
MAPP Centre, Aarhus University, Aarhus, Denmark

Abstract Cultivated meat offers an alternative way to produce animal-based protein and has been positioned as a possible solution to produce meat in a more sustainable manner. As a novel production method with uncertainty about the characteristics of the final product, cultivated meat needs to gain both societal and consumer acceptance in order to succeed in the market. In this chapter, factors influencing consumer and social acceptance are discussed as well as reflections on the future acceptability of cultivated meat.

Graphical Abstract

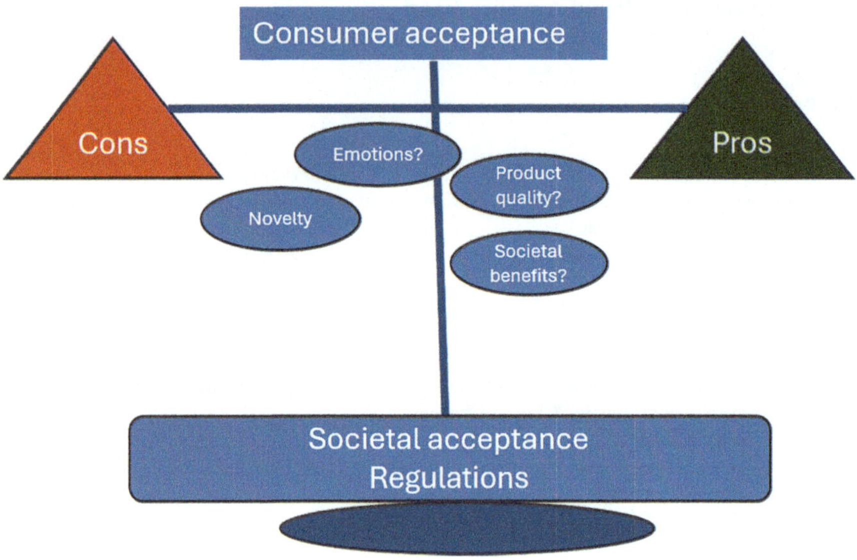

Keywords Consumer acceptance, Cultivated meat, Food innovation, Neophobia, Novel food, Social acceptance

1 Introduction

Moderate meat consumption has been recognised as an essential part of planetary healthy diet that provides adequate nutrition to everyone while being able to stay within the resource boundaries of the planet [1]. Food-based dietary guidelines in many countries use the planetary healthy diet maximum of 300–500 g of meat per week as a target [2], which in many countries requires reducing the meat consumption radically [3]. Finding alternatives to conventional meat as a protein source is recognised as one of the major challenges in transforming the food systems towards sustainability. Cultivated meat has emerged as an option to produce high-protein food in a new way that can overcome some of the issues related to conventional meat production without the need of having to give up meat in the diet.

Cultivated meat has been linked to bringing both health and sustainability benefits for consumers and society. Production of cultivated meat can reduce land use, improve animal welfare and may even modify the nutritional composition of the product towards more beneficial fat composition than conventional [4]. The new production method has also been linked to a possibility to feed the increasing world population as the production of the cultivated meat is not dependent on land area and allows production in locations where keeping cattle is limited. Yet, there are a number of unanswered questions about how sustainable the production of cultivated meat will be, how much energy the production will require and what are the raw materials and production aids that are needed in the large-scale production of cultivated meat.

While the identified benefits of cultivated meat are mainly those that relate to social and environmental aspects such as being climate-friendly and promoting animal welfare, there is great uncertainty about what the benefits are to individual consumers who will use the products. The product not being widely accessible means that only a limited number of consumers have had direct experience with the products and therefore consumers need to form their opinions and expectations based on information. Lack of direct experiences gives rise to uncertainty around two core reasons people have in food choices, namely price and sensory quality of the products: how much it will cost and what they will taste and feel like in the mouth when eating the product [5].

To understand the factors that influence the acceptance of cultivated meat, we need to think about what kind of novelty these products are from the consumers' and society's point of view. Framing the cultivated meat as meat creates expectations of products that are meat-like, and thereby the products as such would be familiar to the consumers, even if the production has moved from animals to bioreactors and industrial production facilities. When the novelty is in requiring new technology in the production, consumers may regard the products less natural and have worries about the safety of the products [6]. This adds a layer of uncertainty to consumer acceptance, especially when the new production method requires special expertise and does not resemble the processes we normally use in home kitchens. Consumers' trust can be built with transparency, but explaining the cultivated meat production process to consumers can be a challenging task due to the high level of novelty in production without a clearly defined final product.

Although the promised benefits of producing cultivated meat are at societal level, the new production method creates also social resistance, especially among those who emphasise cultural values of foods or have concerns about the future of farmers who keep animals and produce meat. Similarly, cultivated meat is a suspect to destroy traditional values of local food cultures. Meat has a strong position on plates in many cultures providing the central piece to the commonly eaten meals, especially evening meals [3]. The traditional and cultural values together with farmers' livelihood have been used as arguments to support the initiatives to ban the cultivated meat from the markets.

Consumer perceptions and social aspects of cultivated meat have been studied widely in the last 5 years with several recent comprehensive review papers coming out [7–12] as well as reviews on regulatory requirements [13, 14]. This chapter uses these review papers as a backbone of current knowledge complemented with some original research papers. Studies on consumer acceptance of cultivated meat are many, but they have explored the cultivated meat at a conceptual level due to lack of real products that would enable the direct product experience. In this chapter, the results on different factors influencing the acceptance of cultivated meat products and production will be introduced in more detail based on existing literature. We start with introducing the challenges that cultivated meat as a novelty has in food products. The next section gathers perceived direct product-related benefits and concerns consumers may have when interacting with products, followed by a section on societal aspects influencing the acceptance of cultivated meat. The fourth section is about policy and regulatory aspects influencing the market access of cultivated meat by providing few examples on how the concept of cultivated meat has been tackled in different countries. The final section gives a short overview with considerations on the future of cultivated meat as a mainstream protein alternative.

2 Consumer Acceptance of Cultivated Meat as a Novelty in Food Domain

Novelty in food can always contain a risk of being something we do not like or being even harmful to us [15, 16]. Cultivated meat is food that is produced in a novel way, but the final product is not clearly defined. The final product has been described as meat, burger, nuggets or just an ingredient that can be used in combination with other ingredients. The major novelty is in the technology used in the production of cultivated meat, but the final novelty experienced by consumers depends on whether the products mimic meat or are something that diverts from traditional meat products standing on their own as a new product category. Thus, the expectations and perceived novelty related to final products can vary widely.

Perceived familiarity has supported acceptance in most studies on cultivated meat, but there are also findings when familiarity has had no impact on acceptance [7]. Yet, as described above, in the case of cultivated meat, the degree of novelty and degree of familiarity depend on the decisions that are made beyond the technology, although the production technology may need to be tailored according to the desired target product. Due to the limited number of physical products so far, there is uncertainty about the novelty hurdles cultivated meat needs to overcome in consumers' minds. Even if the sensory experience would match the conventional counterparties, it is still important to consider the novelty-related production technology as well. Tackling the fear of novelty related to product technology in consumers' minds may require different approaches than tackling the fear of novelty related to sensory experience once the products are introduced.

In principle, we can describe the acceptance of novel technologies at product level as depending on how consumers perceive the benefits and risks in new applications [17]. The rational decision-making can be described as weighing benefits against the risks, and then depending on which ones weigh more, the decision is either for or against accepting the products. We know that benefits outweighing the risks are crucial for the acceptance, but the decisional balance is strongly influenced by emotional reactions as well [18]. The novelty per se represents risk and uncertainty about what is the possible outcome of eating a product, and being cautious is a rational mechanism to avoid negative outcomes that ingesting food may have [19].

Consumers vary in their food and food technology neophobia, but these two types of neophobia are not highly correlated [20]. Whereas food neophobia is related to worries about the actual sensory experience with the food related to how it will taste and what is the texture of the food [15], the food technology neophobia relates to worries about the safety and other negative consequences technologies may have [21]. For technology neophobia, lack of trust on the technology itself can contribute to the fear of novelty as well as use of technology destroying the naturalness of the food. These neophobia tendencies vary between individuals, some of us being open to trying out new foods, whereas others avoid them due to uncertainty about the qualities and consequences of eating the product. In cultivated meat, both of these neophobia aspects are relevant. Although the new food is called cultivated meat, the actual product can be very different from meat as we know it. Producing meat in a bioreactor instead of keeping animals is a new technology that may be hard for consumers to understand and easily regarded as not natural. Thus, as a novel product, cultivated meat has this double challenge in winning consumer acceptance which is not helped by the fact that the food neophobia and food technology neophobia are two distinct concepts [20].

Novelty is a potential risk, but it also provides a possibility to experience something new and thus provide excitement and variety to the eating, which can be appealing to those consumers who seek experiences. According to the Roger's innovation adoption theory [22], consumers can be categorised according to their willingness to adopt new products and technologies. The consumers most open to novelty, so-called innovators and early adopters are expected to cover about 15–20% of the population, and they are likely to be the first ones to try novel products. The openness to novelty may be domain-specific and vary between product categories [23]. In food, lack of neophobia and domain-specific innovation may define the early adopters who are the first ones to try out novel cultivated meat products. From the promotion point of view, the first challenge is to identify these early adopters, and the second challenge is to offer products that have desired product qualities so that first trials result in repeated choices and thereby establish the true product adoption to the market.

3 Consumers' Perception and Attitudes Towards Cultivated Meat Products

When asked about the reasons behind their food choices, consumers typically mention taste, price and health as the major reasons with convenience, environment, sustainability and animal welfare appearing to be lower in the importance ranking [5, 24]. The expectations related to the top three reasons are governed by uncertainty about what the actual cultivated meat product solutions will be like. Novelty as such can be tackled by increasing the awareness and familiarity about the cultivated meat as a concept and providing exposure to the products once they are available, but in novel foods based on new technologies, decrease in perceived naturalness can be a challenge. Many of the suggested benefits in cultivated meat products are credence attributes that need to be conveyed to consumers via information [25], and communicating the novelty and benefits of the new options plays an important role in building consumer acceptance. Consumer trust in information given by companies and regulatory bodies behind safety assessment is crucial for acceptance as well. Below are short summaries of what we know about these factors in relation to consumer acceptance of cultivated meat.

3.1 Sensory Quality/Taste

When consumers refer to taste, they often talk about the overall sensory experience and hedonic evaluation, with flavour, mouthfeel and texture included as well. Uncertainty about sensory quality is one of the open questions in consumers' minds that will be finally solved once the products will be available. As taste is a highly important driver of acceptance [26], it is important that products that enter the market fulfil the expectations consumers have for the products. Positioning of the products sets the expectations, and unrealistically high expectations may result in disappointment with the products if the actual product experience does not fulfil the expectations. Therefore, placing cultivated meat directly as a meat replacement can be regarded as a risky option, because cultivated meat products currently differ from conventional meat in appearance, aroma, tenderness and juiciness.

The current findings on the role of taste are mostly based on concepts and hypothetical products. Thus, consumers' concerns about the possibility of low product quality are not based on direct experience [27]. There are some studies that suggest that it is possible to create cultivated meat products that mimic the colour and texture of their conventional counterparts and can be cooked in the same way as meat [28]. The scientific advancements with a focus on improving flavour and texture can be the path to ensure consumer acceptability [29]. The final sensory eating quality of the products will be crucial for the adoption of cultivated meat, but this quality aspect is product-specific and thereby the quality goals vary as well. To reach the desired taste and texture is partly dependent on the final product development stages, but the ability to develop versatile final products will depend on technological developments in the cultivated meat production. The range of suggested products has varied from structured steak analogues to using meat cell mass as an ingredient in products that contain other ingredients, such as plant-based scaffolding to give the structure to the product.

3.2 Price

Price is another factor that is uncertain in future cultivated meat products. Currently, the production of cultivated meat is expensive, and the first products are expected to have a price premium. One option to reduce the price is hybrid products where cultivated meat cells are combined with conventional meat or other components, including plant-based ones. Currently, the expected high price can be regarded as a major barrier for acceptance and adoption of cultivated meat [30, 31], particularly for price-conscious consumers in rural areas [32] and Eastern Europe [26]. The first producers of cultivated meat are small companies with limited production resources and high development costs. High production costs and low capability to produce large amounts raise the question of what is the best positioning for these products: Should they be positioned as specialties with high premium and target the wealthier markets [33]? There are some consumers who are willing to pay a premium price for cultivated meat [28], but to gain acceptance among large consumer segments, the price needs to be competitive in comparison to meat or any other product that is seen as the direct alternative in positioning of the product.

3.3 Health

Cultivated meat has been linked to both potentially positive and negative product attributes. Health-related benefits can significantly increase the acceptance of cultivated cells [11]. To gain consumer acceptance for cultivated meat as a meat replacement, a key challenge is to ensure that the product contains same essential nutrients such as vitamin B12 and iron, which are primarily obtained from conventional meat [34]. As cultivated meat production is a controlled process, improving the fat composition by reducing the amount of saturated fat is a possibility, and micronutrients can be added to improve the nutritional quality of the product. The end product can mimic the meat composition or be tailored to deviate from it according to what is desired by the producers. Cultivated meat has the potential to produce meat products that have versatile nutrient profiles that may outperform the conventional options, but this may add complexity to the product and not be the first priority in developing the production technology. However, consumers have also expressed concerns about the long-term health effects and potential nutritional deficiencies as a result of eating cultivated meat [35]. Thus, the nutritional composition of cultivated meat may provide a consumer benefit, but consumers also feel uncertainty about long-term consequences of eating cultivated meat as there is no history of use to rely on.

3.4 Perceived Naturalness

Consumers perceive natural meat as inherently positive, something pure where the touch of human hand has not reduced the value of the product [36]. Although naturalness of food does not come on the top of ranking when asked about what the most important reasons behind food choices are, perceived lack of naturalness is often mentioned as a reason to avoid some foods or components, such as food additives or processed foods. Naturalness is perceived as inherently good and is often linked with perceived healthiness as well [37]. Cultivated meat has the double burden of being not only novel and thereby unfamiliar to consumers but also requiring technology and processing that are linked to reduced naturalness perception. Explaining the production method can increase familiarity and reduce the "strangeness" of the process, but at the same time, it increases the awareness of novelty of technologies required in preparation. The familiarity in this occasion may not breed acceptance; instead, it may raise new kind of concerns related to the product. Tackling the perception of cultivated meat being an artificial and not natural as a product may be one of the major hurdles in promoting the acceptance of these products.

Production of cultivated meat happens in bioreactors with methods that are different from what consumers themselves use in food preparation, although the use of bioreactors in food production is not uncommon in other processes such as fermentation and production of sour milk products, beer or tea. For cultivated meat, making references to these well-accepted familiar products that are produced with similar, although not identical, methods could move the perception of naturalness to positive direction. The naming of the technology and outcome products is important as calling it lab-grown, artificial, or synthetic meat contains the idea that the human hand has modified the product into something that is distant from the nature. When the names for the cultivated meat have been studied, acceptance ratings of "clean meat" concept have been shown to lead to better purchasing intentions compared to other terms used [32].

3.5 Disgust

The major challenge for cultivated meat is the potential strong negative emotional response to the concepts and the actual products as well. In addition to the earlier described fear of novelty, disgust sensitivity frequently emerges as a factor that has a negative impact on the acceptance of cultivated meat [11, 19]. Whereas food neophobia is a tendency to avoid food due to fear of the product's properties [15], food disgust is an emotion that has high arousal and negative valence [16]. Rejection of the cultivated meat can be due to high disgust towards its origin as well as to the expected sensory characteristics. Similarly to food or food technology neophobia, disgust sensitivity varies among people: both are more common towards animal-based foods than plant-based foods making cultivated meat vulnerable to disgust reaction. Whereas food neophobia and food technology neophobia can be reduced by increasing familiarity of the products by exposure or information, food disgust may be more difficult to tackle as it may be immune to information and be a high hurdle against getting first-hand sensory experience.

3.6 *Trust and Safety*

We expect foods that are sold in the grocery stores to be safe, but novelty in food is always a potential risk to consumers as there is no direct experience of consequences of eating the food [33]. In order to make an assessment on risks and benefits of new products, consumers have to rely on information they get about food products and production methods from manufacturers or other information sources. In novel foods, trust in different actors in the food system is crucial as consumers do not have direct experience of the products and their knowledge and capability to assess the safety of the product or accuracy of given information themselves are limited. Thus, consumer distrust towards cultivated meat may stem from fear of novelty [31] or from perceived lack of regulatory efficiency and a reduced sense of control over the origin [38]. In a ten-country study, the authors found that trust in food sector actors increased the perceived naturalness of cultivated meat in all countries and reduced the evoked disgust, and trust could even additionally directly contribute to the acceptance of cultivated meat [16]. As many of the possible benefits of the cultured meat are credence attributes related to sustainability and animal welfare, consumers' trust is crucial in conveying the information in an effective manner. Yet, providing too much technical information about the production process can decrease acceptance and negatively affect respondents' opinion [11, 19] similarly to giving the production method a technical name as discussed earlier [32]. However, some studies suggest that information alone may not be powerful enough to change consumer behaviour [26].

4 Societal Aspects in Acceptance of Cultivated Meat as a Food Production Method

In addition to consumer acceptance of cultivated meat products, there needs to be societal acceptance towards the concept and used technology by creating an environment that is open to the introduction of new products and supports the innovation activities. The societal acceptance depends partly on consumers' perceptions of benefits and risks associated with adopting new technologies in the society. Consumers can make these assessments from citizen perspective that goes beyond the acceptance of novel products per se. Cultivated meat can be regarded as a future new food that has the potential to support sustainability goals in society by reducing climate impact and improving animal welfare. Citizens' opinions are important in societal acceptance, but there needs to be support from other stakeholders as well: those developing the products, funding the development and authorities responsible for the regulatory issues. Cultivated meat as a novel production technology requires changes in the cultural traditions of how meat is produced and who are the main actors behind the production. When changes are implemented in food systems, there will be some actors who are likely to benefit and those who will lose. For example, consumers may regard the cultivated meat products as a possibly interesting addition to their daily diets, but consider them unacceptable due to the anticipated negative consequences to the beef producers. Similarly, there may be a good support towards cultivated meat as a product that has a positive environmental impact, but due to perceived unnaturalness, the products are not accepted to one's own diet. New products and production methods can gain success in the market if they manage to gain both societal and product acceptance.

In the following, the societal benefits and possible risks are discussed in more detail starting with environmental benefits and animal welfare followed by reduction of hunger and possible negative effects on farmers.

4.1 Environmental Benefits/Climate Change

Meat production and especially beef production require a lot of energy, water and other resources and as output produce harmful greenhouse gas during the production [4]. Cultivated meat as an alternative way of producing meat has been regarded as a solution where meat production can become more sustainable. Environmental concerns are a key motivator for some consumers to reduce their meat intake, but the majority of consumers in high-income countries with high meat consumption still remain reluctant to change their behaviour, even if they are aware of the environmental impact of meat production [39, 40]. The most positive consumers towards reducing meat are typically young, female, and ecology-oriented and are more likely to live in Europe and Asia than in the USA [39]. However, the true environmental impact of cultivated meat will depend on how the new production methods develop and where the needed energy and other resources for producing the cultivated meat will come from [34]. From a consumer perspective, the environmental benefits will appeal to those consumers who are environmentally conscious.

4.2 Animal Welfare

Improvement in animal welfare is regarded as one of the strongest benefits of cultivated meat [34]. As animals are sources of cells, they are still needed in the production of cultivated meat but in smaller numbers than for the production of conventional meat. The cells may even be harvested without a need to slaughter animals from living animals. Thus, the "meat" can be positioned as a "slaughter-free" option, which appeals to consumers who are against killing animals in food production [35]. Less animals and less killing of them can result in better animal welfare, but there are open questions on how the production of animal cells will be organised and what the role of animal farming in the process is. This has created a concern about the future of animal farmers and their livelihood resulting in strong lobbying against the cultivated meat as a production method in some countries. Although consumers are concerned about animal welfare in meat production, the animal welfare does not rank high when asked about the reasons behind food choices [24], and willingness to pay a premium for meat produced with high animal welfare standards is low.

Issues about animal welfare as well as environmental concerns can be regarded as pragmatic problems that cultivated meat production can solve, at least partly. However, these issues can also be based on deeper ethical concerns about doing the right thing in relation to the animals and saving the planet for future generations. The ethical aspect can cause feelings of guilt about eating meat and regardless of concerns about still preferring meat to legumes and other plant-based alternatives [34]. For these devoted meat eaters, cultivated meat is likely to be particularly appealing as it enables eating meat while taking the moral concerns into account [26, 32].

4.3 Reducing Hunger and Responding to Increasing Meat Demand

As a new way of producing high-quality meat protein with lower input resources including less need for land area, cultivated meat has been suggested as part of the solution to reduce world hunger. Despite the recommendations to limit the meat, and especially red meat consumption in food-based dietary guidelines and EAT Lancet Commission's planetary healthy diet [1, 2], meat consumption is predicted to increase in the near future especially in many middle-income countries due to increasing income levels [41]. Cultivated meat can provide a solution to respond to the growing demand without compromising the use of land resources to grow food.

4.4 Future of Farmers

Introducing new production methods may also include risks and not all benefits are advantageous to all actors in society. Future of meat-producing farmers has been discussed as one of the changes induced to the food system if the cultivated meat production becomes the dominant meat production method. The change not only affects farmers but also the whole ecosystem around meat production and livelihood of areas highly dependent on meat production. In addition to the farming sector, consumers can have concerns about the potential negative impacts on the agricultural sector. For cultivated meat, these concerns relate to farmers and employment in animal production [27, 33, 34]. To gain acceptance among different actors requires thinking how the possible negative impacts can be mitigated and clarifying the relation between conventional and cultivated meat production and the possible models to integrate the two.

5 Who Is Willing to Adopt Cultivated Meat Products?

Consumer adoption of new technologies and solutions depends on what benefits the cultivated meat products can bring them, but the value of different benefits varies among consumers. These benefits can be directly linked to the product, such as taste, price and convenience, or to the benefits that are aligned with one's values. Consumers can be segmented according to sociodemographic variables, but often, psychographic or behavioural segmentation based on value-related factors such as identities or practices is better in explaining differences in consumers' acceptance responses. The consumer acceptance has so far mainly been measured in relation to cultivated meat as a concept rather in relation to real product alternatives.

5.1 Sociodemographic Factors

In their review, Hanan et al. [10] gathered the knowledge so far on sociodemographic factors linked to perceived acceptability of cultivated meat. The results are somewhat mixed and vary between different studies, but younger respondents appear to be more positive towards cultivated meat than older respondents, although some studies show the opposite results. The majority of studies have found that men and frequent meat eaters are more positive towards cultured meat than women. Those hoping to reduce their meat consumption also have a positive view about cultivated meat. Cultivated meat acceptance is linked to high education in many studies. Although there is a large variety in methods, locations and selection criteria of respondents in reported studies, there seems to be a rough alignment that cultivated meat is likely to be better accepted by younger age group, men and those with longer education, especially if they are frequent meat eaters [10]. However, due to variability in findings, consumer acceptance needs to be verified when planning the market approach for any specific product.

In a recent review, consumers in different countries varied in the way they perceive cultivated meat, but these differences were also study and question dependent [11]. The used acceptance measures have a hierarchy depending on the level of commitment in the question: willingness to try is typically higher than willingness to buy or willingness to pay. As willingness to try gives the highest acceptance rates, they may give an unrealistic picture of acceptance. Yet, the willingness to try varied between 31% and 77% of respondents depending on the country and information given [11]. As the products do not yet exist, these findings suggest that there is a reasonable number of consumers who are interested in trying cultivated meat products and thereby potential to create familiarity with the products over time. One possible explanation to differences in expressed acceptability indicators is religion as Lewisch and Riefler [9] in their review suggest that individuals accepting cultured meat are less religious than other consumer segments. The differences can also reflect culture and the role of meat in daily diets. Countries where meat and dairy products are widely used may have less interest in cultivated meat as an option compared to densely populated countries like Singapore, where cultivated meat has already been introduced [34]. Cultivated meat may be more appealing as a concept in countries where the meat consumption is rising [11].

5.2 *Psychosocial Factors*

Acceptance of cultivated meat can also be linked to the consumers' values and attitudinal background. In addition to curiosity and interest in new innovations and low neophobia of food and food technology, consumers differ in their environmental awareness and health consciousness, which may promote the acceptance of cultivated meat as the more sustainable and possibly healthier option than conventional meat. Those consumers who seek for options to replace the meat in their diet may be the first group to adopt cultivated meat. Cultivated meat may appeal especially to those consumers who do not wish to abandon meat eating. Meat attachment as a strong positive attitude towards meat has appeared as a barrier towards acceptance of plant-based alternatives [42, 43], and cultivated meat may provide an option that can be tolerated by those who are unwilling to give up the meat but still would like to choose options that are more climate friendly [43].

When introducing cultivated meat products to the market, building capability to use the products is important as well. Lessons from plant-based meat alternatives suggest that consumers may find it difficult to prepare food from new alternatives that do not behave similarly to conventional meat product when cooking [44, 45]. One option to overcome these issues is to target restaurants and institutional kitchens with the products as these have professionals who can prepare the products in a manner that is appealing to consumers. At the same time, this increases familiarity towards the products, which promotes acceptance of cultivated meat if the experiences are satisfactory.

6 Responses to Introduction of Cultivated Meat in Different Countries

Cultivated meat can be regarded as a transformative innovation that enables a novel way to produce food, and thereby it can challenge the current meat production. As cultivated meat is a novel food, most countries require a regulatory safety assessment before the products are allowed to enter the stores in order to protect consumers against possible risks related to novelty. Consumer studies find a potential market for cultivated meat in many countries [11, 46], but a clear and transparent regulatory framework is essential for building public trust [37], and trust is important in the acceptance of cultivated meat [16]. If consumers trust that regulatory systems are able to protect them, they are more open to novel options.

The regulations vary between countries, but their main purpose is to guarantee consumer safety [14]. The differences in regulatory systems relate to how much other criteria, such as broader social, environmental and ethical implications, are used in assessing the new cultivated meat options. The requirements made by regulatory frameworks have been criticised of slowing down the process of new products accessing the markets, as well as not considering the wider social perspectives and concerns of different actors in food systems, including those of consumers [14, 47]. Building the social acceptance among different stakeholders requires time and lengthens the duration of the evaluation process, which has been seen as a problem, although acceptance building may be crucial in building the future success of the technology.

The regulations of different countries have been described in the study by Johnson and Monaco [14] together with the situation of accepted and submitted applications of culture meat products as in 2025. Currently, very few cultivated meat products have been gone through the assessment successfully; there are two products in Singapore, two in the USA and one in Israel. A few applications have been submitted to the process, but the availability of products to the consumers is still limited with one product out in a retail store in Singapore and one in the USA available on special events. Currently, only very few consumers have direct access to the products and experience with them, consequently raising awareness about cultivated meat, which is based on information about the potential products that will appear in the stores in future.

Instead of solely relying on animal farming, the cultivated meat production is moved into food manufacturing plants with bioreactors requiring new type of expertise to produce the food. As in all truly new innovations, the roles of actors in the food systems will change, and this can affect livelihoods of actors as well. Cultivated meat production is likely to still require animals but in smaller numbers.

This has raised concerns related to the role and livelihood of farmers in meat production, if cultivated meat will become the main production channel of meat in the future. If there are fewer animal farms in the future, this will have an impact on the life and landscape in rural communities. These changes will challenge the traditional food culture that relies on rural production of food. The producers of cultivated meat may need expertise that comes from fields that core actors in current food systems do not have at the moment.

Whereas some countries regard cultivated meat as a new option in our diet and an opportunity to develop new food production methods that may be beneficial from sustainability and animal welfare perspective, others have started to react strongly creating barriers against the new innovation. Singapore has adopted a proactive approach becoming the first country in the world to approve the sale of cultivated meat and establishing clear labelling regulations, allowing terms like "cultured" and "cell-based" in food products [13, 48–50]. The USA was the second country to allow commercial sales of cultivated meat [51]. The regulatory framework is a cooperative effort between the Food and Drug Administration (FDA), the US Department of Agriculture (USDA) and Food Safety and Inspection Service (FSIS) where FDA regulates the initial production phases and the USDA oversees processing, packaging and labelling [13, 14, 50].

Although cultivated meat is accepted in the USA, the acceptance is not unanimous throughout the whole country. Thirteen US states have enacted laws that prohibit the use of traditional meat vocabulary on the labels of alternative or cell-based meat products [52]. Moreover, in 2024, Alabama and Florida passed laws banning the manufacture, sale or distribution of cell-cultured meat. Other states, including Tennessee and Arizona, have also fielded bills concerning cultivated meat [53].

In Europe, the views on cultivated meat are divided among countries. From the regulatory perspective, in the European Union (EU), cultivated meat falls under the EU's Novel Food Regulation (EU 2015/2283) that is based on precautionary principle putting high emphasis on safety [48]. To gain acceptance to novel foods requires a comprehensive risk assessment by the European Food Safety Authority (EFSA) at every stage of the production chain. As an authority, EFSA has an advisory role giving opinions, but the in the final phase, the decisions are made by the EU Commission who may deviate from EFSA's recommendation or consider other factors beyond safety as more important when making decisions. These can be related to political, cultural or ethical considerations that are separate from scientific safety evaluations. The EU's diverse policies and reliance on the precautionary principle create a complex environment for market entry and commercialisation of cultivated meat. This is further compounded by political and institutional disagreements among member states, which prevent the formation of a unified regulatory system [13, 54].

Countries differ in their openness to cultivated meat in Europe. For instance, the Dutch Government has successfully developed a "code of practice" that allows for tasting in controlled environments in partnership with cultivated meat companies. In contrast, Italy became the first country to ban the production and sale of cultivated meat through a politically motivated law (172/2023), with the aim to protect human health and the country's gastronomy [55]. This ban was not based on scientific consensus [47, 48] and lacked legal effect due to procedural flaws [55]. Moreover, the strong reaction creates a perception that cultivated meat can be unsafe and may hinder investment and research in the field [47]. This suggests that regulatory support that involves all stakeholders from the public to regulators [56], combined with a receptive cultural environment, will be key to public acceptance of cultured meat.

Australia, New Zealand and Korea have similar frameworks like the EU and US systems [13, 14]. Japan and Israel have also integrated the promotion of alternative proteins into their national strategies for decarbonisation, food security and economic growth [50].

7 Conclusion and Future Perspectives

7.1 Acceptance of Cultivated Meat

Cultivated meat creates a promise of finding an alternative, more sustainable way to produce a traditional food that is under pressure from sustainability and health point of views. While the goal is to manufacture meat as a familiar and prestigious food, production-related radical novelty can become a hurdle in gaining consumer acceptance to cultivated meat. The final product requires acceptance at two levels: acceptance of the technology required in production and acceptance of the product itself. The social acceptance of cultivated meat relies on acknowledging the benefits to environment, animal welfare and health that the new production method can provide. The social acceptance endorses different actors in the food system to support the development of cultivated meat as an alternative high-protein food option. However, for consumer acceptance, the product quality becomes crucial for adopting the products. Consumers may well be aware of the societal benefits, but their willingness to sacrifice product quality is known to be low. Sensory experience, price and nutritional composition of the products need to be on par with other options including conventional meat as well.

The benefits of cultivated meat are still to an extent promises and possibilities rather than benefits that have fully been realised. Cultivated meat is produced by growing animal cells in bioreactors that bypass the need of land area and reduce the number of animals needed in meat production. However, the technology still requires further development in order to be an efficient way of producing products in quantities that can make a substantial contribution to the food supply with an affordable price. In addition to the technological and production-related hurdles, there are also several regulatory and market-related hurdles to overcome. The regulatory acceptance of cultivated meat has been regarded as one of the major challenges that can slow down the acceptance and development of these new products in some part of the world. Country and region differences in regulatory acceptance processes can result in large diversity in how companies in various geographic areas invest in the development of cultivated meat as well as how much access consumers will get to these products. While strict scrutiny to the safety of new foods is an important part of protecting consumers against harmful products, the slowness and uncertainty related to outcomes of the process can act as a barrier to development due to the high risk and cost of resources required.

The social acceptance enables and supports the development of cultivated meat products, but the final acceptance of products depends on their eating quality and affordability. At the moment, the quality of final products is still unsure: which meat products they will be and how they will taste and feel in the mouth. Meat products as a category cover a wide range of foods that also vary in the meat content, type of meat used and how the meat has been processed during the product manufacturing process. Thus, the cultivated meat products can vary in shape, size and taste from minced meat to steaks and from burgers to sausages. The existing few applications suggest that first products will be some kind of hybrids that combine the bioreactor-produced meat cells with other components in the final products. They may mimic products that are close to burgers and nuggets which are at the lower end of prestige value among meat products. However, the future will show how the product variety of cultivated meat products will develop.

7.2 Building Social Acceptance Towards the Concept of Cultivated Meat

To gain social acceptance, increasing the awareness of the benefits related to environment and animal welfare that cultivated meat can offer to the society is important. Instead of just informing about the benefits, awareness-raising approach should offer an opportunity to open a dialogue between different actors in food systems including civil society organisations and consumers. Increased awareness and possibility to engage in discussion about the benefits and possible risks related to the cultivated meat production create transparency on the pros and cons of the new way of producing meat and are thereby likely to build trust in the production method as well.

The companies and other actors in the cultivated meat food systems should consider where their shared interests are and combine their forces in promoting the cultivated meat as a product concept. Although the products from different manufacturers will compete in the market, in the precompetitive stage, the cultivated meat producers share common interests in creating consumer awareness, promoting feasible regulatory practices and even running shared campaigns to promote the concept together with other actors. Regulatory demands related to labelling the products are another topic where different actors can find shared interests.

Developing a harmonised nomenclature for cultivated meat should be a shared interest to all actors in the food systems. There should be commonly agreed definitions of terms used when talking about cultivated meat, what is required from a product to be categorised a cultivated meat product and what a product should and should not contain in order to be called cultivated meat. The lack of agreement on use of terms may confuse not only the consumers for whom the whole technology is strange but also the actors working with the supply side and regulatory issues. Agreement of the terms would benefit the whole field and build understanding and trust among actors and support companies in their own product-related communication as well.

Social acceptance is also dependent on the factors that go beyond the cultivated meat. As a new production method, it may replace partly the existing production of meat and have an impact on conventional farming of animals and thereby on local economies around meat production. The examples of strong reactions from Italy and the USA banning the cultivated meat illustrate that the new technology causes concerns among communities. Justification of these concerns is not based on sound evidence-based assessment; instead, they are often driven by emotional reactions. Part of societal benefit and risk assessment should be a thorough analysis on who will benefit and who will be harmed by adoption of cultivated meat production. The findings may support the concerns, but they may also disperse them. These assessments should be considered as a standard part of the innovation process, and if done at the early stages of development, there is time to plan how to mitigate the possible harms. These considerations should be based on evidence and include all relevant actors in discussions to find the possible mitigation solutions. The wide inclusion of actors in these discussions builds transparency, trust and shared understanding of what would be future developments.

7.3 Product Quality Will Be Essential

As the products are not yet in the market, building awareness and creating realistic expectations among consumers are important so that when the physical products enter the market, consumers' interest and acceptance of the production method will be in place. At the same time, avoiding overpromising and keeping the product-related promises at a level that can be achieved in product application should be the target. Transparency is important in relation to how the products have been manufactured and how much cell-grown ingredients are there in the final product.

For producers of cultivated meat, framing of the products will be one of the crucial aspects of future acceptance. The term cultivated meat implicitly positions the products as meat, but it is likely that the first products will be hybrids containing other raw materials such as plant-based scaffolding to create the desired texture to the product. Most of the currently available cultivated meat products contain a minor part of meat cells with the rest being other components. Meat is a food that has high prestige value, and positioning the new products as meat suggests that product quality is comparable to meat. Open positioning of the products as hybrids can lower expectations as the reference point moves away from conventional meat.

Harmonising the use of terms that relate to components used in final products could benefit the whole industry in the long run.

Even if consumers are aware of cultivated meat as a production method, the products need to tackle the novelty aspect. Consumers vary in the concerns related to novelty as well as in their interest in detailed information about how the food has been produced. Yet, having this information easily available for those who wish to know more is important. The regulatory requirements for labelling foods produced with cellular technologies are not yet fully established, but making additional information available via links in product packaging may be an option to increase transparency. The ability to link the information to the package will also enable the more space-demanding options to convey the desired information with audiovisual materials in addition to plain text. These techniques enable the use of storytelling techniques to associate the novelty of products with familiar options and thereby decrease the concerns related to novelty in consumers' minds.

Price of the cultivated meat products will be a crucial factor to market success. The development costs of cultivated meat products are high, and having the first products as specialties with high premium may be the most feasible strategy to enter the market. However, to respond to the promise of using less resources in meat production and improving animal welfare require that the products will be able to replace the conventional meat or fulfil the increasing global demand of meat in countries where the consumption is growing due to increased incomes. Currently, there is no unified view on whether the products in the long run will be more or less expensive than conventional meat, as it depends on what the final products are and how the production technologies will develop.

The final acceptance of cultivated meat will depend on what the products in the market will be like and how they are positioned. Emphasising the societal benefits of environment and animal welfare requires conveying these benefits with effective information approaches, and they may appeal to specific consumer segments. Yet, the sensory quality of the products will be an essential factor in repeated choices and building long-term adoption of cultivated meat. Tailoring the quality of the products towards the high end of the market may be a challenge, but the eating quality of the food has to be at the level of competing alternatives.

References

1. Rockström J et al (2025) The EAT–Lancet Commission on healthy, sustainable, and just food systems. Lancet 406:1625–1700. https://doi.org/10.1016/S0140-6736(25)01201-2
2. Cámara M et al (2021) Food-based dietary guidelines around the world: a comparative analysis to update Aesan scientific committee dietary recommendations. Nutrients 13:3131
3. Font-i-Furnols M, Guerrero L (2025) An overview of drivers and emotions of meat consumption. Meat Sci 219:109619
4. Zão ACAF, Nogueira WV, Rondan FS, Scaglioni PT (2026) Cultured meat: a multidimensional review of technological, nutritional, ethical, and regulatory advances (2020–2025). J Food Sci 91. https://doi.org/10.1111/1750-3841.70915

5. Steptoe A, Pollard TM, Wardle J (1995) Development of a measure of the motives underlying the selection of food: the food choice questionnaire. Appetite 25:267
6. Etale A, Siegrist M (2021) Food processing and perceived naturalness: is it more natural or just more traditional? Food Qual Prefer 94:104323
7. Raverta P, Sandi I, Martin B, Loera B (2025) Unfamiliar familiarity: a scoping review on the role of familiarity in consumer acceptance of cultivated meat. Appetite 211. https://doi.org/10.1016/j.appet.2025.108000
8. Rana RL, Bux C, Tommaso V (2026) The sustainability nexus of cultured meat: integrating environmental, social, and technological-economic insights. Food Res Int 227. https://doi.org/10.1016/j.foodres.2025.118245
9. Lewisch L, Riefler P (2023) Cultured meat acceptance for global food security: a systematic literature review and future research directions. Agric Food Econ 11. https://doi.org/10.1186/s40100-023-00287-2
10. Hanan FA, Karim SA, Aziz YA, Ishak FAC, Sumarjan N (2024) Consumer's cultured meat perception and acceptance determinants: a systematic review and future research agenda. Int J Consum Stud 48. https://doi.org/10.1111/ijcs.13088
11. Szendrő K (2025) Consumer perceptions of lab-grown cells: awareness, barriers, and the power of information. A review. Czech J Anim Sci 70:203–222. https://doi.org/10.17221/31/2025-CJAS
12. Lin JWX, Maran N, Lim AJY, Ng SB, Teo PS (2025) Current challenges, and potential solutions to increase acceptance and long-term consumption of cultured meat and edible insects – a review. Future Foods 11. https://doi.org/10.1016/j.fufo.2025.100544
13. Bakhsh A et al (2025) Cell-based meat safety and regulatory approaches: a comprehensive review. Food Sci Anim Resour 45:145–164. https://doi.org/10.5851/kosfa.2024.e122
14. Johnson H, Monaco A (2025) Global developments in the regulation of cultivated meat: a comparative study of the EU, Singapore, US and Australia and New Zealand. Rev Eur Comp Int Environ Law 34:496–511
15. Pliner P, Hobden K (1992) Development of a scale to measure the trait of food neophobia in humans. Appetite 19:105
16. Siegrist M, Hartmann C (2020) Perceived naturalness, disgust, trust and food neophobia as predictors of cultured meat acceptance in ten countries. Appetite 155:104814
17. Bearth A, Siegrist M (2016) Are risk or benefit perceptions more important for public acceptance of innovative food technologies: a meta-analysis. Trends Food Sci Technol 49:14–23. https://doi.org/10.1016/j.tifs.2016.01.003
18. Giacalone D, Jaeger SR (2023) Consumer acceptance of novel sustainable food technologies: a multi-country survey. J Clean Prod 408:137119
19. Monaco A et al (2024) Consumers' perception of novel foods and the impact of heuristics and biases: a systematic review. Appetite 196. https://doi.org/10.1016/j.appet.2024.107285
20. Fantechi T, Marinelli N, Casini L, Contini C (2026) Exploring alternative proteins: psychological drivers behind consumer engagement. Br Food J 128:939–956. https://doi.org/10.1108/BFJ-10-2024-1066
21. Cox DN, Evans G (2008) Construction and validation of a psychometric scale to measure consumers' fears of novel food technologies: the food technology neophobia scale. Food Qual Prefer 19:704–710
22. Rogers E (1983) Diffusion of innovations. Free Press, New York
23. Huotilainen A, Pirttilä-Backman AM, Tuorila H (2006) How innovativeness relates to social representation of new foods and to the willingness to try and use such foods. Food Qual Prefer 17:353–361
24. Onwezen MC, Reinders MJ, Verain MCD, Snoek HM (2019) The development of a single-item food choice questionnaire. Food Qual Prefer 71:34–45
25. Fernqvist F, Ekelund L (2014) Credence and the effect on consumer liking of food – a review. Food Qual Prefer 32:340–353. https://doi.org/10.1016/j.foodqual.2013.10.005

26. Onwezen MC, Dagevos H (2024) A meta-review of consumer behaviour studies on meat reduction and alternative protein acceptance. Food Qual Prefer 114:105067
27. Tsvakirai CZ, Nalley LL, Tshehla M (2024) What do we know about consumers' attitudes towards cultured meat? A scoping review. Future Foods 9. https://doi.org/10.1016/j.fufo.2023.100279
28. To KV, Comer CC, O'Keefe SF, Lahne J (2024) A taste of cell-cultured meat: a scoping review. Front Nutr 11. https://doi.org/10.3389/fnut.2024.1332765
29. Tomiyama AJ et al (2020) Bridging the gap between the science of cultured meat and public perceptions. Trends Food Sci Technol 104:144–152
30. Verbeke W, Sans P, Van Loo EJ (2015) Challenges and prospects for consumer acceptance of cultured meat. J Integr Agric 14:285–294. https://doi.org/10.1016/S2095-3119(14)60884-4
31. Akinmeye F, Chriki S, Liu C, Zhao J, Ghnimi S (2024) What factors influence consumer attitudes towards alternative proteins? Food Hum 3:100349
32. Siddiqui SA et al (2022) Consumer behavior towards cultured meat: a review since 2014. Appetite 179. https://doi.org/10.1016/j.appet.2022.106314
33. Bryant C, Barnett J (2018) Consumer acceptance of cultured meat: a systematic review. Meat Sci 143:8–17. https://doi.org/10.1016/j.meatsci.2018.04.008
34. Rasmussen MK et al (2024) Critical review of cultivated meat from a Nordic perspective. Trends Food Sci Technol 144. https://doi.org/10.1016/j.tifs.2024.104336
35. Pakseresht A, Ahmadi Kaliji S, Canavari M (2022) Review of factors affecting consumer acceptance of cultured meat. Appetite 170:105829
36. Rozin P, Chen R, Scott SE, Cusimano C (2024) Americans believe in the benevolence of nature, and this belief is not lower in people who have experienced natural disasters. Judgm Decis Mak 19:e18
37. Hartmann C, Furtwaengler P, Siegrist M (2022) Consumers' evaluation of the environmental friendliness, healthiness and naturalness of meat, meat substitutes, and other protein-rich foods. Food Qual Prefer 97:104486
38. Kantono K et al (2022) Consumer acceptance and production of in vitro meat: a review. Sustainability (Switzerland) 14. https://doi.org/10.3390/su14094910
39. Sanchez-Sabate R, Sabaté J (2019) Consumer attitudes towards environmental concerns of meat consumption: a systematic review. Int J Environ Res Public Health 16. https://doi.org/10.3390/ijerph16071220
40. Hartmann C, Siegrist M (2017) Consumer perception and behaviour regarding sustainable protein consumption: a systematic review. Trends Food Sci Technol 61:11–25
41. FAO (2025) Meat market review. Overview of global market developments in 2024
42. Graça J, Calheiros MM, Oliveira A (2015) Attached to meat? (Un)willingness and intentions to adopt a more plant-based diet. Appetite 95:113–125
43. van Dijk B, Jouppila K, Sandell M, Knaapila A (2023) No meat, lab meat, or half meat? Dutch and Finnish consumers' attitudes toward meat substitutes, cultured meat, and hybrid meat products. Food Qual Prefer 108:104886
44. Schösler H, de Boer J, Boersema JJ (2012) Can we cut out the meat of the dish? Constructing consumer-oriented pathways towards meat substitution. Appetite 58:39–47
45. Wendler M, Halkier B (2024) Dietary transition requires work: exploring the practice-transition processes of young Danish meat reducers. Food Cult Soc 27:1358–1376
46. Bryant C, Barnett J (2020) Consumer acceptance of cultured meat: an updated review (2018–2020). Appl Sci (Switz) 10. https://doi.org/10.3390/app10155201
47. Fino MA et al (2024) Cultivated meat beyond bans: ten remarks from the Italian case toward a reasoned decision-making process. One Earth 7:2108–2111. https://doi.org/10.1016/j.oneear.2024.11.002
48. Lanzoni D et al (2024) Cultured meat in the European Union: legislative context and food safety issues. Curr Res Food Sci 8. https://doi.org/10.1016/j.crfs.2024.100722
49. Stevens H, Ruperti Y (2024) Smart food: novel foods, food security, and the smart nation in Singapore. Food Cult Soc 27:754–774

50. Miyake Y, Tachikawa M, Kohsaka R (2023) Policy frameworks and regulations for the research and development of cell-based meats: systematic literature review. Food Res Int 167. https://doi.org/10.1016/j.foodres.2023.112599
51. Kirsch M, Morales-Dalmau J, Lavrentieva A (2023) Cultivated meat manufacturing: technology, trends, and challenges. Eng Life Sci 23. https://doi.org/10.1002/elsc.202300227
52. Pitkoff J (2021) Note state bans on labeling for alternative meat products: free speech and consumer protection. NYU Environ Law J 29:297–348
53. Gerber S, Bae H, Ramirez I, Cash SB (2025) Publicly tasting cultivated meat and socially constructing perceived value politics and identity. npj Sci Food 9:94
54. Lähteenmäki-Uutela A, Rahikainen M, Lonkila A, Yang B (2021) Alternative proteins and EU food law. Food Control 130:108336
55. Monaco A (2025) A perspective on the regulation of cultivated meat in the European Union. npj Sci Food 9. https://doi.org/10.1038/s41538-025-00384-0
56. Petetin L (2014) Frankenburgers, risks and approval. Eur J Risk Regul 5:168–186

The manufacturer's authorised representative in the EU is Springer Nature Customer Service Centre GmbH, Europaplatz 3, 69115 Heidelberg, Germany. If you have any concerns regarding our products, please contact ProductSafety@springernature.com

Printed and bound by CPI Group (UK) Ltd, Croydon, CR0 4YY
16/09/2026
02217783-0008